THE
LOST
FAMILY

THE
LOST
FAMILY

How DNA Testing
Is Upending Who We Are

LIBBY COPELAND

Abrams Press, New York

Library of Congress Control Number: 2020944995

ISBN: 978-1-4197-4793-9
eISBN: 978-1-68335-893-0

Printed and bound in the United States
10 9 8 7 6 5 4 3 2

Abrams books are available at special discounts when purchased
in quantity for premiums and promotions as well as fundraising
or educational use. Special editions can also be created to
specification. For details, contact specialsales@abramsbooks.com
or the address below.

Abrams Press® is a registered trademark of Harry N. Abrams, Inc.

ABRAMS The Art of Books
195 Broadway, New York, NY 10007
abramsbooks.com

For Dan, always

CONTENTS

PROLOGUE

This is the story of a woman named Alice, but it is also the story of a man named Jason, and a woman named Jacqui, and a whole host of people you haven't yet met but who could be you. This is about the countless people who have seen ads for at-home DNA tests, and—eager to learn about their ancestral heritage, perhaps—decided those tests looked like fun, and got more than they expected. It could be that there's a fundamental fact about you, or about your family's past, that you don't yet know, and that will change the way you think about truth and family and who you are. The only thing standing between you and the discovery of this story is that DNA test.

If the hundreds of Americans I've spoken with are any guide, before you take such a test, you're unlikely to think there's any such disruptive fact in your family history. Even if you know there's a statistical possibility that you'll discover something surprising, you're unlikely to expect the statistics will apply to you. The problem is, it is impossible to say, in the absence of the inkling of any such surprise, whether or not you'll be better off for knowing this information. And once you know it, it's too late.

I came to think of the community of people I encountered researching this book as seekers. They are people obsessed with figuring out just what's in their genes. Not everyone who orders a direct-to-consumer DNA test qualifies as one, though a lot of folks who enter into this casually find themselves unexpectedly captivated by questions they never thought to ask, questions posed by the DNA itself. At-home DNA testing is sometimes called "recreational," to distinguish it from genetic tests that are ordered by doctors. But its implications can be far more profound than the term "recreational" implies.

The seekers generally fall into one of three categories. Some started out as avid genealogists. Before they ever ordered a DNA kit, they were already subscribers to the vast stores of genealogical records at Ancestry.com. Some had already flown out to the Mormon-run Family History Library in Salt Lake City, Utah, and spent a week there, camped out, up to their ears in old church records on microfilm. For these people, DNA testing through a company like Ancestry or 23andMe was a logical extension of this deep genealogical curiosity. Indeed, they probably tested in the early 2000s, when the tests available were much cruder and could tell you much less about your immediate forebears. Some of them, DNA junkies, have tested over the years at a variety of companies, including some that no longer exist, spitting and swabbing and scouring their results for meaning, in a grand history project of the self. These days, they canvass relatives willing to test, and return from family cruises with their kin's samples. They try to trace their family trees back hundreds of years, using DNA combined with genealogy to suss out the identity of a particular nineteenth-century ancestor or to figure out how exactly they're related to a distant cousin. If elderly relatives have difficulty generating enough saliva to spit into a vial, these seekers know the answer is a slice of lemon. A few, thwarted by the unexpected timing of events, have even resorted to swabbing the insides of their parents' cheeks after death.

The second category of seekers is marked by a more immediate and pressing puzzle—brought on, say, by lingering suspicions that the man they called "Dad" might not be their genetic father. Or they might be looking for evidence that they're not related to a family they never felt at home in. Some find their suspicions validated and others are disabused of them, and some find something else entirely, like a previously unknown half-sibling. Many seekers are adoptees looking for their biological relatives, or people conceived via sperm donor seeking the stranger who contributed half their genetic material. Finding genetic cousins in a DNA database can make it possible for a seeker to trace his way to his biological mother even if that mother never submitted her saliva for analysis. These seekers join Facebook groups dedicated to the mechanics of what's known as genetic genealogy, learning to talk of haplogroups and centimorgans and triangulation. They sketch countless family trees, trying to figure out where they might fit into them. They call their hobby an addiction. They read tips on how to phrase letters of introduction to their fathers, a topic so fraught and fascinating for what it reveals about the vagaries of human nature, and the complications of the human heart, that it deserves a dissertation or three.

The third category of people have no inkling they're about to find anything surprising, and this is often the most disruptive way in which DNA testing plays out. These people do not start out as seekers. Instead, they test because they want to find out where their forebears came from. They test to find out if that old family story of a Native American ancestor is true. They test because they want to know whether they're like that guy in the AncestryDNA commercial, who thought he was one thing and turned out to be another: "I traded in my lederhosen for a kilt!" he declares, happy in his certainty. They test because they got a kit as a gift for Christmas—and then Easter dinner becomes awkward, because the test has revealed something strange. Perhaps a stranger has appeared in the DNA database who might be an aunt or a niece or a grandparent or a half-sibling, based on the shared genetic material, and now everyone is asking questions. Perhaps a family's understanding of its heritage is muddied by a pie chart offering one's "ethnicity estimate," and the question "Why is it saying I'm half Greek?" becomes a curiosity, which becomes a nagging doubt, which becomes a family conversation no one will ever forget. Then, these people, too, become obsessives, trying to solve the mystery, finding possible relatives through Google and Facebook searches, and then offering to buy and ship DNA kits to those people if they'll agree to test.

Well, some of them do. Some of them, presumably, back away slowly, like a hiker who spies an old grenade from a long-ago war on a beach. Those people don't tell their stories. Who could blame them? Those people don't *have* stories; they have DNA results they'd rather not have seen, and which they may prefer not to believe.

The seekers filled my inbox and my ears over the last few years with their questions. What sense could they make of a secret kept for so many decades? How much of the past did they need to go back and footnote in light of this new knowledge? How many conversations were, in retrospect, lies of omission? If old taboos—about being the child of an extramarital relationship, or the child of adoption, or being conceived via sperm donor—were a thing of the past, why was it that the shame surrounding these origins felt so palpable even now? And how *now* to treat these strangers who'd come abruptly into their lives, these strangers with whom they shared genetic material? Are these people "family," and if not, do we need a new category to describe them? Is it always better to know these genetic relatives exist? Is it always better to know the truth? Should they invite the truth over for Thanksgiving dinner?

Secrets, we are all discovering, have a propulsive power all their own, and time and complicity only make them more powerful. Once you decide to keep a secret, the secret maintains a circular logic, even when circumstances change. Many seekers say the fact of the secret is the thing that nags at them, more than the nature of the secret itself.

DNA testing has brought the past forward to the present, forcing us to grapple with decisions made long ago in different, often desperate, circumstances. It forces us to think about the people whose truths have been hushed up for decades—the teenager consigned to a home for unwed mothers, the medical student who contributed his sperm, never dreaming that sperm would become a person knocking on his inbox five decades later. Some seekers are in their eighties and nineties, just discovering siblings they've spent their lives without. Some are left with only questions, and there is no one left to ask.

"I had no clue that I was adopted," says a seeker named Linda who discovered this blunt fact at the age of fifty-one, through a DNA test. When Linda figured out the identity of her biological mother using puzzle-solving techniques well-known to seekers, she learned the woman had just passed away, and that she would never have answers to certain things. "Did she ever think of me? Did she ever look for me?"

In researching this book, I toured the first American company to offer at-home DNA testing for ancestry, in Houston, Texas, and the biggest, in Lehi, Utah. I tested my own DNA at three companies, and looked for relatives, and learned the imprecisions of ethnicity predictions while comparing the results the various companies gave me. I spent a day at the biggest genealogical library in the world, where history was a breathing thing, alive in the records. Meanwhile, police began making end runs around the limitations of government-run DNA databases, instead using genetic material gathered by consumer testing services to solve horrific cases that had been cold for decades—and igniting fierce privacy debates. And the databases of people who'd undergone at-home DNA testing went from eight million by the summer of 2017 to nearly thirty million just two years later. Over the next year, those numbers continued to rise, albeit more slowly, and by 2020—on the twentieth anniversary of the creation of the industry—the four major companies had sold at least 35 million kits. The sheer girth of those numbers means that even if you don't choose to send away for a kit, it increasingly doesn't matter. Especially in the United States, where DNA testing is more popular than anywhere else, all of us are already drawn in by the decisions of other people who share our genetic material—people who, in many cases, we've never met. As bioethicist Thomas H. Murray told me, "You don't get to opt out."

This means you may become a statistic—someone who discovers a genetic fact that upends your sense of self, forces you to rethink what you know about race and religion, about your place in your family and your role in the world. Statistics are heartless that way. The more I reported for this book, the more seekers I talked to, the more I came to feel that we are embarking on a vast social experiment, the full implications of which we can't yet know.

This is the story of the seekers, who are grappling with essential questions about identity. What makes us who we are? It is the story of Alice, who did not know she would become one of them. That is, one of us.

1

ALICE IS NOT ALICE

The first thing Alice Collins Plebuch told me when she opened the door to her house was that I was tall. I am not tall.

Alice is very short. Five feet if she's lucky, she likes to say. She's in her early seventies, and her grandkids call her Grandma Nerd for her love of technology. Before we walked to her office-slash-sewing-room, she warned me not to take my shoes off, because sewing pins were scattered all over and I was likely to get them embedded in my feet. And then she led me back to her busy, treacherous, creative space, filled with bits of fabric and piles of books and genealogical research folders

Alice Collins Plebuch

arranged so delicately that if you had the bad sense to reach for an object that looked interesting, a whole stack of things might come tumbling to the floor, and Alice might or might not notice, and might or might not say *don't worry*, she'd get to it later.

She gave me a tour of her advanced sewing technology, so unlike standard sewing machines as to be an entirely different species. Atop one desk was something called THE Dream Machine 2 by Brother, a sewing, embroidery, quilting, and crafting machine so complex I'd been confounded each time Alice had tried to describe it to me over the phone. It was huge, with a ten-inch high-definition screen to choose an embroidery image and the capacity to scan and to plug in a memory stick so that you could add more. (It had been retailing for the cost of a Kia when she'd bought it a few years before, though Alice being Alice, she did not pay anywhere near that much.) Nearby were several other serious-looking contraptions, including something called a serger for executing what are known as overlock stitches, and something called an AccuQuilt, which I read is supposed to help you "reclaim your quilting joy."

Then Alice showed me the things she'd made on these machines, the beauty of which I could comprehend. By her desk was an early effort saluting a book series she devoured long before HBO made it into television: a direwolf from *Game of Thrones* with the phrase "Winter is coming." She showed me pictures of the fancy, full-body burgundy *Star Trek* costumes she'd made for herself, her eldest son, and his children, refashioning the complex rank braid each time her son told her it wasn't quite faithful to the on-screen version. On the back of a denim jacket she'd crafted an elaborate, plaintive-looking green-and-gold dragon, studded with red and silver crystals, its one sad eye outlined in shiny nuggets of gold. Just now she was working on removable pillow covers for the cushions of her couch, using a patterned turquoise fabric she'd found on deep discount. She had plans and measurements sketched out, and throwaway cloth for playing with, and as she showed me her plans, she improvised, folding the cloth this way and that, looking for the design that offered the right fit and the right give. For Alice, sewing is not a tribute to domesticity; she tends to resent domesticity, given the way she was raised, the eldest girl in a family of nine. Rather, sewing is nerdy, technical, challenging. Alice loves a challenge.

Over the years, I've thought a lot about Alice's brain—how for decades she honed certain skills involving technological efficiency and problem-solving, so that when the time came for her to answer the most important question of all, she dug in like she'd been training for it. When she was growing up, her family was not wealthy—no matter how many overtime shifts her dad worked, there were still seven kids to care for—so she put herself through college, washing dishes and tutoring kids in math and sewing her own clothes to save money. In time, she worked for a professor as a "keypuncher," putting information onto computer cards to be fed into a mainframe computer, and quickly demonstrated what a college friend describes as "a natural feel" for crunching data on those finicky machines. She got a degree in political science at the University of California, Riverside, specializing in statistical analysis, and nearly finished a master's degree in political science. In 1971, she married Bruce Plebuch, who studied philosophy in grad school and later became a lawyer. The same year, she got a job working as a clerk for the University of California, Santa Barbara.

There, Alice stumbled on a problem in need of solving. Faculty members' records were on card files, and a small staff spent an inordinate amount of time tracking and updating their salaries and academic advancements. Alice had the idea to put this information onto early computers. This set a pattern, and became the focus of her career in a series of positions in the field of information systems and data processing throughout the University of California system, where she worked with personnel files, payroll systems, and benefits enrollment, facets of working life that most of us notice only when they're clunky and broken. Early on, Alice was part of a rising cadre of women in a male-dominated environment. She asked her boss for a title and salary to match her responsibilities—"You know if you had a man doing this you'd be paying him a lot more"—and got the first of many promotions.

And she kept making suggestions for how things could be better. *It sure would be a lot nicer if we had these automated.* And, when the Internet

came along: *It sure would be cool if people could access their benefits online.*
All that data could be put to better use if only someone were willing to
tame it, and Alice was just the woman for the job. "I'm not intimidated
by lots of data," she told me. She'd always been good at math and science; as a kid, if a problem stumped her, she'd go to sleep and wake up
with the answer. She can see the big picture, but she also notices the
details. She told me it used to piss off one of her programmers, the way
she could scroll through acres of his work and glimpse the one error that
would mean a whole program wouldn't work. She could do it quickly,
by picking out the "nits."

Alice has been retired for many years. She stopped working in her
late fifties, the result of financial planning and good retirement benefits.
But she remained an early adopter of new technologies. In 2007, before
anyone had an iPhone, she flew to the Macworld Expo in San Francisco
to cradle the prototype in her arms. She had the first model the very
day it became available, and still does with every new iPhone. And when
the thing happened that set everything into motion and changed Alice's
life, this was also the result of her love of new technologies. Alice being
Alice, she could not leave a puzzle unsolved, no matter how difficult it
was or how troubling its implications.

She could not.

* * *

Alice had long had questions about her family. Her mother, who was
also named Alice, was into genealogy and kept an old family bible from
the 1840s with birth, death, and marriage notations that traced back her
English roots. Alice the elder had encouraged her daughter to embrace
her hobby. But sometimes daughters want to do the opposite of what
their mothers tell them, which is why Alice the younger refused to
indulge whatever nascent interest she had in the topic until after her
mom died in 1992.

Then, she dove in. Alice found her mother's line easy to document,
even in the years before lots of genealogical records were online. Her

mom, Alice Nisbet Collins, was descended from Irish people on one side, and on the other from Scottish and English people, some of whom had been on this side of the Atlantic as far back as colonial America. Alice was able to follow one line of her mother's ancestors back to 1500s England. Alice's father, Jim, lived for seven years after his wife died, and as Alice updated him on everything she was finding along her mother's line, she began to feel guilty. "Because my dad had nothing—he had no history," she says.

Alice's father, Jim Collins

Jim Collins, the son of Irish immigrants, knew little of his parents, one of whom he barely remembered and the other not at all. His mother had died when he was a baby, and his father had given him and his older siblings away to a Catholic orphanage. For a long time, Jim didn't even know what year he was born; sending away for some vitals as a young man, he'd discovered he was a year younger than he'd thought. So, in her father's waning years, Alice dove into the project of Jim. She had just one image of him as a child, taken in 1914, shortly before he and his siblings were sent away. In the photo, curly-haired Jim sits on his father's lap, clad in a white baby dress, with his tiny sister and brother standing on either side. He is the only one smiling.

The Collins children, Kitty, Jim, and John,
with their father, John Josef

Some things Alice already knew. She knew that life in the orphanage had been difficult. She knew her dad was probably malnourished there, because a doctor later told Jim that this likely explained his small stature. She knew that Jim had left the orphanage as a young teenager, and that he'd lived some rowdy years he liked to describe as his "misspent youth," before he joined the Civilian Conservation Corps and then the US Army Corps of Engineers, and met and married Alice's mother. She knew Jim had loved his sister, who died as a young woman, and had not been close to his brother.

Alice and her siblings sensed that Jim's Irishness and his Catholicism were important to him; they were what was left of his identity when so much else was taken away. He cooked a wicked corned beef on St. Patrick's Day, and liked to brag to one of his granddaughters about the time someone had described him as a "thoroughbred Irishman." Jim and Alice the elder had five boys and two girls, and the family went to church on Sundays, and much of the kids' schooling was at Catholic schools, and it was "important to him that the kids be raised Catholic," says Alice's sister, Gerardine Collins Wiggins, who goes by Gerry. But for all that, Jim's knowledge of his roots was shallow. He knew his dad was from County Cork. Before traveling to Ireland in 1990, Gerry had sent away for Jim's birth certificate, hoping it would help her trace his roots. But when she arrived in Ireland with the document and the few family names her dad was able to give her, the office she'd planned to consult for historical resources had closed down, and Gerry couldn't search the old records. Looking back, there were several moments like that: moments when questions were asked but answers were not forthcoming, when the paucity of information available to Jim fastened riddles into tight knots.

Here's another. Jim was not terribly knowledgeable about his extended family, and they seem not to have embraced him, either in childhood, when he might have needed them most, or later. But in adulthood, Jim tracked down an aunt while traveling in the Southwest, and went into the hotel she was running to inquire if he could visit with her. He could hear her informing someone that she was taking her nap and that her nephew would have to come back later. He left, hurt, and never met with his aunt, and any questions he might have had about the mysteries of his childhood and the parents he barely remembered and the extended family who didn't take in those three kids bound for an orphanage—those questions went unasked.

When Alice embarked on her genealogical quest on behalf of her father, she knew it would not be nearly as easy as it had been for her mother, but that was OK. Alice likes research. She sent away for Jim's parents' death certificates and headed over to the National Archives in San Francisco, where she was living at the time, to comb through microfilm, looking for her grandparents in the 1910 census. She was surprised by how little she knew about Jim. At one point, she realized she didn't even know his mother's proper first name.

She was able to glean just the barest outline of her father's parents. When Jim was born, his family lived in a working-class neighborhood in the Bronx. His father, John Josef Collins, was listed as a "driver" and later a longshoreman. His mother, Katie, died at thirty-two, when little Jim was just nine months old. When Alice got her grandmother Katie's death certificate, she strained to read the words by *cause of death*, which were written in the kind of loopy, inscrutable script that is the bane of every genealogist's existence. She thought it read, in part, "cranial softening" or "cerebral softening"; she showed it to a pathologist friend, who thought Katie might have had encephalitis. The last word, though, was absolutely clear. It said "insanity." How, Alice wondered, does a person die of insanity?

Before Jim died in 1999, just before his eighty-sixth birthday, Alice wasn't able to tell him much more than he already knew. And in the years that followed, she continued to be stymied by the paper trail. Searching census records, she managed to figure out the name of the orphanage in Sparkill, New York, where Jim had spent his childhood. It had been called St. Agnes Home and School, though the orphanage itself no longer existed by the time Alice found it. She wrote the religious sisters located there, asking for information on her dad and his parents, and got back a letter with the barest of facts: the date Jim and his siblings were admitted, and the varying dates when he and his siblings were discharged.

Before many Americans were aware of the way the Internet was transforming genealogical research, making it possible to find ancestors in old records from all over the world, Alice was already decades deep into this hobby and had subscribed to Ancestry.com to access its vast trove of historical records. And this is why she found out as soon as the company started offering an at-home DNA test in 2012.

These days, Ancestry's database is stocked with the genetic data of more than eighteen million customers, but back then, the kind of test the company was making available to subscribers—autosomal DNA testing—was relatively new and not widely known about. Alice's scientific curiosity was piqued. If the paper trail could not help her, she thought perhaps the genetic trail could. The timing was fortuitous, as Alice thought she'd homed in on the village her father's father had emigrated from, and was considering a trip to Ireland. She got on the company's waiting list for the test.

There are two ways that home DNA testing companies collect your genetic material. One method is to have the customer swab the inside of her cheeks, and another is to have her spit in a vial. AncestryDNA uses saliva. When the company's box arrived at Alice's house, she opened up the vial and spit and spit, about ten times, until the bubbles of saliva cleared a line on the side of the tube. She closed the top, which released a preserving liquid into her saliva, put the vial in a package, and sent it back to the company. And then she waited.

* * *

Before autosomal DNA testing came on the scene, two types of genetic testing dominated the consumer market for those interested in tracing their ancestral histories. One, called Y-DNA testing, examines the Y chromosome a man inherits from his father's father's father . . . all the way up what's known as the patrilineal line. Only men can take the test, since only men inherit a Y sex chromosome along with their X (women, of course, inherit two X's). The other test looks at mitochondrial DNA (mtDNA). Whereas our twenty-three pairs of chromosomes sit inside the nucleus of the cell, the tiny mitochondrion sits outside the nucleus with its own DNA, and both men and women inherit mitochondrial DNA from their mother's mother's mother . . . all the way up the matrilineal line. Both of these kinds of testing can tell a person about one particular genetic line, but not about their maternal or paternal side as a whole, nor about the genetic heritage they inherited from their other parent.

Because mitochondrial DNA and Y-DNA are inherited more or less intact, geneticists use mutations, which happen slowly over time, to trace historical migrations and track branches on our huge family tree of humanity. Y-DNA and mtDNA testing can tell us where some of our ancient ancestors lived, but typically can't give us precise information about our immediate family trees, nor can they offer predictions about a person's overall ethnicity. Although they can reveal that you and another person share a common ancestor, it can be difficult to tell how closely related you are to that other person. However, there are instances where these kinds of DNA can be helpful for genealogical purposes. Because the Y chromosome passes from father to son relatively unchanged, for instance, it can be used in combination with a surname, which also typically passes from father to son, to trace the genealogy of a male line.

As it happens, some of Alice's family had already begun dabbling in genetic testing. Her sister, Gerry, who was also interested in the Collins family's history, had recently had her mitochondrial DNA tested, showing her deep ancestry along her matrilineal line, and Gerry and Alice had convinced one of their brothers to test his Y chromosome, in hopes of getting a sense of their father's side. But the AncestryDNA test, when it became available, was a wholly different animal, offering much more recent and comprehensive genealogical information. AncestryDNA's test looked at the twenty-two pairs of chromosomes that don't determine a person's genetic sex, known as autosomes. It considered the genetic material contributed by both parents and by their parents before them. It could reveal recent genetic ethnicity and show relatives along both the maternal and paternal sides, assuming they were in the company's database. It promised, in other words, to help Alice unravel her father's genealogical knots.

Alice waited over a month to get her results back. Finally, she got a notification they were ready. She logged on—and was utterly perplexed.

She was just 48 percent British Isles, AncestryDNA informed her. She should have been more like 100 percent, given her dad's Irish ancestry and her mom's Irish, English, and Scottish mix. Instead, the test results suggested that the other half of Alice's ethnic makeup was a mix of what the company called "European Jewish," "Persian/Turkish/

Caucasus," "Eastern European," and "uncertain." Alice emailed her sister in Minnesota, characterizing these findings as "kind of surprising." Her tone was moderate, although she did not feel moderate. She felt confused and, as the days went on, increasingly annoyed at what seemed like flaws in this new technology. "Keep in mind," said a little note at the top of her AncestryDNA estimate, "your genetic ethnicity results may be updated. As more DNA samples are gathered and more data is analyzed, we expect our ethnicity predictions to become more accurate." *More accurate?* Did that mean the estimate was mistaken? Alice tried to imagine how it could possibly be correct. Perhaps these Eastern European results indicated ancestry going back tens of thousands of years?

"I know we pick up ancient stuff," Alice wrote her sister, but if so, why was it that so many of the modern-day customers who showed up as predicted genetic relatives to her in AncestryDNA's database shared this European Jewish ancestry? "Since we know the ancestry that far back on Mom's side, does that mean it comes from Dad? Or is Mom's side not what it appears?"

This was a new puzzle, and Alice threw herself into it. She couldn't be sure whether the mistake was in the company's test or in the story she'd long believed about herself, and while she was inclined to dismiss the test as flawed, not knowing made her uneasy. The closest matches that showed up within Ancestry's database were third and fourth cousins, and she pored over their names and realized she had no idea who any of them were. She looked at her relatives' family trees, hunting for a common ancestor who might explain how her family hooked up with theirs. She got nowhere.

Ancestry had message boards, and Alice began to lurk where she saw people complaining. A lot of testers were saying their AncestryDNA results seemed to overestimate their Scandinavian heritage—results that must surely be wrong, testers wrote, because they had no paper trail, no family stories, no nothing to indicate this much Scandinavian. This was simultaneously reassuring and discomfiting to Alice: On the one hand, it seemed to indicate that the company was capable of making mistakes. On the other hand, no other customers were complaining about her precise situation: a false positive result for "European Jewish," "Persian/Turkish/Caucasus," and "Eastern European."

There was a woman on the Ancestry message boards named CeCe Moore who seemed to know more than everyone else and was answering questions. Alice googled her, found she helped run a Yahoo group called DNA-NEWBIE, and began to follow it. At first, what Alice read was above her head, but she watched and learned, observing how adoptees used their DNA results to figure out their biological families, and following links to videos showing how DNA gets passed down through the generations, refreshing what she'd learned decades earlier during a temporary stint as a biology major.

The bulk of our DNA sits in our twenty-three pairs of chromosomes, which carry the blueprints for our existence, about half inherited from each parent. Our parents, in turn, have each inherited from their parents, so Alice could expect to get about 25 percent of her genetic material from each grandparent, on average, though because of the way chromosomes are randomly inherited and because DNA is shuffled through a process called recombination, the exact percentage varies. The amount of autosomal DNA coming from each ancestor would get progressively smaller each generation back, with some ancestors failing to transmit DNA altogether as they became more distant.

The funny thing is, about 99.5 percent of our DNA is identical to that of every other human being on the planet. It's only in that remaining 0.5 percent that we get the distinctions in eye color, skin color, height, and other things we can see, as well as plenty of things we can't, and it's only a portion of that 0.5 percent that companies like Ancestry are looking at when they examine our DNA sequences. Four chemical "bases" form the building blocks of our DNA—adenine (A), cytosine (C), guanine (G), and thymine (T)—and they are arranged into pairs along the double-stranded helix structure of our DNA. (A and T pair, and C and G pair.) Testing works by reading these letters. But Ancestry wasn't looking at all of them, since whole genome sequencing involves looking at all three billion or so of a person's complementary "base pairs," and was too expensive for the direct-to-consumer market. Instead, using an approach called genotyping, Ancestry was looking at about seven hundred thousand locations along the genome—called single nucleotide polymorphisms, or SNPs (pronounced "snips")—where the bases are known to vary among human beings.

Alice's crash course in DNA taught her a lot, but it did not tell her how accurate Ancestry's test was. The Collins kids are spread out all over the country, but it so happened that within days of Alice getting her results, her brothers Jim and Brian were both in the Vancouver area. When they visited, Alice told each in turn about her findings. *Crazy*, thought Jim, dismissing the results. *Interesting but impossible*, thought Brian, dismissing the results.

But Gerry's reaction was different. "Whoa!" she'd written back to Alice's email, in huge letters, intrigued by the mystery. The sisters started discussing what to do next, and in between, they emailed about the trip to Ireland that Alice was thinking of taking. Gerry offered Alice advice from her 1990 trip, explaining the hazards of driving on the other side of the road, and telling Alice not to be surprised that most Irish people she'd encounter would be taller than their family.

About a week later, the Y-DNA results came back for their brother Jim (we'll call him Jim the younger), showing his ancient lineage along his father's father's father's line. It listed his paternal haplogroup (a genetic population sharing a common ancestor) and his subclade (a smaller group within that haplogroup). They seemed to suggest the Ancestry results might be much ado about nothing. "Today, roughly 70 percent of the men in southern England" belong to that haplogroup and subclade, the explanation from the National Geographic Society's Genographic Project read. "In parts of Spain and Ireland, that number exceeds 90 percent." This made sense to Alice—this test was tracing her father's line, after all, and everyone knew Jim Collins was Irish.

The nuances of this finding were not yet clear to the Collinses. For a host of reasons, a finding like this wasn't necessarily inconsistent with Alice's AncestryDNA results, assuming they were correct. For one thing, the Y-DNA test did not track anything from the Collinses' maternal side, so if the mystery lay with their mother, this test did nothing to address it. And the particular results Jim the younger had received actually couldn't tell him definitively whether or not he had Jewish heritage on his father's side. But in any case, Alice's brother started calling himself "Irish Jim." Back and forth the siblings went, discussing their assorted DNA results and what they all meant. Gerry pointed to her mitochondrial results, explaining that her maternal haplogroup appeared to be of Western European lineage. "I see it as a fun mystery to be solved and nothing to be worried about," Alice wrote her siblings about the whole affair. The AncestryDNA test was still in beta testing, and this made Alice especially dubious of its accuracy. She found a feedback form and sent the company what she'd later describe as a "nasty" note: Her genealogy, family stories, and life experience all indicated she was Irish, Scottish, and English, she told them, and Ancestry's science clearly wasn't up to the task yet.

In some sense, Alice was right that the ethnicity estimates of the major testing companies are flawed, in ways both practical and existential. They are estimates for a reason, based on data and techniques that are imperfect and evolving. They are flawed in another sense, because there are big questions around what precisely "ethnicity" is and whether we can ever find such a thing in a person's DNA sequence. Those were big issues, and in time, Alice would grapple with them.

But for the time being, despite her nasty note to Ancestry, she still had a queasy feeling that she might be missing something. This should go without saying, but the notion of being ancestrally Jewish was not what disquieted Alice; indeed, she found this idea intriguing, except for the implication that she didn't know important aspects of her family's history. She had traced her genealogy along both sides as exhaustively as she could. She knew—or at least thought she knew—who her parents were. None of it made sense. She couldn't drop the matter without investigating, though perhaps some people might have. She and Gerry began to speculate about what could explain the test's finding, on the off chance it was correct.

Alice knew she'd chase this question till she had an answer, but she had no way of knowing how this would become the thing she did all day, every day, taking over her retirement, teaching her an expertise in a kind of genetic detective work that a growing number of Americans are learning. She had no way of knowing how difficult it would be to solve this mystery, or how very long it would take, and how many people she'd pull into the effort—family and strangers and genetic relatives she'd never known before. She had no way of knowing that she'd be joining a community of seekers, each one of them with questions of his or her own, and with stories about the strangeness of the human condition at this, the very dawn of the consumer genomics age.

She could not know then, in the summer of 2012, that in coming days she would start to rethink that trip to Ireland. She could not know that eventually, she'd be moved to email Ancestry's feedback team with an addendum to her earlier note: an apology.

2

CRUDE BEGINNINGS

Roberta Estes got into genealogy in the late 1970s, when she was pregnant with her first child. There were a lot of unanswered questions about Roberta's father's side of the family, just as for Alice's. Back then, genealogy was as fast as polishing rocks. You couldn't sit at your computer, log onto Ancestry.com, and figure out your great-great-grandfather's name in a few hours. Oh, no. You asked family members for their stories. You wrote letters, actual letters sent through the mail. You traveled to small towns and chatted with clerks and went to libraries to scroll through microfilm till your eyes were red. Genealogy was for the few, the dedicated, the serious family historians. It required leisure time and, ideally, enough money to travel. It was not for the idly curious.

"The slowness of the process was part of it," Estes told me the first time she talked to me from her home in a rural part of Michigan outside Ann Arbor. She had long silver hair and was wearing a T-shirt that said, "I'm a genealogy rebel—I research past my bedtime."

It can be hard to grasp how profoundly the world of genealogy has changed from then to now, with the advent of the Internet, social media, and consumer DNA testing. The bar to entry is so much lower than it was when Roberta Estes started. Since the first genetic genealogy tests became available in 2000, DNA testing has become much cheaper and much better at finding close relative matches. Now, for ninety-nine dollars (and less during flash sales), the idly curious can stumble into genealogy; within weeks of submitting saliva they'll have one of those ethnicity estimates like the one Alice got, as well as a list of perhaps thousands of relatives who match them in a company's database. They can quickly find out more about those relatives through family trees, Facebook, Google, LinkedIn, and people-search websites. If they want to learn more about their ancestors, they need not write letters and travel to libraries but can see billions of the world's genealogical records instantly via a subscription at Ancestry and through the free resources at the Mormon-run FamilySearch.org. And they can join groups, mainly on Facebook, for all the flavors of genetic genealogy seekers: adoptees seeking their biological families; African Americans and people of Jewish descent, who face unique roadblocks when they search; those who've discovered that the men they thought of as their fathers aren't their fathers as far as biology is concerned; those conceived by donor sperm. There are groups for those whose ancestors come from County Donegal in Ireland, and for Asians adopted into the United States, and for Americans descended from Germans who lived in Russia.

Which means that whereas in the past, you had to work years, even decades, to solve a mystery—and Roberta eventually found she had one, too—nowadays, the answer can often be found within days.

America has become a nation obsessed with genealogy. The mere existence of so many genealogical materials digitized, indexed, and searchable online, and our communal drive to find them, comes from a suite of personal and cultural motivations, as well as a complex history around the search for lineage. In his 2013 history of American genealogy, *Family Trees*, historian François Weil traces how the American impulse toward genealogy has often been in tension with itself. In the early days of the new American republic, Weil writes, the idea of establishing one's family line was associated with the British aristocracy's obsession with social rank, and viewed with suspicion by a society that saw itself as more egalitarian and forward-looking. Why would one be driven to document one's ancestors, if not to prove some connection to better birth and station?

But over the course of the nineteenth century, that shifted, enough that by 1879 the *New York Times* could declare that "we are becoming the most genealogical nation on the face of the earth." Weil writes that American genealogy transformed into a respectable middle-class endeavor as Americans began to justify and sanctify the activity within the context of family, which came to be viewed as an almost holy thing. The family "was viewed as a refuge from the outside world in an ever-changing environment," Weil writes, and genealogy became a mechanism for remembering and solidifying that unit. Besides, some Americans came to see the process of learning one's family history as a moral endeavor—a person could learn much from what her ancestors had done right or wrong. Reframed within the context of republicanism and democratic ideals, genealogical inquiry could become the means to celebrate not just the richest and most titled of forebears, but even the humbler sort. One 1850s Pennsylvanian went so far as to boast of his family's "mediocrity." The practice of keeping one's family history in a household bible had long been popular; now, middle-class New England families augmented those bibles with wall hangings of family registers and embroidered family trees.

Yet for some, genealogy was also a means of forging identity by relating to some people and excluding others. Even before the Civil War, there was "lineage consciousness" among those descended from elite colonial families, who used their descent from "high" birth to justify and enforce their higher social rank. (Some nouveau riche bought fake family pedigrees and had family crests painted on their carriages.) Americans also became intrigued by the idea that the country boasted different "stocks"—there were the Puritans, the Southern Americans, the New Yorkers. And in the wake of the war, Weil writes, genealogy became a vehicle for many whites to bind "ancestor worship, nationalism, and racism" into one big ugly package. By the early 1900s, these notions were well entrenched, so that presidents general of the Daughters of the American Revolution could talk of the importance of the country's Anglo-Saxon roots, of the "purity of our Caucasian blood," and the importance of "the white man's standard of living."

Of course, there were other forces at play in this transformation, including some key scientific breakthroughs that would form the basis for the field of genetics—as well as a theory that, in attempting to explain the vast social, racial, and economic divisions that ran through society, served to reinforce them.

* * *

Gregor Mendel died in obscurity. The priest and naturalist whose work helped us understand genetic inheritance was only recognized well after he died. Born in Moravia (now part of the Czech Republic) in 1822, Mendel made his breakthrough in the 1850s and '60s, breeding garden peas in the monastery garden at Brno. In a series of innovative experiments, he examined things like height, seed shape, and pod color, trying to track the traits of plant hybrids through the generations. Years of study led him to develop a theory of dominant and recessive traits that explained how things like wrinkled seeds can seem to disappear for a generation or more, and then reemerge intact.

Mendel's intention was not to establish laws of heredity, nor to anticipate the concept of the gene, but when his 1866 paper on those pea experiments was rediscovered in 1900, he came to be seen as a prophet because of all that had happened in the sciences in the decades in between. In 1859, Charles Darwin had published *On the Origin of Species*, describing his astonishing theory of evolution by natural selection. *Origin of Species* explained how selection drives change by way of inherited variations within populations, helping organisms with traits better suited to their environments to survive and pass their adaptations on to their young. The book was seen as intriguing, radical, implausible; copies flew off the shelves. Although Darwin's ideas weren't immediately accepted whole cloth by his peers, they sparked larger conversations about the concept of evolution. There were also lively debates about the mechanism through which heredity might work, though none of those theories—including Darwin's, involving something called "gemmules"—would prove right.

Origin of Species had a big influence on Darwin's cousin Francis Galton, an explorer, statistician, and all-around polymath who was fascinated by the topic of heredity. Galton became deeply interested in studying the relative contributions of what he would later coin "nature and nurture," helping frame a debate that has fascinated generations of psychologists (and would, over a century later, entrance a six-year-old named Anne Wojcicki, who grew up to co-found a company called 23andMe). Galton was a data guy—known to methodically count yawns during lectures—and he set about quantifying what made human beings what they were, pioneering methods that would become influential in human genetics research. He had the innovative idea of using twin studies to examine the influences of nature versus nurture. He used pedigree charts to try to study how traits ran in families. He set up an "Anthropometric Laboratory" at a major exhibition, charged visitors three pence for admission, and collected information on all sorts of things about them, like "keenness of sight and of hearing," "force of blow," and "breathing power." He developed a system for classifying fingerprints and studied photographs of criminals, hoping their facial features would illuminate their characters.

Ten years after Darwin published *Origin of Species*, Galton published *Hereditary Genius*, attempting to demonstrate through the genealogical study of eminent men that character, intellect, and ability ran in families. (He dismissed the idea that these judges, statesmen, and commanders might have benefited in a big way from social and economic advantages.) And he offered his theory of how to make humanity better: Just as it was possible to breed horses and dogs for speed, people could be bred through a series of "judicious marriages." Later, he coined the word "eugenics" to describe what he called the "science of improving stock . . . to give the more suitable races or strains of blood a better chance of prevailing speedily over the less suitable."

All of this—as well as a better understanding of cells and new thinking about the nature of heredity—created a new context for the rediscovery of Mendel. His paper on pea experiments, seen as proof of the existence of paired hereditary particles, set the stage for more research establishing the classical concept of the gene as a material unit in the chromosome. Of course, Mendelism and the early era of genetics tell only the beginnings of what we understand about heredity today; the rise of molecular genetics and a century of research were to follow. Mendel's pea plants could not account for the full complexity of inheritance patterns—for the role of environment, or the way that multiple genes play parts in influencing most traits and diseases. Scholars like Gregory Radick, a historian of science, have pointed out that the emphasis on the work of Gregor Mendel in contemporary science curriculums may encourage belief in "the primacy of the gene" at the expense of a more accurate modern understanding of the complex roles that genes play in shaping traits. Such simplistic thinking can contribute to genetic determinism, a sense that genes are fate.

We know this danger because we've seen it before. The rediscovery of Mendel's laws coincided with political and social movements advocating this kind of genetic fatalism—movements influenced by Galton, whose work lent itself to a hereditarian worldview. In the early 1900s, the theory of eugenics, coupled with the principles of Mendelian genetics and the reform-minded zealotry of the Progressive Era, became a frightful cultural force, the basis for a supposed biological justification for racism and discrimination that spread through England, America, and much of Europe, and still echoes in far-right corners of the Internet today. At the First International Eugenics Congress in 1912, lecturers showed pedigrees appearing to document families doomed to imbecility and lunacy. Germany, whose scientists referred to eugenics as "race hygiene," was ominously well represented at the exhibition. In the United States, Indiana enacted the first compulsory sterilization law in 1907, with more than thirty states passing similar laws over the coming decades.

To modern eyes, eugenic efforts can appear by turns horrifying and comically unscientific. The concept of "eugenic love," which promised to scientifically modernize marriage by encouraging what one editorial called "love in the right direction," became popular among women's rights reformers. At state fairs across the United States, Better Babies Contests emerged, where babies could be examined and tested, with points taken away for defects like scaly skin, delayed teething, and slow reactions. Meanwhile, psychologists administered IQ tests to US Army recruits and concluded that "race" was linked to intelligence; conveniently, the racial hierarchy they discovered perfectly conformed to preexisting prejudices. Eugenics influenced the Immigration Act of 1924, which restricted the numbers of people coming from "socially inadequate" areas like eastern and southern Europe.

In other words, eugenics was no mere fringe movement; its assumptions shaped what both scientists and reformers thought they were seeing. There was panic about the degradation of the human race by the unfit. "It is difficult to find many early-twentieth-century American biologists who were not advocates of eugenics in some form," observes historian Mark Largent. In 1927, in *Buck v. Bell*, the US Supreme Court upheld a state's right to forcibly sterilize a woman after experts testified that not only was she "feebleminded," but her mother and infant daughter were, too. "Three generations of imbeciles are enough," Justice Oliver Wendell Holmes Jr. famously wrote.

In Germany, where the Nazis explicitly borrowed from American eugenicists, the movement reached its terrible apogee, with forced sterilizations, marriage restrictions, euthanasia of the mentally diseased and disabled, and, finally, genocide. The Supreme Court's decision in *Buck v. Bell* was actually invoked at the Nuremberg trials in defense of Nazi sterilization experiments. German atrocities provoked revulsion in America, but even so, Largent argues, it took decades for the movement to fade, and legislation to limit immigration—justified by eugenics—persisted. Involuntary sterilizations also continued, about sixty thousand between 1907 and the 1970s, when the laws authorizing them began to be repealed by states. Yet as these ended, historian Alexandra Minna Stern told me, a different set of circumstances enabled another wave of unwanted sterilizations affecting women of color and poor women. Reporting and a state investigation revealed that California was unlawfully sterilizing female prisoners as late as 2010.

Historian Nathaniel Comfort argues that eugenics is not an aberration of history. He hears its echoes in the promise of medical genetics, in thinking that frames "heredity as the essence of human nature," in our impulse to perfection via the genes. The legacy of eugenics is just part of the much larger story of how genetics has helped us understand where we came from and how we're made, but this legacy matters when we look at all the questions that ancestry testing and genetic genealogy bring to the fore: How do we define what we are, and what role should genetics get to play in that definition? How much of what we read in our genes and in our lineages is a product of the very specific culture we swim in? Today, some scientists debate whether "race" as a proxy for genetic variation and difference is useful, if imperfect—or so flawed that it's not a worthwhile scientific concept. (We'll return to this later.)

We look for ourselves in our family histories and in our genes, but such things alone do not make identity. We human beings are the meaning-makers, each of us a product of a particular time and place, with ideas about what we value and, indeed, what we hope to find when we look.

* * *

I went to look, too. Partly, I was interested in how genealogy informs how we think about ourselves—and what it is that made people like Roberta Estes and Alice Collins Plebuch seekers so many years back, before there was such a thing as DNA testing or online census records, when genealogy required patience and spare time. I went where seekers go. I went to Salt Lake City, to the Church of Jesus Christ of Latter-day Saints, also known as the Mormon Church.

"The greatest responsibility in this world that God has laid upon us is to seek after our own dead," Joseph Smith, the founder of the Latter-day Saint movement, said shortly before his own death in 1844. The charismatic figure experienced his first religious vision at the age of fourteen on his family's farm in western New York, and founded a movement that eventually moved to Utah. Smith was aware that the young religion had a problem when it came to the souls of loved ones who'd died before it was established: How could these people be saved if they'd never had the chance to accept the restored gospel?

As recounted in the LDS Church's history of its genealogical efforts, *Hearts Turned to the Fathers*, Smith had a vision in which the Lord told him that death was no barrier to salvation for the soul of a person who would have received the Gospel when alive. After that, he preached the doctrine of baptism for the dead. In fact, the Church came to believe a series of practices were necessary for salvation, and that it was the sacred duty of church members to perform them on behalf of their ancestors. These included not only baptism by proxy but also the enactment of temple "sealing" ceremonies that create eternal bonds within families. It matters deeply to members of the Church of Jesus Christ of Latter-day Saints who their ancestors are because this knowledge allows them to carry out a grave responsibility: to save souls and ensure the sanctity of the family unit in the afterlife. At times, the sense of mission among members have of the LDS Church has gotten out of hand, as when members have submitted the names of Jews, including Holocaust victims, for posthumous baptism.

That genealogy is not a mere hobby for the Church helps explain the enormous resources it has poured into the gathering and dissemination of old family records. In 1894, the Church created the Genealogical Society of Utah, which is now known as FamilySearch International, the Church's nonprofit genealogical arm. By the late 1930s, the Church had begun microfilming archival records all over the world, and in 1965, it completed a storage facility called the Granite Mountain Records Vault, meant to protect all those records from decay and disaster under more than seven hundred feet of stone. Weil writes that it was also around this time, during the latter half of the twentieth century, that genealogy in general began to shed some of its racism and nativism, helped along by the civil rights movement and by Alex Haley's novel and miniseries *Roots*. The pursuit of family history became a means for a wider, multicultural swath of Americans to understand who they were and how they got here. In more recent decades, the personal computer and the Internet made family history research easier and far more accessible to the average American, as well as an increasingly big business.

FamilySearch, though, is not a business. It is a massive project free to everyone, dedicated to the idea that we're all better off if we know our ancestors. Across the world, it maintains more than five thousand family history centers, what journalist Christine Kenneally calls a "sacred municipal library system." It also runs the website FamilySearch.org, where you can search more than five billion records.

But there are some things you can find only if you come to the Church's Family History Library, which is the biggest genealogical research facility in the world. The place is astonishing. Inside, different floors house regions of the world. There are records from more than 120 countries, in more than 170 languages, including the largest collection of Chinese genealogies outside mainland China. There are hand-drawn pedigree charts from the Polynesian state of Tonga, and old and historical books with names like *Rural Cemeteries, New Canaan, Connecticut.* There is special seating for the deaf and people with disabilities, and brochures in fifteen languages, and rows and rows of microfilm readers. Over the course of a week, the library hosts as many as four hundred volunteers, in addition to paid staff. Some are retired and on multi-year missions, having paid their own way from their homes to live nearby and help out here. One library researcher is an expert in old-world synagogue records, while a volunteer expert is skilled in deciphering Gothic German script, a disappearing wisdom valuable to visitors who show up with their family secrets encoded in old German church documents. And what's here is just a fraction of the LDS Church's effort to preserve the past. FamilySearch has teams in more than forty-five countries working on digitizing records, though making them searchable is a whole other ball of wax, requiring the help of hundreds of thousands of people to read through old documents, decipher them, and type in salient details like name, date of emigration, date of death.

Every morning, seekers line up outside before the Family History Library opens at 8 A.M. They come with dolly carts and heavy bags filled with family photographs and document files. Paul Nauta, the library's PR director, calls these the homesteaders, for the way they set up camp at particular computers and squat all day. They may be hobbyists or pros; they may travel as groups of genealogical societies, the better to swap stories and resources. They may come from far away—Canada, France, England, New Zealand, all over the United States—and park at the library every day for a full week. Sometimes, people planning to do just a little research stay far longer than they meant to, as if they fall into some kind of wormhole that alters time. This place can do that to you.

The other major force on America's genealogy scene is Ancestry, located just half an hour away from the Family History Library. Ancestry makes money offering subscription access to old records, many of which not even FamilySearch has. Even though these two organizations are motivated by different things—the one pursues profit, whereas the other sees its work as a godly endeavor—they often team up, though they have no financial relationship. More and more, FamilySearch has been working with commercial genealogical partners like Ancestry, as well as Israeli-based MyHeritage and UK-based Findmypast, to share resources and data "because even collectively, we can't do it all," Nauta told me.

This revolution of information about the dead is breathtaking and makes you think about all the ways we're giving up our privacy not only in this life but in the hereafter. Nauta told me he was bullish on the amount of information we citizens of the Internet share online—he figured it would make the lives of future genealogists so much easier.

"Can you imagine having just a week's worth of Pinterest posts from your grandma?" he asked, and I shuddered inwardly, thinking of all the things I'd posted on social media over the years. But the concept of privacy has radically changed for all of us, even over the last decade, from what we share online to how faithfully our movements and shopping habits are tracked by corporations, to, yes, the very secrets of the genetic material curled into chromosomes in the center of our cells. Why should the privacy of the dead be any different?

Members of the LDS Church I spoke with at the library had a profound sense of connectedness to the past. The nineteenth-century notion of knowing one's ancestors as a moral endeavor was alive in them. Nauta told me that DNA testing had lowered the age of people interested in family research, and brought in adherents through a different door—they tested first and then became interested in genealogy. That was the case for me. I'd tested at 23andMe in 2017, after my dad gave me a kit for the holidays, as so many Americans do these days. The test made the past accessible in a way it never had seemed before, even if I found the results only mildly interesting: I wondered how I was related to a marvelous fantasy and sci-fi writer who showed up among my relative matches, and pondered the lost ancestor who must explain my 1 percent Korean. I was at once disappointed and relieved not to find any big surprises in my results. By then, I'd begun to talk to other testers, and to understand that surprises can cut both ways, and that boring results can be a blessing.

When I met with Nauta, I brought a piece of paper with the names of my eight great-grandparents. He practically rubbed his hands in excitement over my paternal side—"We have the most comprehensive records for Sweden of anywhere outside the archives of the country itself!"—but I told him what I really wanted was help finding the paper trail of my mom's mom's mom, Sare Kahn. My mom is descended from turn-of-the-nineteenth-century Eastern European Jewish immigrants, whose documents are far more difficult to access than those of many of my dad's great-grandparents, who include Brits and Swedes who benefited from better record-keeping. Plus, my mom's family alternated between Hebrew, Yiddish, and English versions of their names, and when they came over, many said they were from Russia, a vast area that on a modern map might be a number of different countries. The problem of using a paper trail to find your ancestors is that it favors the side less affected by war, famine, dislocation, and persecution—in other words, the people who were already privileged by history are privileged again by the records, as African American genealogists know well. In some ways, then, DNA testing is democratizing for modern-day genealogists. It can give testers access to new cousins whose family stories and family trees can provide an end run around those historical gaps.

But if there were records to be found, I knew, this was the place to find them. A library manager showed me around the Family-Search resources, and I planned to do an hour's worth of work before driving back to my Airbnb. Yet a funny thing happened after I was seated at a computer, with access to all those records right at my fingertips. Time seemed to collapse. Three volunteers guided me through all the different resources, teaching me how to follow Sare through the records like a hound following the scent of a hare. They saw patterns

I didn't, skipping over records I would have wasted time on to zero in on ones that at first seemed unpromising. They were patient. One studied the hundred-year-old scrawl of a ship's manifest until she could decipher a "J" that looked like no "J" I'd ever seen. Sometimes my great-grandmother was Sare and sometimes Sara and sometimes Sarah, and sometimes Kahn and sometimes Cohen, until she married and became a Jaffe. I found images of her marriage and death certificates, giving us for the first time her parents' names and her husband's parents' names, and I found a record of the birth of her first child, a sibling of my grandmother's who must have died young because my mom never knew that child existed. I thought I had a sense of what it must have been like for Alice and Roberta in the early days of their genealogy hobbies. I could understand what hooked them—the tantalizing hope that somewhere in the next file was a record that could explain who their ancestors were, what their lives were like before and after arriving in America.

I was so engrossed I didn't care when dinnertime passed, and at 8 P.M. I looked up in a panic and realized the garage I'd parked in might have closed. I gathered my things and rushed to the lobby, where a friendly volunteer directed me to the exit. It had been five and a half hours.

A wormhole indeed. This is how seekers are made.

* * *

The questions started for Roberta Estes when she was eighteen months old, though she wouldn't know it till later. That's when her father was in a car accident, and the hospital called his wife, and then the hospital called his other wife. Both wives showed up at the hospital with babies, which is how they found out about each other. Estes's mother dumped her father.

Roberta did not unravel the full story of her father's other family till she was an adult, and the questions sent her digging. It was well before the DNA age, so all she had was word and paper: her reticent mother's accounts, interviews with family, troves of family letters. She wanted to find her half-brother, the little boy she'd crossed paths with in the hospital when both of them were too young to know what was happening. His name, she discovered, was David Estes. She tried city directories, obits, marriage records, and divorce records and eventually consulted a private investigator, who found a police record for her half-brother and, through that, located the man's address. Roberta wrote him a letter.

David Estes turned out to be a long-haired, tattooed long-haul truck driver, and Roberta was struck both by his rough-around-the-edges demeanor and by his loyalty to those he loved and those who needed his love. The first time they spoke by phone, his deep voice was everything to her, and at first, she could barely speak. They became close. Dave was the kind of guy who rescued dogs that had been abandoned at truck stops, and once saved a Rottweiler puppy by slugging the guy who was abusing it (hence, that police record). He lived in Ohio, about three hours from Roberta, but he would take the worst runs Idaho potato runs, which required traveling through the mountains—just because they took him near her house. Roberta loved it when David called her "Sis." She was his only biological family aside from his children. He did not talk about feelings easily, which made it all the more meaningful when he told her he loved her.

Over the years, Estes gained a lot from genealogy. She learned she was descended from a white man who owned a slave named Harriett and fathered children with her, and she found descendants of Harriett, distant cousins with whom she grew close. She learned that, through her paternal side, she had ancestors from sub-Saharan Africa. And she learned a lot about her father that was difficult to reconcile: how he drank, how he squandered his intelligence, and how he deceived two families. As a child, Roberta had idolized him, much to her mother's annoyance, and as an adult, she had to honestly look at the whole man. But genealogical research also allowed her to peel back the context for his life: The son of an impoverished bootlegger who was fed alcohol when he was hungry, he went into the hospital at least twice in an attempt to stop drinking. She wonders whether his pain may have driven him to take his own life in a second car accident, when Roberta was seven.

"Both David and I unquestionably knew that man loved us," she told me. "I have really good memories with my father, and so did David. There is just no black and white; there's just a really big range of gray."

Genealogy was an incredible gift to Roberta Estes. It solved mysteries; it gave her empathy and understanding for life's grays. And most of all, genealogy gave Roberta her brother David.

3

SOMEBODY OUGHT TO START A BUSINESS

In the summer of 2012, not long after she got her strange DNA results, Alice and her sister Gerry came up with a plan. If the Ancestry results were wrong, the surest way to disprove them was to test through another company. Alice researched and learned about 23andMe, a company that offers DNA testing directly to consumers, with a focus on genetic diseases, traits, and ancestry. She sent away for a spit kit.

Gerry decided to test at 23andMe, too. She was curious about her own results, and besides, if she matched Alice, that would eliminate some possibilities that Alice needed to consider—that Alice had been adopted or that Jim Collins wasn't, biologically speaking, Alice's father. These were contenders for an explanation, even if Gerry told Alice she was insane when Alice suggested their mother could have had an affair.

Their mom had had "eight live births in ten years," Gerry pointed out—there are seven living Collins kids, plus little Nancy Jane, who would have made three girls in the family if she'd survived past the age of two months. Gerry tried to imagine her mother carrying on something torrid in between being pregnant, giving birth, and caring for the endless stream of infants. "There was no time."

In truth, Alice thought the scenario unlikely, too. For one thing, she thought, she looked like her siblings. All the Collins kids seemed to have their father's small, hooded eye shape, and plus, all were short, like both of their parents. (They are a family for whom each inch counts; once, when Gerry declared herself five foot one and a half, Alice suggested she was using her hair to gain a height advantage.) For another, Alice could not imagine her difficult and rather unglamorous mother ensconced in an affair. When she mentioned the theory to her husband, a taciturn man with a dry sense of humor, Bruce just laughed and laughed. But the possibility could not be ignored. Jim Collins served in the army for more than twenty years before retiring at the rank of chief warrant officer 2, in 1957. He served at West Point and in Guam, Virginia, Germany, and Texas. For the most part his family traveled with him, but still, there were times when it was possible, at least in theory, that her mother could have been with somebody else.

Besides, Alice knew, the simplest explanation is the most likely one. She was noticing, when she shared her strange results with people she knew, that they tended to ask if she was certain her father was biologically related to her. It was "the most probable likelihood," her best friend, Kathy Long, observed. Alice's methodical mind could not abandon considering such an obvious solution, if only to prove it wrong. Same for the adoption theory, which seemed even more far-fetched but was still within what Alice considered "statistical possibility."

If the Ancestry results were correct, there were other theories, too—enough theories, in coming days and weeks, to fill a season's worth of *The Young and the Restless*. Some, more far-fetched, were nonetheless more palatable than the idea that Alice the elder had hidden that Alice the younger had come into the world by way of an affair or an adoption. For example, Gerry wondered whether their father's mother, Katie Collins, had had an affair. She studied pictures of her paternal grandfather, John Josef, whose picture hung on her wall, and thought the man looked nothing like their family. In truth, she'd long thought that. But if grandmother Katie Collins *had* had an affair, the mysterious lover should account for only about a quarter of Alice's ethnic makeup. AncestryDNA was telling Alice that half of her ancestral heritage was unaccounted for. Perhaps, Gerry speculated, spinning out the theory, Katie Kennedy Collins was not really Irish, but an Eastern European woman with ancestry from the Caucasus who had an affair with a Jewish immigrant. Alice tried to make the theory work. *Maybe she was adopted by an Irish couple*, she wrote in an email. *Maybe she changed her name when she came to the United States.*

It was a weird, liminal time. The sisters stewed in the questions, in the unknowns, in the riddles posed by a test that Alice would later say she'd entered into "blithely" and was thoroughly unprepared to consider.

What if she wasn't related at all to her siblings? What if everyone but her had Jim as a father—and her father was some stranger? It is a strange thing to find a mystery sitting inside you. It is a strange thing to look in the mirror at the face you've grown old with and find you don't quite recognize it.

Meanwhile, Gerry attended an Irish wedding for a friend's daughter. There were Irish songs, and there was an Irish blessing, and the priest's homily went on and on about Irish pride. All her life, Gerry had connected with her Irishness, seen it as part of her very being. As she sat through the happy occasion, listening to the singing, Gerry felt a sadness overtake her.

* * *

Genealogy had been Bennett Greenspan's thing since he was a kid. Actually, genealogy and a certain entrepreneurial streak were two of his defining characteristics when he was growing up in Nebraska, though it was only decades later that they'd come together. Greenspan is in his sixties now, but his fascination with ancestors began when he was twelve and his grandmother died. Relatives gathered for the Jewish mourning period known as shiva, and he went around asking family history questions and writing it all down. When I visited Greenspan in Houston during the summer of 2018, he showed me a family chart he'd written in 1967 or so, when he was a teenager, attempting to trace four generations back. There were corrections of misspelled names, dates filled in with different pens over time, and question marks for mysteries he could not solve. I was struck by the care he'd taken with this document, and the fact that he'd saved it. It's a rare fifteen-year-old who takes an interest in his Hungarian great-grandparents.

Around the same age as he caught the genealogy bug, young Bennett figured out how to game the junior high chewing-gum business. His father knew a wholesaler, which meant he could buy a case of twenty packs of gum for less than four cents a pack. He sold them for a dime a pack, or two for fifteen cents, making a profit even while cutting a deal. He showed up each morning at school with his pockets and socks stuffed full of gum.

Greenspan grew up and became an entrepreneur who pursued genealogy in his spare time, putting it on hold whenever life got in the way or he hit a mystery the paper trail couldn't solve. By the late 1990s, he was in semi-retirement, having recently sold his photographic supply company, and when he kindly offered to rearrange his wife's pantry—the various tomato products, he observed, were nowhere near one another on the shelves—she informed him that he needed to get a hobby.

So it was back to genealogy, and just a matter of time before Greenspan hit another question he could not answer. Genealogists refer to such frustrating experiences as "brick walls," and there are countless books and blog posts dedicated to figuring out ways around them. Greenspan's brick wall materialized when he came across a potential relative with an unusual last name he recognized from family records; the guy lived in Argentina and claimed to hail from the same Ukrainian village as Greenspan's ancestors. Greenspan was in touch with this man, but none of the historical documents he painstakingly located—an exit passport, an old marriage contract—could confirm whether they were actually related. The paper trail had failed him. And that's when he started thinking about whether science could help.

By then, a few academic papers had been published on genetics and genealogy that intrigued Greenspan. One used DNA testing along the Y chromosome to study Jewish men who believed themselves to be descendants of the first Israelite high priest through their father's line. It found that many of these men, who often carry variations on the last name Cohen, shared certain genetic markers. Greenspan realized he might be able to use the same technology to figure out if the man in Argentina was really a long-lost cousin. But he couldn't do it himself; he wasn't a scientist. He found the name of a University of Arizona geneticist, Michael Hammer, atop the Cohen paper, located Hammer's home phone number online, and called him up on a Sunday. Greenspan was excited, but Hammer seemed less than enthusiastic about taking this unexpected call on his day off.

"Look, all I want to do is buy a couple DNA tests from you," Greenspan told Hammer. Hammer told him the university didn't offer such tests to the public, so Greenspan asked him who did. The answer was nobody did.

Greenspan paused, unsure what to say, and in that pause, he told me later, something magical happened. Hammer offhandedly remarked that somebody ought to start a business, because he got so many phone calls from crazy genealogists asking him the same question.

Greenspan recognized the opportunity instantly. A split second later, he asked if Hammer might be interested in perusing a business plan that he was planning to write up—a plan to have Hammer's lab supply and analyze the tests and have Greenspan market them. Hammer said he was willing to take a look. Greenspan wrote the plan and then stalked Hammer, calling his office again and again till he could get the scientist on the line and find out if he'd read it yet.

This was the start of what would eventually be FamilyTreeDNA, the oldest American genetic genealogy testing company. "Who knew it was going to become this huge success?" Hammer remarked when we spoke. He confessed that he initially thought the company such a long shot, and was himself such a crummy businessman, that when he agreed to be involved, "I made a very poor deal with Bennett."

Greenspan and his co-founders launched in the spring of 2000, rolling out tests analyzing Y-DNA and then mitochondrial DNA. It was a heady period. At the same time as Greenspan was opening his company, multibillionaire Mormon philanthropist James LeVoy Sorenson was launching the Sorenson Molecular Genealogy Foundation, which had as its stated goal the aim of connecting the world by gathering genetic samples from around the globe and showing how similar we are to one another. Just a few months after FamilyTreeDNA launched, President Bill Clinton announced that a rough draft of the human genome had been sequenced as part of the Human Genome Project, a massive, multinational effort to decipher the more than three billion base pairs that make up the blueprint for a human being. "Today, we are learning the language in which God created life," Clinton said.

So much has changed since 2000. When FamilyTreeDNA launched, Greenspan's tests cost several hundred dollars each, and they could not tell you anywhere near as much as they can now. The company kept improving its tests as technology got better and cheaper. In 2005, the National Geographic Society launched a nonprofit anthropological effort called the Genographic Project, soliciting genetic samples to trace the ancient migration routes of people's ancestors. Two years later, 23andMe—named for the number of chromosome sets in the human genome—launched its "Personal Genome Service," offering its first kits for the whopping price of $999 and hosting so-called spit parties to gin up interest. But the real breakthrough for genealogists came in 2009, when 23andMe announced that it was able to use autosomal DNA testing to help people find their genetic cousins. These days, FamilyTreeDNA's most popular offering is autosomal testing, but it is the only major company that still offers separate mitochondrial and Y-DNA testing, a niche market.

Greenspan's first customers were serious genealogists like himself. He told me that in the beginning, that's all he imagined his market to be. He saw this venture merely as another advance for hobbyists, akin to the conversion of paper records to digital ones. At genealogy conferences, he struggled to explain the relevance of his new company to the people walking by. Sometimes, he told me, he'd literally strong-arm potential customers. He'd shake their hands and walk backward, pulling them toward his booth, talking all the way.

* * *

When Alice and Gerry sent away for kits from 23andMe, that was only part of their approach to solving the mystery of Alice's strange AncestryDNA ethnicity estimate. Thinking ahead to getting her second set of genetic results, Alice knew that 23andMe might conflict with Ancestry's findings—or it might replicate them. If the second test turned up the same unexpected ethnic mix as the first, she hoped that the company would give the same results to Gerry, because that would eliminate the theories that Alice was the product of an adoption or an affair. But if it did, the sisters would have a new problem: They would not know whether their maternal or paternal line was contributing the unexpected ancestry, and would have no clue which side of the family to focus their efforts on. Dad seemed the likelier candidate given several factors, including how little the sisters knew about his family, and the fact that Alice thought she was seeing DNA cousins from her mother's old New England side in the AncestryDNA database. But they could not know for sure. Gerry's mitochondrial DNA results couldn't rule out their mom. The solution, they realized, was to test both sides of the family.

The sisters decided to approach two first cousins—their mother's nephew and their father's nephew—and see if the men would be willing to undergo DNA tests themselves. As it happens, they had only one first cousin on their father's side. His name was Pete Nolan, and he was the son of their dad's sister, Kitty. And the crazy thing is—for as much as Alice came to love him, and as important a role as Pete would play as the sisters unraveled their mystery—he very nearly hadn't existed for them.

Pete lives outside Charlotte, North Carolina, and we spoke by phone one day. He put me on speakerphone, and I struggled to hear him through the phone's echo, and he struggled to hear me because he's hard of hearing, so his wife listened in, translating when one of us couldn't understand the other. Pete is in his upper eighties, and he shares a birthday with Alice, except that he was born fifteen years before her. The Collins kids grew up hearing about their cousin, but never having known him, because by the time they were born, he had disappeared.

Pete grew up in the Kingsbridge section of the Bronx. He was reluctant to delve into his childhood, but the paucity of what he took from that period in his life was moving to me. He was five when his mother got sick with what he'd later learn was bone cancer, seven when she died, and "the family came apart when I was nine," he said. "I was sent into foster homes." He grew up with little knowledge of his mother's side of the family, and few memories to draw upon when he grew older and wanted to find out more about it. But he remembered his mother's maiden name was Collins, and he remembered the address in the Bronx where he'd spent his early years, because his mother had made him memorize it. He told me he also remembered two uncles—his mother's brothers—who used to visit him.

Pete grew older, went into the Air Force, got married, and moved on with his life. But after he retired from his work at IBM, he tried to learn more about his mother and her family. He found his old childhood address on a map of the Bronx neighborhood, and traced a route he'd taken as a child—down the hill, cross the street, turn right—in order to figure out the name of the church where he'd attended Catholic school. He contacted the church and got information on his baptismal record from 1933, which listed his godfather and uncle, James "Jim" Collins. Searching for Jim, he wrote to the Department of Veterans Affairs and the Social Security Administration and even cold-called some men by that name. He got nowhere until, through a chance conversation with a high school classmate he reconnected with at his fiftieth reunion, he managed to figure out the name of the former orphanage in Rockland County, New York, where his uncles had spent their childhood, and in 2001, he wrote a letter to the religious sisters running the site, asking for more information.

As it turned out, this was a fateful letter.

Unbeknownst to Pete, Alice had also written to the former orphanage months earlier as part of her genealogical research. She'd been trying to find out more about her dad's childhood, and had gotten back a letter from the archivist with the barest of biographical details. But whoever was managing the mail realized that two different people were writing to inquire about the same family. Pete got back a reply letting him know about Alice. And that was how Pete found his mother's family, and came to know seven first cousins who, for most of his life, he hadn't known existed.

Alice was astonished to get a letter from Pete. He had figured in Collins family stories like a ghost. Growing up, she was reminded of her first cousin every year on her birthday, when her father would tell her his nephew, Pete, had been born fifteen years earlier on the same day. Alice's parents had even met because of Pete—Alice the elder was friends with Pete's mom, Kitty, and babysat for Pete. Jim was fond of his sister and he spoke with affection of her in later years. She was tall and willowy, Alice told me, showing me a picture of a beautiful brunette in a smart outfit. After Kitty died young, Jim fell out of touch with his sister's widower and with his nephew. Alice felt that the mystery of Pete's fate weighed on her father. She remembered him speaking about his nephew in 1999, on the day he went in for a hospital procedure that he did not survive.

"I wonder what happened to Pete," Jim had said.

Now Alice knew the answer, and both she and Pete were grateful for this connection between long-lost families. Over time, they got to know each other, and shared photographs and stories from their lives. They met in person, and Alice was surprised by how tall Pete is. (Granted, many people seem tall to Alice.) "She is the best cousin I ever had," Pete told me.

Were it not for those crossed letters, Alice and Gerry would not have had a paternal first cousin to ask for DNA, and without that DNA—well, things would have been very different. In August 2012, Alice emailed both of her first cousins and told them she was trying to answer questions raised by a DNA test. In an email to Pete, Alice used the nicknames they reserved for each other—Pete was "Elder," for elder cousin, and she signed off as "Younger." "Just when you think the world is solid," Younger wrote to Elder, "something happens to shake things up." She laid out some of the possibilities she was hoping her cousins' DNA could resolve, including the theory that their shared grandmother had had an affair. Most likely, it was all much ado about nothing. "The test interpretations may need refinement," she wrote.

The year 2012 was not like now—everybody and his brother hadn't taken DNA tests, and plenty of people didn't know what they were. It was not a certainty that the two cousins would agree. But Younger included a link in her email to Elder to some videos explaining DNA testing, and said she and Gerry would pay for the test.

"What do you think?"

Elder wrote back that yes, he would take the test, and so did Alice's maternal first cousin. Now the investigation could begin.

* * *

For a long time, Bennett Greenspan was right about the niche market for consumer genetic tests. The early years were slow going for awareness of what DNA testing for genealogy and ancestry was. It wasn't till about 2013—thirteen years after Greenspan's launch—that consumer genetics companies like his collectively tested their one millionth customer.

FamilyTreeDNA was for years primarily a company for people like Roberta Estes, the silver-haired genealogist, who heard about at-home DNA testing well before most Americans, in the early 2000s. (This was shortly before she managed to track down her long-lost half-brother, Dave, from her father's "other" family, through genealogical research and a private investigator.) Estes was curious about the Y-DNA testing Greenspan was offering for male customers. If she got some male relatives to test, she wondered, could this genetic analysis help her trace the Estes line? She called up Greenspan—he was not only the kind of entrepreneur to call a scientist at home, but also the kind of entrepreneur to get directly on the phone with a customer—and asked him how the science worked. Greenspan recognized another seeker when he met one. He talked to Estes for about ninety minutes, explaining how his Y-DNA test could help her.

Estes was convinced, and she began the first of twenty surname projects, using DNA combined with traditional genealogical methods to understand the lineage of her own family and others. She had worked in technology for years, had an MBA and a master's in computer science, and her mind bent naturally toward data and logic and information systems, much like Alice's. Eventually her skills would prove immensely valuable in a field—genetic genealogy—that she, along with other citizen-scientists, would help create. In time, she also consulted for Greenspan.

When I visited Greenspan, he told me he started his company because of his interest in people and where they came from. "I didn't start this to make a lot of money," he said. But as it happened, FamilyTreeDNA has turned a profit every year since it has existed. It is a small, closely held company of about 150 employees, which Greenspan told me is privately owned by himself and his business partner, Max Blankfeld. They could not have imagined back in 2000 that at-home DNA test kits would blow up, that one day they'd be building their own lab, and then expanding it, and taking on contracts for other genetic testing companies. They could not imagine how busy the lab would become over the holidays, when DNA testing kits become the quintessential holiday gift for the person in your life who already has everything. In the company's first year, when FamilyTreeDNA analyzed a mere three hundred consumer kits, they could not imagine their company's kits would someday be available for sale all over the world.

Greenspan is a talker. He is friendly, occasionally brusque, a born salesman, happiest when he can unfurl a good yarn, and sometimes prickly when he can't control the narrative. His company occupies several floors in a bland brick building off a freeway in Houston. On one wall of his office is an assortment of framed political buttons going back to President William McKinley, and on another, a sheet of white paper at least six feet long shows his family tree going several generations back. When I saw it, I asked Greenspan about the genealogical brick wall that had started his odyssey, and his company. Had the man in Argentina with the unusual surname actually turned out to be a cousin? Greenspan smiled, and walked me through the lengths he'd had to go through to find out, using Y-DNA testing because autosomal DNA testing hadn't yet come along to make genealogists' lives so much easier. He had to send two kits: one to the fellow in Argentina, and one to a male cousin on his mother's side who lived in California. And? Greenspan walked me over to his family tree wall chart and pointed to the Argentinian cousin's name.

For lunch, we drove to a Mexican restaurant where the portions were so large we shared a platter of fajitas. Some weeks before, I had paid for an autosomal test through FamilyTreeDNA, emailing Greenspan to see if he'd be willing to walk me through my results when I was in Texas. I had questions about methodology and definitions, and about why the company's estimate of my ancestral mix differed from the results I'd gotten from 23andMe. Now, as I scooped chicken and onions, Greenspan surprised me by rattling off his company's assessment of my ethnic percentages from memory. He remembered the last name of one of my relative matches (Mendoza), and the number of relative matches I had in his database (eleven thousand), and he recalled even the most granular details of the ancestry report his company had issued for me—I was possibly as much as 2 percent Middle Eastern. I didn't remember seeing much of this, but later when I checked online, I saw he was right about all of it. He'd paid more attention to my results than I had.

When we got back to Greenspan's office I sat by his computer while he signed in and showed me his own DNA results. Examining his relative matches, he knew how certain fourth cousins were related to him through the same set of great-great-great-grandparents. He recalled which distant cousin's father had discovered a relative in Mallorca. "Here's another guy on my dad's side—it's my dad's father's sister's grandson," he said, pointing to the screen.

I probably don't need to say this but: Most people can't do this. Maybe we lack the memory or organizational capacity in our brains. Maybe we simply don't care that much. If America is turning into a nation of seekers—so many of us doing DNA testing on a lark, then backing into family discoveries without quite meaning to—it has taken the passion of longtime genealogists like Greenspan and Roberta Estes and the volunteer researchers at the Family History Library to get us there. They cared about the past before it was easy, before it was sexy and high-tech and quite so commercial, back when it was associated with colonial dames and retirees.

And because many of these original seekers got to genetic genealogy sooner, they learned early on how it can delight—and how it can disappoint.

* * *

For nearly eight years, Roberta Estes had her half-brother, David, in her life. The potato runs, the rescued dogs, the foul mouth, and the steady love. Then David got sick.

Many years before, when David had served as a marine tail gunner in Vietnam, he'd been shot down and received a blood transfusion; almost three decades later, he developed Hepatitis C from that transfusion and then liver cancer.

Roberta was a genealogist, so it made sense that she'd tested David's Y-DNA over the years as part of research into the Estes line. But it wasn't till shortly before David died in 2012 that she ordered autosomal DNA tests for him and for herself, and understood why his Y chromosome didn't match those of other Estes men she'd studied. She and David were not biologically related. Her father—the man who'd had two families—was not actually David's biological father, a fact her father likely never knew himself. David's mother had had a relationship with someone else. Roberta tried to talk to David about what she'd discovered, but he made it clear he didn't want to know the results from the DNA test, so Roberta never told him. The knowledge would have wounded him, and he was sick. And besides, David *was* her brother, in all the important ways.

At David's funeral, the room was filled with his trucker friends, who left their big rigs parked around the funeral home with engines running in a tribute to David. His wife was sobbing. The preacher didn't show. Roberta realized there was no one else to give the eulogy. So she did it.

For years after that, Roberta searched for David's family, wanting to give David a kind of closure, wherever he was. She vowed to find his biological father. Just as she'd learned to decipher patterns of genetic relatedness using Y-DNA, now she learned to decipher autosomal DNA. She was still administrator of David's 23andMe account, so she searched among his matches for close relatives—David's DNA was still working for him years after his body was cremated.

In time, Roberta figured out who David's father had been, and posted about it on her blog in a series titled "Dear Dave." "Dave, meet the man we believe is your father," she wrote, below a grainy black-and-white image of a man with a narrow face and deep-set eyes, who looked exactly like his son. "Dave, meet your half-sister Helen," she wrote, posting pictures of herself with Helen, the two women smiling and hugging.

Funny thing about Roberta and Helen. Because they were both David's sisters—though one woman never had the chance to meet him in life, and the other loved him even though they weren't genetically related—it seemed to Roberta as if they were each other's sisters, too.

When I spoke to Roberta Estes, her depiction of David was so real, I felt like I knew him—his rough edges, his warm heart. She told me that once, when he came by her house for an impromptu visit along a trucking run, he saw some medical paperwork spread out on the table—Roberta thought she might have cancer, though she'd later turn out to be fine.

He confronted her: "Were you going to tell me?" If she was sick, David said, he'd sell his house and come take care of her. She could tell he meant it.

"You can't do that," she said.

"You can't tell me what I can and can't do," her brother replied.

Roberta cried when she told me this story. "I don't know if I've ever had anybody in my life love me like that," she said.

The story of David was not an indictment of DNA testing, Roberta told me. Genealogy had given David to her; DNA could not take him away. The legacy of her father—his brilliance and his dysfunction—had taught her that genetics only goes so far. "Your genes neither guarantee you nor condemn you to anything," she said. And genetics is not the only measure of family. DNA is just DNA—it could not account for her and David.

"It's not love," she said.

* * *

When the 23andMe kits arrived at Alice's house in Vancouver, Washington, and at her sister Gerry's house, in Forest Lake, Minnesota, they spit into the vials, sealed them up and sent them back. Elsewhere in the country, two elderly men—Alice and Gerry's first cousins—were also spitting into vials, just the first of many people who would eventually be drawn into Alice's quest. In 2012 this wasn't as evident to many Americans as it is now, but when one person spits into a vial or swabs her cheek, her whole family is implicated. Because we share DNA with others, one woman's genetic mystery can never be hers alone—which meant that in one way or another, Alice's test results could inevitably prove to have implications for many other people.

For Alice, the waiting was fraught this time, because she felt what she was waiting on, really, was a verdict on herself. If the 23andMe test were to replicate the findings in the AncestryDNA test, it would mean that one of her parents wasn't who he or she was supposed to be. And that would mean Alice wasn't in some sense who she thought she was—a bald fact about herself that Gerry might or might not share, depending on whether they had the same parents. Alice wrestled with this idea: She *was* still herself, of course, but if AncestryDNA was right, it meant some important facts she'd long thought about herself and how she'd come to be were wrong. The context for her life was wrong. And that's a pretty big thing to be confronted with when you've taken a DNA test, as Alice had, "just for fun."

4

YOUR TRUTH OR MINE?

It makes sense that so many adoptees turned to genetic testing when companies like Ancestry unveiled their services. It was adoptees, and the volunteer "search angels" who helped them—a largely female band of citizen-scientists, some of them adoptees themselves—who pioneered many of the techniques that seekers turn to when they search for kin. Within the adoptee community, there is much talk of how deeply frustrating and destabilizing it is to lack basic information about the circumstances of one's own birth because of state laws restricting access to records: the names of one's parents, one's own birth name, important medical information. Not to mention getting answers to other, more profound, questions: Why was I given up? Does my biological mother want to know me? Before the age of genetic testing, an adoptee could spend decades looking for her biological family, and piles of money hiring private investigators, and still get nowhere.

In the pre-genomic era, the prospects for Jacqui Ochoa were bleak. Her longing was palpable. Jacqui wanted things many of us take for granted: She wanted answers to her questions, and she wanted a beginning to her personal narrative, an origin story. Most of us grow up with this story, and it's as familiar to us as the sight of our own face in the mirror: the look of our parents when they were young, as glimpsed in old photos; the story behind how they planned for us (or didn't plan and were happily surprised); their description of how the sky looked and the wind blew on the day our mothers checked into the hospital. How we emerged fat or scrawny, squalling or silent, sick or hungry.

Jacqui wanted her beginning. And she wanted family—perhaps that most of all.

Jacqui Ochoa was a foundling, which is a term for a child abandoned and found, a term that made Jacqui think of a little bird when she first heard it. In October 1965, she was a four-day-old baby tucked into a basket and left in front of a pastor's home in Van Nuys, California. She was adopted, and grew up with four other adopted siblings, and with parents who showed her unconditional love. They made her feel wanted. They told Jacqui her mom must've loved her so much. Jacqui grew up knowing that when she was given away, she'd been bundled in a basket, with a woolen blanket and a soft bunting bag—plus, as she later learned from a news clipping about her own case, diapers, "one undershirt, one pair socks, baby oil and powder."

But despite her upbringing, Jacqui grew up with a feeling there was something wrong, something fundamentally unlovable about her. She described it as a hole in her gut. She thought about the woman who'd given birth to her.

"How could she walk away?" asks Jacqui, who is now a school-teacher in Orange County, California. "That's what I thought about for thirty-five years."

Jacqui wanted answers. She wanted to learn the identity of her mother, and to understand what led her mother to give her up, and to know her own medical history, but her adoption records were sealed. She began searching at fifteen, writing letters to the local county agencies, petitioning the county court. Over the years, she managed to speak to the police officers who were called when she was found, and to read their police report, and to see the house where she was abandoned. As the mother of two boys, she tried to imagine the desperation that would lead a woman to walk to that doorstep and leave that basket, and then walk to a pay phone and call the pastor's home and inform the voice on the other end to go to the front door. She wondered if her mother might be out there, ready to be found but thwarted by the fact that Jacqui did not know how to find her. She placed classifieds in the *Los Angeles Times*. She searched the Internet when that came along.

When Jacqui first became aware of autosomal genetic testing, she knew it could help her, but the price was too high for a single mom. In the absence of a DNA test, or the unsealing of her adoption records, she needed dumb luck. She needed that one-in-a-thousand, needle-in-a-haystack breakthrough that was the stuff adoptees dreamed of in the pre-genomic age.

Which is what she got.

One day in 2016, almost four decades into her search, Jacqui got a letter from a social worker at the Los Angeles County Department of Children and Family Services. They'd been in contact with her siblings. Would Jacqui like to meet them? Jacqui burst into tears. Yes, by god, after thirty-five years of searching, she'd be thrilled to meet someone from her biological family. This felt like everything, the consummation of so many years of looking and waiting and wanting.

* * *

The landscape of the consumer genomics market now would have been barely recognizable a decade ago. One study by scholar Andelka Phillips, then at the University of Oxford, found that as of January 2016, at least 246 genetics testing companies across the globe were selling their wares directly to customers online. These tests are generally lightly regulated, with the exception now of health-related testing such as that offered by 23andMe, which is more regulated in the United States by the Food and Drug Administration. Not all DNA testing companies offer services related to predicting ethnicity and finding relatives; indeed, the spectrum of services they offer is dizzying, and their usefulness and accuracy sometimes dubious. They range from the paternity tests you can pick up at Walgreens to tests that look specifically for African or Native American ancestry, to others promising DNA-based matchmaking services. Phillips's survey placed consumer DNA tests into a long list of categories that included "child talent," "nutrigenetic," and, most ominously, "surreptitious." The catchiest company names I've seen are "She Cheated" and "Who'zTheDaddy?"

"Have you found an article of clothing with a suspicious stain?" asks the website of one Florida-based company called All About Truth DNA Services, which informs readers that "aprrpoximately [sic] 60%

of husbands and 40% of wives will have an affair at some point," and recommends consumers wait for their "suspicious item" to dry and then send it in for testing. Also accepted: cigarette butts, toothpicks, hair.

While the lion's share of DNA companies cater to questions of ancestry, health, paternity, and relatedness, much of the emerging consumer genomics market falls into lifestyle and fitness categories, encompassing products *The Atlantic*'s Sarah Zhang has likened to horoscopes: "vague, occasionally informative, sometimes amusing." Their claims, and the science used to back them up, are of varying quality. Some tests, met with horror by a wide swath of researchers, promise to offer insight into children's athletic ability. Another company offers an "Inborn Talent Genetic Test" for children, to help with career profiling—the better to maximize "the chances of them becoming an elite in life." Marketing "faux scientific authority," these kinds of tests aren't just harmless entertainment, warns one scholarly paper; they threaten to diminish consumer confidence in the clinical genetic tests that doctors order to guide medical decisions.

The landscape is confusing for the average consumer, and it can be hard to tell what genetic tests to take seriously. Large ancestry-testing companies, like AncestryDNA and 23andMe, may be characterized as "recreational," but they employ teams of scientists and rely on robust data to understand genetic relatedness and to track patterns of ancestral heritage (even if the latter is imperfect and constantly being refined). On the other hand, when 23andMe announced it was teaming up with a health-coaching app, and allowing customers to integrate their genetic results to help generate personalized diet and exercise advice, a number of geneticists were skeptical, concerned the company was getting ahead of research. What is a consumer to believe? A few years ago, Helix, originally a spin-out of genomics giant Illumina (which makes many of the chips and machines used to analyze DNA), unveiled a "DNA app store" allowing third-party companies to sell products off its DNA testing. These have included the Mayo Clinic GeneGuide, a test that requires the sign-off of a physician and, with the help of Mayo Clinic professionals, interprets your genetic material for insights into things like disease risk and carrier screening, but they also included the "Vinome Wine Explorer," which claimed your genetic data could help predict what wine you'd like, a concept one geneticist described as "completely silly." Helix has since announced a shift away from this "consumer-initiated" model, but there is still a lot of confusion over what genetic testing can and should be able to tell us.

One spring day, I found myself watching an ad for a special partnership between 23andMe and Lexus, which promised to find cars optimized to people's genetics. It's a credit to how out-there some DNA testing claims have become that it took me a few seconds to realize this was an April Fool's joke. As ludicrous and playful as that ad turned out to be (the driver licks the steering wheel to start the engine), it hit on a deeper message rooted in a suite of cultural messages we get about our genes. Consider the marketing campaigns that consumer genetics companies actually do run. During the 2018 FIFA Men's World Cup, for which the US team failed to qualify, 23andMe urged people to root for a team "based on your genetic ancestry"—they called this campaign "Root for your Roots." AncestryDNA has partnered with Spotify to create custom playlists based on the ancestral regions that customers hail from. "Solidify a true connection to the motherland," suggests one ad by a company called African Ancestry. "Know who you are"—as if DNA might know us better than we know ourselves, might act as a kind of historical id, reminding us of cultural affinities forgotten over generations but remembered in our cells.

These efforts are targeting—and reinforcing—a deep-seated belief that if we peer closely enough, we'll be able to decipher nearly everything about ourselves, our likes and loves, from the ACGTs along the strands of the double helix of our DNA molecules. It is an idea we've held for decades. In *The DNA Mystique: The Gene as a Cultural Icon*, which was published in 1995, sociologist Dorothy Nelkin and historian M. Susan Lindee warned of the rise of genetic essentialism. Examining portrayals of the gene in mass culture, they found a tendency to point to genes as the explanation for "obesity, criminality, shyness, directional ability, political leanings, and preferred styles of dressing. There are selfish genes, pleasure-seeking genes, violence genes, celebrity genes . . ." In popular portrayal, good genes and bad genes lead to good and bad traits.

We're such believers in genes that a recent Stanford University study found that informing people of their genetic predispositions for certain traits—rather, *misinforming* them, by telling them whether they had certain gene variants associated with exercise capacity and obesity, regardless of their actual results—influenced their actual physiology. Those told they had low-endurance versions of a gene variant did worse on a treadmill test, with poorer endurance and worse lung function (even if they didn't actually have that gene variant). Those told they had a variant that made them feel easily sated felt fuller on average after being given a meal, and tests revealed their bodies had produced more of a hormone that indicates feelings of fullness. By believing they were genetically destined for something, these subjects appear to have made it true.

In their book, Nelkin and Lindee looked back at the eugenics movement earlier in the century and saw thematic links between the 1990s obsession with genetics and those old notions of heredity. Yesterday's "Better Babies" were "still a highly desired reproductive commodity," while yesterday's feebleminded women, forcibly sterilized so as not to pass their problems on to the next generation, had been transformed into contemporary welfare mothers said to be birthing tomorrow's poor and criminal classes. "Ideas about heredity have as much to do with social meaning as they do with scientific research," they wrote.

Beneath all this, the authors argued, lay the mystique of the genes: "DNA has assumed a cultural meaning similar to that of the biblical soul. It has become a sacred entity, a way to explore fundamental questions about human life, to define the essence of human existence." Like the soul, DNA in this reading has a moral meaning and has implications not only for a person's sense of identity but for her place in society. Twenty-five years after that book came out, we still talk about DNA in quasi-religious terms—"the language," to harken back to Bill Clinton, "in which God created life." And perhaps there's something deeply human about this. Cultural psychologist Steven J. Heine has written that "in every society that has been investigated, there is clear evidence to show that we are predisposed to think of the world as emerging from hidden underlying essences"—whether that be blood or chi, humors or souls. Essentialism "is one of the most persistent and widely documented psychological biases."

Heine writes that we tend to think of essences as being "deep down and eternal," as defining and dividing categories within the world— black from white, dog from cat. When it comes to the genetics of traits and disease, Heine says, we are prone to a related outlook he calls "switch thinking"—the tendency to assume that particular genes control these things in a direct and fatalistic fashion. Yet it is not true that there is an "infidelity gene," as some outlets have reported, for instance, that makes a partner's cheating inevitable. "The vast majority of genetic influences on our lives operate in immensely complex ways and are not at all accurately captured by our essentialist biases," Heine writes.

Kristen V. Brown, a Bloomberg journalist who covers the intersection of technology, business, and health, told me she blamed some of this essentialist thinking on the Human Genome Project, "because part of the way that the Human Genome Project was sold to the masses was this idea that your genome explains everything." She added: "And then we decoded most of the important parts and were like, 'Shit, this *still* doesn't explain everything.' . . . But that was the marketing message and it was a good one, and it stuck."

So when companies urge people to root for a soccer team based on genetic heritage, or promise an exercise plan based on their DNA, this homes in on an idea that already holds great currency in the popular imagination. We are eager to know more about ourselves, and within the consumer space, ancestry testing appears to be driving the market for self-discovery. An Ancestry ad during the 2018 holiday season showed Kelly Ripa ordering biscotti in imperfect Italian, since she had just discovered she was "74 percent Italian!" What fun! Who wouldn't want an excuse to expand their diet of baked goods? Seemingly less fraught than health testing, ancestry testing is, as Brown puts it, "the killer app."

The market may change over time, of course. As our understanding of genetics improves, things like pharmacogenomics (the relationship between drugs and genes) and nutrigenomics (the interplay between nutrition and genes) may become much bigger forces. What's now considered "recreational" health testing may become more clinically relevant, and the genetic health market in general may prove to be bigger than that for ancestry and genealogy. But for now, as University College London researcher and genetic genealogist Debbie Kennett points out, the largest genomic dataset in the world isn't in the hands of governments, pharmaceutical companies, or research organizations. "Instead," she writes, "it is the ancestry companies which have been accumulating most of the genetic data."

Today, four companies lead the pack in terms of database size for autosomal tests: AncestryDNA is in the lead, then 23andMe, then the increasingly popular MyHeritage DNA, and, finally, FamilyTreeDNA. The bulk of these tests is being bought in the United States. That they are big business goes without saying, though the different companies operate on different models. Ancestry and MyHeritage have primarily focused on combining DNA testing with subscription-based access to genealogical materials, while 23andMe has carved out the health risk and traits market alongside ancestry testing. Meanwhile, in addition to selling consumer tests, FamilyTreeDNA's lab takes on work for other genetics companies, universities, and hospitals. And there are relative newcomers trying to break into the market.

In the fall of 2018, I attended the New York State Family History Conference, where a big session was led by an up-and-coming UK-based company, Living DNA. One of the brand's ambassadors did a series of cartwheels up the long aisle of the huge ballroom, eliciting gasps from the audience, and the company's co-founder David Nicholson made a strong pitch for the serious genealogists in attendance to buy his company's test. He asked if there was anyone in attendance who *hadn't* yet taken a DNA test; out of the rows upon rows, only four people raised their hands. Yet Nicholson was counting on the fact that he was offering something so unique that not only DNA newbies but even people who'd already tested at two, three, or four other DNA ancestry companies would want to test at his, too. He explained the strengths of the company's forthcoming relative matching system, and showed the granularity of the company's ethnicity estimate, explaining that Living DNA could pinpoint DNA from twenty-one regions in the British Isles alone, even distinguishing between genetic heritage from the county of Devon and that from the neighboring county of Cornwall.

But if Living DNA wanted to catch up to the other companies, especially in the United States, it would have to hustle. It had launched in 2016, only four years after Ancestry started offering autosomal DNA tests, but those four years had been key in the business of DNA testing. Its biggest rival already had more than ten million people in its database by then.

* * *

News outlets outdo each other trying to quantify the hugeness of AncestryDNA's database of genetic data. It's "the world's largest collection of human spittle, numbering in the hundreds of gallons," McClatchy observed a few years back. The approximately 1.5 million kits the company sold between Black Friday and Cyber Monday of 2017 added up to "2,000 gallons of saliva—enough to fill a modest above-ground swimming pool with the genetic history of every person in the city of Philadelphia," wrote *Wired*. The pace of the company's growth is fairly astonishing: It debuted its autosomal DNA test in 2012, several years after 23andMe. Yet when I visited its headquarters in the summer of 2018, it was just about to announce that it had double the number of DNA customers that 23andMe had.

Ancestry as we know it today began with two companies coming together. John Sittner founded a genealogical publishing company called Ancestry in 1983; it produced genealogical reference books, as well as the *Ancestry Newsletter*, which later became a magazine. Meanwhile, in 1990, two Mormon entrepreneurs named Paul B. Allen and Daniel Taggart founded a company called Infobases to sell religious and educational texts, first on floppy disks and then on CD-ROM. In 1995, Allen and Taggart moved into genealogy, licensing some of Sittner's reference books and packaging them with family tree software and other resources into something called the LDS Family History Suite, which made a million dollars in less than half a year. Allen told me later that this was a breakthrough moment for him: To be sure, Mormons were more interested in genealogy than the average American, but still, they made up just a tiny fraction of the US population. What was the potential for the genealogy market if he could broaden his audience?

Realizing they needed a genealogically oriented brand, Allen and Taggart started buying into Sittner's company, eventually coming to own it outright. They launched Ancestry.com in 1996, offering genealogy hobbyists free access to the Social Security death index for their research needs. Allen started surveying online visitors to the site, asking them questions about what they'd spent on various genealogical endeavors over the course of a year. How much on reference books, family tree software, genealogical magazines, travel? What about on postage and photocopies? When he crunched the numbers, he realized his average visitor was spending more than five hundred dollars a year on this hobby. "This is a multibillion-dollar industry and nobody's noticed it yet," Allen realized. "Why don't we digitize everything they're spending money on and make it possible to do it all through one subscription?"

And so they did.

The company dabbled in genetic testing over the years, offering Y-DNA and mitochondrial DNA testing, but these tests had limited appeal. The game-changer was the 2012 debut of autosomal DNA testing, the one that Alice Collins Plebuch tried out in beta, after Ancestry purchased the DNA assets of the nonprofit Sorenson Molecular Genealogy Foundation, and used them to help build its reference panel and make its ethnicity estimates. Suddenly, all of Ancestry's strengths could be brought together into one lucrative venture. The company had gone public in 2009, and in 2012 it was bought by an investor group led by a European private equity firm. At the time, Ancestry was valued at $1.6 billion; when more investors bought stakes in the company four years later, its value had gone up by a billion dollars, and in 2020, the investment giant Blackstone Group announced it was acquiring a majority stake in the company in a deal worth a hefty $4.7 billion.

Compared to FamilyTreeDNA, which has an understated presence in a building it shares with other companies, Ancestry's headquarters can't be missed. The headquarters in Lehi, Utah, completed in 2016 at a cost of $35 million, is a gleaming, modern structure—two connected buildings housing about half of the company's 1,600 employees. The lobby windows look out on two majestic mountain ranges, barely visible through a haze on the day I was there. It had been difficult to get permission to tour; access to the company's headquarters was tightly controlled, and visitors had to sign confidentiality agreements just to get past the lobby. Ancestry is almost omnipresent in this area; in addition to the headquarters, its customer service operation is twenty minutes away in Orem, while its team of professional genealogists, who are available for hire, are located forty minutes away in Salt Lake City.

Once I was inside, I was met by Jennifer Utley, the company's director of research and its longest tenured employee, who had been at Ancestry for over twenty years. When she started, the company had just established an Internet presence, and Utley's work was editing genealogy reference books and *Ancestry Magazine*. Now, clocks on omnipresent video screens throughout the headquarters showed the current time in the company's many offices throughout the world, including in San Francisco, Dublin, Munich, and Sydney. About eighty paid interns were eating in the cafeteria, just off the lobby, getting ready to do "speed dating" with executives so that they could learn what it's like to work there. Utley took me upstairs, showing off artwork with a genomic theme: hanging pendant lamps inspired by the double strands of the

DNA helix, and a massive, colorful bar graph representing the fifteen principal biogeographical ancestries of eighty-four human populations. Large portrait photographs of some of the company's employees were mounted on the walls alongside images of their ancestors.

Utley introduced me to two members of the content acquisitions team, who told me they travel all over North America to work with archival facilities, mainly at the state level, digitizing and indexing old records. In exchange for this, the company is given permission to place the records on its website so that its millions of paying subscribers can access them. The company has a tiered subscription model offering access to things like birth, marriage, death, and census records, international and military records, passenger lists, and old newspapers, depending on how much you pay. Based on the subscription you require and the length of time you need it, this could run into several hundred dollars for a year's worth of access. A customer who is simply trying to solve a few basic questions about her great-grandmother could conceivably pay twenty dollars for a month of basic access and never spend any more—assuming she's not converted into a seeker once she gets access to all those astonishing records.

Paul Allen, who left the company in the early 2000s, told me that one way to think about the brilliance of Ancestry's model is to frame its genealogical product as the opposite of breaking news. "The value of the records *increases* over time; they don't decrease over time," he said. "Every year that goes by, more people are entering middle age, where they start to be interested in genealogy, and then they have children and grandchildren." This means the birth record of a single ancestor from 1850, for instance, "is of interest to more people every year just because of aging and population growth."

Utley used her employee pass to unlock a door into a special room into which people were discouraged from bringing pens or open containers of liquid. (An exception was made for my pen and pad.) Ancestry has more than twenty billion genealogical records, and this room was one where the collection grew. There were microfilm and microfiche readers, and machines for scanning. A temporary archive housed the physical objects while they were being digitized—records of yearbooks and city directories, probates and wills. It represented an astonishing effort, but also an endless one—even if it were possible to scan, digitize, and index every document of genealogical interest in the world, human beings keep making more.

"It's not just about capturing images, it's also about preserving them," Utley told me. "We've been in archives where they had to prop things up on toilets" to protect them from water damage. "When records disappear, they're gone forever." The folks at the Family History Library spoke of a similar race against time, mentioning places in the world with so little funding for archives that precious records were kept on dirt floors. It occurred to me throughout my time in Utah that the work of both entities, the corporate one and the church-funded one, must at times feel Sisyphean: trying to save the records of the world before they give way to the ravages of Mother Nature and the forgetting of time. Nearby, a desk held a stack of metal canisters the size of personal pizzas. I glanced at the top one: It contained reels of the *Ottawa Citizen* newspaper, May through October 1972. Who knew what secrets were preserved there, laid down in mundane detail as part of the quotidian duties of putting out a daily paper? Amid all the lines of text in those reels might be an obituary or a marriage announcement containing a single name that was the clue one seeker needed, if only that detail could be ferreted out.

Elsewhere in the building were other operations: marketing, emerging businesses, mobile apps, human resources, and people from the DNA product and features team. Utley showed me a cubicle where several different versions of the company's DNA kit were displayed, customized for the more than thirty countries in which they were being sold. DNA testing at Ancestry works hand in hand with the rest of its business. If you spit into a tube, you'll get your ethnicity estimate and relative matches, but only with a subscription will you gain access to most of your matches' public family trees, or to genealogical records that might help you figure out where your mysterious Italian heritage is coming from. And what's more, sales beget sales.

AncestryDNA becomes a more appealing place to test as its database of customers grows, because people are more likely to find close genetic relatives there.

We walked over to the area housing Utley's own unit, which is responsible for researching and publicizing the genealogical discoveries made possible by the company's resources. (It was Utley's team that uncovered that Emma Watson, who played the witch Hermione in the Harry Potter films, was in real life distantly connected to a woman convicted of witchcraft in 1592, and that Benedict Cumberbatch, who plays Sherlock Holmes on TV, was very-very-slightly related to the author of the Holmes series, Sir Arthur Conan Doyle. If you're willing to extend the definition of "relative" far out enough, the world of genealogy is full of kismet like this.) As we walked, Utley pointed out people who'd worked at the company for a decade or more. She told me she loved her work for the sense of mission it gave her. The company was hosting eighty million family trees on its website; customers' sense of the past informed who they were in the present. She could be selling shampoo or something and that would be fine, but this work gave her the sense she was actually changing people's lives for the better. "There's a real sense of purpose," she said.

At the same time, Ancestry's size and mission can't help but draw scrutiny. A few months before I visited, McClatchy ran a multi-part series on AncestryDNA. One article was headlined "Ancestry wants your spit, your DNA and your trust. Should you give them all three?" The company's entire operation—the DNA testing, the family trees, the incredible treasure trove of genealogical records—is built upon reams and reams of incredibly intimate data—data that tells the stories of people's ancestral backgrounds and genetic traits, not to mention the sacrifices, triumphs, and scandals of their ancestors.

There's a tension inherent in being such a big company managing the genetic information of millions of people. There is a tension between the idea, promulgated by the Sorenson Molecular Genealogy Foundation, that we are all the same, and the idea that our differences can be parsed into pie charts. There's a tension between my right to spit into a vial and have its contents analyzed by a private company—and the right of my brother or mother not to be seen.

* * *

Starting in 1989, Arizona State University researchers joined forces with a small group of Native Americans known as the Havasupai, who were living in a remote area of the Grand Canyon accessible only by foot, horseback, or helicopter. Many Havasupai were afflicted by diabetes, and the researchers hoped to figure out if there was a genetic component to the disease by conducting genetic testing among members of the community.

But many years later, the relationship soured. The Havasupai said they found out in 2003 that the researchers were using their genetic material to look into topics to which they said they didn't consent, including schizophrenia, inbreeding, and the community's ancient population migration. One paper based on the tribe's blood samples suggested the Havasupai's ancestors may have migrated from Asia over the frozen Bering Sea. The Havasupai's origin story was that "it had originated in the canyon and was assigned to be its guardian," as the *New York Times* described it, and this belief was the basis of its sovereign rights, the tribe's vice chairman explained. Therese Markow, the geneticist overseeing the research, maintained that tribal members gave their consent for more far-reaching studies, and that different lines of scientific inquiry can be interrelated—understanding a people's ancestral origins may help in understanding a disease. But in 2010, the Arizona Board of Regents reached a settlement to pay $700,000 to forty-one of the tribe's members and return the blood samples.

This was a decade ago, long before a genetic genealogist and a detective teamed up to use DNA samples in a public genealogy database to solve the Golden State Killer case, and started a new trend for solving long-cold cases, while also driving home that a handful of people can effectively give up the privacy of hundreds or thousands of relatives when they choose to give up their own. It was also before Bennett

Greenspan's FamilyTreeDNA decided to give law enforcement access to its database without telling its customers, angering many genetic genealogists. It was before then-prospective presidential candidate Elizabeth Warren tested herself for Native American ancestry, causing controversy over just what DNA gets to say about Native American identity. It was before Alondra Nelson, Columbia University sociologist and author of *The Social Life of DNA*, observed that genetic testing arguably presents minorities with a Faustian bargain, "where, in order to get information that's been denied because of historical trauma, one has to enter into a conversation with thinking about race in a biological way"—in a way that, throughout history, has often reinforced so many of the racist stereotypes, so much of the reductive thinking, that have oppressed people of color.

And yet the Havasupai case presaged so many of the thorny questions we're only just beginning to think about, including the question of what "informed consent" means. Kim TallBear, a University of Alberta professor who wrote *Native American DNA: Tribal Belonging and the False Promise of Genetic Science*, told me that Native peoples are the "canaries in the coal mine" when it comes to issues of consent in genetic research. Can you truly consent if you can't even imagine what use your genetic material or information might be put to in years to come? What if you can't anticipate how someone else's interpretation of your genetic material will materially impact your sense of identity, your idea of family, or your spiritual beliefs? Bioethicist Thomas May points to the research done suggesting that the Havasupai community had come from across the Bering Sea. "You could take the position, 'Well, that's the truth,'" he said. "You could also take the position that some things aren't open to scientific verification." For science to use someone's body to attempt to disprove something sacred to that person—is that the uncovering of truth or a violation?

TallBear says that reducing the Havasupai story to a clash between science and religion doesn't capture this dispute, nor is it accurate to say that tribes like the Havasupai are anti-science. Origin narratives speak to how Native peoples see the world and their place within it. These narratives are important because they are guides to living, offering moral lessons that inform how Native peoples ethically care for the land and for one another. Some non-Natives "insist on viewing indigenous origin narratives as literal, and if we can't prove them, they're false," TallBear told me. "There are different kinds of truths in the narratives we tell."

The rise of consumer genomics poses questions about the emphasis we put on genetic identity, and what we do when DNA test results come into conflict with the narratives we've long believed about ourselves. How much of your sense of yourself should scientists and algorithms be allowed to dictate? We all hold certain truths to be sacred—they may be truths about where our people came from or about a sibling we love. What we do with genetic information, how we incorporate it into our story, is the tricky part—the part that Alice and Roberta and so many of us are just now figuring out.

It seemed like Jacqui Ochoa, the foundling abandoned on a pastor's doorstep in 1965, had the good luck to sidestep the problems of genetic testing coming into conflict with identity. She didn't need to depend on complex techniques to trace her way to her family, nor did she need to ask herself questions about whether her siblings would want to be found. Instead, she had Vicki, Eric, and Katrina, the siblings who'd been looking for her. As children, after their mother's suicide, Vicki and Eric had learned that their mother had had a little girl they'd never known, and Vicki came across an old news clipping of an abandoned baby she believed might be that little girl. Years later, Vicki and Eric and their half-sister, Katrina, filed paperwork to find her. Staff at the Los Angeles County Department of Children and Family Services matched up the two files—that of Jacqui the foundling and of the three siblings looking for their long-lost sister—according to articles from a few years ago covering their tearful reunion. "It's quite unbelievable that we were able to do it," a social worker for the county agency told

ABCNews.com, describing the bringing of these siblings together as a "miracle."

"We reunited and met and fell in love," Jacqui told me. She and Vicki Costa bonded quickly, trying to make up for lost years, with Jacqui flying to Central California to see Vicki, and Vicki coming south to Orange County. They drank wine and talked till the sun came up. Vicki became an aunt to Jacqui's sons. And while in matters of politics and religion they differ, in other ways they are strikingly similar—both are bighearted, emotionally generous, quick to laugh and to cry. There were times, talking to one woman, when I felt as if I were talking to the other. When Jacqui learned of her mother's emotional intensity, she saw herself in those descriptions, and when she learned her mother had committed suicide, she wondered what this meant for her. Could that instability come for her someday, like a kind of genetic curse?

After a year, it occurred to Jacqui that it made sense to do DNA testing. She was completely certain that Vicki was her sister, yet she wanted the confidence and certainty that only science could give her. It was then, when their lives were woven together, that she discovered Vicki and Eric and Katrina were *not* her siblings. (The Los Angeles County Department of Children and Family Services clammed up, telling me that without a court order, it couldn't comment on the specifics of how it matched up two unrelated families.) Jacqui was heartbroken. She still loved them, and they her, and she remains so close to Vicki that the two still call each other "sister." But Jacqui felt she no longer belonged to Vicki and her siblings in quite the same way. She envied them their genetic bond.

"I felt like somebody just stuck me on a raft and pushed me out to sea," Jacqui said. The final blow came when, with the help of CeCe Moore, the genetic genealogist Alice had also stumbled across, Jacqui located her actual biological siblings, five half-sisters, none of whom embraced her. For the half-siblings on her father's side, Jacqui is an unwelcome reminder of a man she is told was neglectful and mentally unstable. On her mother's side, Jacqui's half-sisters cannot believe their mother would abandon a baby.

"Our mother wouldn't do that," Jacqui says they told her, her voice rising and breaking when she repeated their words. She could never shake the sense of having been abandoned when she was born; now, she felt, it was happening again. Jacqui says her maternal half-sisters told her they didn't trust the DNA results, even though the science of relative-matching is quite reliable. She struggled to make sense of their rejection. How could she find such likeness and warmth from genetic strangers, and yet her own blood cast her out? How could human beings be so wonderful and so cruel? "I just can't believe it," Jacqui said, "that all of the time and the effort and the love and the excitement and the joy and the anguish and the fear and the frustration ended up like this. *You don't believe the science?!*"

I tried to imagine what it must be like for Jacqui's maternal half-sisters to be told that their late mother had given up a baby decades before. Was it impossible to incorporate this into their memories of the woman they knew? Did this unwelcome intrusion invite all sorts of revisions of the past and reassessments of her character? Did it cause them to wonder about the limits of her love? Such a fact—the fact of Jacqui's existence—must pose a cognitive disconnect, like being told the sky is red. This stranger's truth was in conflict with theirs, so her truth had to be wrong.

* * *

The first inkling Bennett Greenspan got that this DNA stuff could be life-changing, and not always in a good way, was before he started selling DNA kits. He was doing what he calls "proof of concept" Y-DNA testing within a large group of men he knew. Even within that early cohort, he found a major and unwelcome surprise, the details of which he didn't want me to publish. "Do no harm" was his philosophy, he told me. But in the very near future, he said, the concept of an adoptee who didn't know her biological family simply would not exist, no matter what state laws dictate.

When we sat in his office under his huge family chart, he rumi-
nated on how the legitimate interests of an adoptee seeking health information could bump up against the legitimate interests of a biological
mother who'd given her child away decades before. Perhaps this birth
mother had never told anyone about her child, and her marriage was
built in part on this omission. Perhaps she never took a genetic test, but
her sister did, and that was all it took.

"How did I know in twenty years they were going to come up with
a DNA test that was somehow going to allow you to find me?" Greenspan asked, channeling this birth mother. And on the other hand, he
pondered, was any of this the daughter's fault? Was she not entitled to
know what and who she was made of? Whose rights should predomi-
nate? "I'm not an ethicist," he said. "Look, I think everything we do
should be guided on, let's try not to hurt anybody. OK? But eventually,
probably, somebody's going to get hurt. Either you as the mother or me
as the kid is going to get hurt when I'm happy to see you, you're not
happy to see me."

Greenspan has a slightly mordant sense of humor. He's started calling the phrase "anonymous sperm donor" an oxymoron akin to "jumbo
shrimp." He told me that when he gives talks, he likes to tell audiences
that it is the duty of grandparents to send their grandsons beer money
so that today's young men won't resort to selling their sperm. Otherwise, he can, like a fortune-teller, predict their futures. In twenty years,
those boys will be getting a knock on the door from someone calling
them "Dad."

5

NON-PATERNITY EVENTS

Jason is in his late forties. He lives in the Midwest. If it were up to him, his full name would be printed here; he is done with secrets. But this is his mother's story, too.

Jason was mostly raised by his grandparents, with his mom and a stepfather in and out of his life. His mom wouldn't talk about who his father was, not when she came to his high school graduation, nor when he was getting married. He was too young to understand, she'd say, or now wasn't a good time.

Jason was in his thirties, a father with two young kids, when he decided to ask again. A relative had recently died, and it occurred to him his mother might pass away without ever revealing the mystery of his paternity. In the pre-genomic age, he was at the mercy of what she and any other secret-keepers were willing to tell him. So, he wrote his mom a letter and put some teeth to his request: He told her that if he did not hear back, he'd start asking around; he'd heard some cousins might know some things. It worked. His mother wrote back with a name and little else. "Here is what you wanted, sorry it took so long," she wrote. "I would just as soon leave as is."

But Jason could not leave as is. He wanted to know his father. He was not only acting out of a desire to know about his dad's medical history, even if that's what he told himself at the time. What he wanted, he knows now, was something deeper. He felt a hole inside himself. "Like a yearning," hc says.

He wrote a letter to the guy, who lived a few hours away, and they forged a restrained father-son relationship. The man told him he'd dated Jason's mother in college, and he knew she'd gotten pregnant, but she'd told him the baby wasn't his. The man was kind and polite, but he did not welcome Jason into his life, nor tell his adult children that Jason existed. Instead, once or twice a year, they went golfing together. Once or twice a year for over a decade.

Jason told me later that this arm's-length relationship was what drove him to do a DNA test, in hopes that scientific proof would improve their relationship. You might already know where this is going, but Jason didn't. He never dreamed the man wasn't his father; he only wanted to stop feeling like a secret. So, in 2016, at the urging of a genealogist friend, Jason bought an AncestryDNA test.

That's how Jason discovered he had the wrong man. His purported father wasn't in the database, but that didn't matter. Jason's friend came over and, looking at Jason's list of genetic relatives, swiftly traced Jason into a totally different family, as the likely son of one of three brothers.

In my reporting, I encountered a number of stories like Jason's, stories in which mothers would not or could not talk to their children about how they'd come into the world. Perhaps they did not know who the father was—and how could they tell their children that? Perhaps the circumstances surrounding this conception had been traumatic, had involved coercion or violence. The questions brought back shame or anger or an experience too private to share. They brought back a world in which a pregnant teenager was whispered about and shunned. DNA testing brings old taboos to the fore, secrets that are like the proverbial snowball rolling down the hill, getting bigger and heavier with each passing day. How do you undo a secret?

Only later did Jason look back and see how, in his long relationship with the man who was not his father, he'd been forcing the facts to fit a narrative. He had searched for resemblances between them, and seized on the fact that they were both the same height, both even-tempered and reserved. It was a testament, he realized later, to how badly he wanted this kinship. "You're going to find commonalities with anybody," he says. In one sense, then, Jason's account is a rebuttal to those who believe that DNA testing emphasizes genetic ties over other family bonds. After all, Jason found value in what he thought was a biological bond long before he could prove it with a test. Consumer genomics simply told him he was wrong.

When Jason handed the man a letter, at their next golfing session, twelve years into the relationship, he watched his would-be father's face move through shock and confusion. The man looked up. He told Jason that he was relieved. For twelve years he'd carried the guilt of believing he'd unknowingly abandoned a child. Now he knew he had not. And he said he was sad—"Anyone would be proud to have you as a kid," he told Jason, who cried when he repeated these words to me. The man's words were balm to the sense of shame Jason had felt since he was a child.

Jason went home and once again wrote a letter, with three copies for three brothers, informing them that one of them was his father. The letter had ripple effects—one brother said his wife had learned about it, and it was causing problems in his marriage—but it turned out to be a different brother who claimed Jason. When, at last, Jason met this man at a restaurant, he did not need to search for resemblances. The man's eyes were "like looking into a mirror." The man was warm, gregarious, funny. They sat and talked for hours. The man had asked Jason for an old photo of his mom, and recalled knowing her briefly. He told Jason he already knew they were father and son, though he agreed to do the paternity test Jason picked up at Walgreens. They sat in an SUV outside a post office and swabbed their cheeks together, then sent it off. Within a week, the test confirmed their relationship. The man welcomed Jason into his own blended family, in which no distinction was made between full and half- and step-siblings. Suddenly, Jason had a whole brood of brothers and sisters, several of whom lived nearby, and they all got in touch to tell him how happy they were to meet him.

That first Christmas, Jason's new family insisted he and his wife and kids join them, and it was the beginning of holidays together, and summer weekends, and talking on the phone, and visits to see one another's kids' in their plays and recitals. Jason's wife told me Jason had changed in the years since finding his father. He had become a more confident person. The hole was filled. "My wife said it was a dream come true, and it was—being welcomed," Jason said, thinking back to that first Christmas and starting to cry again. "And nobody was ashamed, and everybody was happy, and it was a great thing."

I could not help but think, hearing this story and thinking of the foundling Jacqui Ochoa, who wanted but did not get the bounty that Jason got, that spitting into a vial in search of family was like spinning a roulette wheel, with no ability to predict the outcome in advance, and the highest of stakes.

* * *

When a person discovers his dad is not his genetic father, this is often known in the world of seekers as a "non-paternity event," or NPE. Because of the popularity of home DNA testing, the term has trickled down from the scholarly confines of genetic genealogy to become an increasingly mainstream acronym. In Facebook groups devoted to finding family through DNA, people are constantly posting about the discovery of NPEs in their families. A nonprofit called NPE Friends Fellowship, dedicated to support and education around this phenomenon, has so far held a conference and a cruise, and offers grant programs, including one to supplement travel costs for those who want to meet their newly discovered biological families. Sometimes, the term "NPE" becomes a way to refer not to the fact of unexpected paternity but to the adult child uncovering it. "I'm an NPE," seekers write, and the term is transformed into a modern and reductive form of identity.

At genealogy events and in academic papers, I came across many other terms to describe NPEs, including "misattributed paternity," "misidentified paternity," "false paternity," "misaligned paternity," "extra paternity event," "surname discontinuity event," and "non-patrilineal transmission." There's the old-fashioned term "cuckoldry," which has a very *Jerry Springer Show* feel, and the rather opaque "extra-pair paternity," a term I found in biology papers on such populations as tree swallows, white-handed gibbons, and human beings. Whatever you call it, questions of paternity are at the heart of most of the DNA surprises America is experiencing right now. "We get a lot of people who found out their dad isn't their dad," says Yoav Naveh, the director of MyHeritage's

DNA team. Some seekers use "NPE" to mean "non-*parental* event" or "not parent expected," to refer to the broader category of cases in which both or either parent is not who the seeker thought.

How common are NPEs? It's hard to say. There is a common misconception that the phenomenon of what one academic study calls "cuckolded fathers" may be as high as 30 percent. But the most reliable recent studies, parsing the question in a number of ways, tend to put the rate in the low single digits, probably between 1 and 2 percent per generation.

More broadly, it is difficult to quantify just how often recreational testers discover significant unexpected things, whether that be instances of misattributed parentage, or surprise half-siblings, or the truth of an ethnic identity hidden from them for years. There are so many ways these surprises can play out. There are cases where adoptees were never told they were adopted, and found out via DNA test. (This is more common among older testers, whose childhoods harken back to a time when more adoptions were undisclosed to children.) There are cases in which people did not know they were conceived by sperm donors, particularly going decades back to a time when assisted fertility was kept secret, and when some doctors were in the practice of mixing a husband's sperm with that of the donor, or encouraging couples to have sex around the time of insemination so that husband and wife could say they never truly knew the genetic paternity of their child. In my reporting, I encountered a number of testers of significant African American ancestry who did not know of this part of their heritage till they tested, for a variety of reasons. Growing up, they were told they were Italian or Native American.

The shared nature of genetic material makes it possible to discover non-paternity events even when a parent and child haven't both tested to compare their DNA. For example, a man doesn't need to be in a DNA database to discover he's not related to a child he thought was his, or that he *is* related to a child he didn't know of; it might be that that child tests and doesn't match a paternal cousin, and things unravel from there. Or perhaps two siblings test, and they discover they share only 25 percent of their genetic material, instead of 50 percent, suggesting one has a different parent.

Ancestry and 23andMe told me they don't gather statistics on how often their customers get unexpected results about their immediate families. Anecdotally, though, Jennifer Utley of Ancestry said that unexpected paternity "happens more than people would expect," though some of this may be because seekers are self-selecting—they choose to test because they have questions. "I think it's impossible to quantify because I see it so much," says CeCe Moore, who, in the years since Alice saw her name on a message board, has become well-known outside the close-knit world of genetic genealogy. Even with a very narrow definition of unexpected findings—if we confine it to consumers who discover one of their parents isn't a genetic parent, or that they have a previously unknown sibling or half-sibling—the number is in the low single digits at least, Moore told me. (Rebekah Drumsta, public relations director for NPE Friends Fellowship, told me she thought the number of testers affected could be as high as 15 percent.) But if you assume just 3 percent of consumers, which is a conservative estimate, spread across thirty-five million testers in databases, you're talking more than a million people—and more every day.

Yet 3 percent doesn't even capture the scope of the phenomenon. It doesn't account for the people who haven't tested and may never wish to but will find out the truth anyway, as a result of a decision someone else made. It doesn't account for the fact that NPE rates increase cumulatively the more generations back you go, which means you may find a surprise in your grandparents' generation if not in your parents'. And it doesn't account for all the people affected by one genetic revelation. Because of the nature of family, each revelation must refract across a series of different people, and in different ways. A man who discovers that an affair decades back yielded a child may also mean his wife discovering this, too, once that child starts contacting him and his other children. A tester who finds out the man who raised him isn't his genetic father is also finding out that the brother he was raised with is genetically a half-brother, a piece of news he may now share with that sibling, while possibly discovering half-siblings he's never met through his genetic father. Every adult who discovers she was donor-conceived may have five or ten or more previously unknown siblings, some of whom will discover this truth in years to come. In a family, everyone is implicated by a secret.

"You see an intact family and you just assume everyone is genetically related, and that's just not always the case," Moore told me. "At first I thought it would just be a trickle—but it turns out to be more like a flood."

Moore has been a consultant for many seasons on the PBS series *Finding Your Roots*, a genealogy show that incorporates genetics to tell celebrities about their ancestors and familial connections. Her background is in television, and she has an intelligence and camera-ready charisma that make her good at explaining not only the technical intricacies of genetic genealogy techniques but also the emotional implications at work in the cases she's seen. I attended a talk she gave in which she explained that she used to try to help eager genealogists in their efforts to persuade reluctant relatives to undergo DNA testing. She understood their enthusiasm—a particular aunt's genetic data might truly be the key to tracing an ancestral line back to colonial America. But, Moore told the audience, she no longer honors these requests, because of the possibility that something else will be discovered.

"People ask all the time at conferences, 'How do I convince my relatives to test?' I say, 'Don't,'" Moore told me. This was both an ethical concern and a pragmatic one, she said; she did not want genetic genealogy tainted by bad press. "If we pressure people who don't want to test and they find out some unexpected surprise, that's not going to go well for our field," she said.

In many ways, the genomic testing revolution represents the end of family secrets. There is a vast army of American men and women searching for their families, and many of them didn't even know they needed to search until they took a DNA test. Was Alice one of these people? Was Jim Collins not the man who'd contributed half of her genetic material, even if in countless ways big and small, he'd made her who she was? And if so, what would she do with that information?

FamilyTreeDNA's Bennett Greenspan told me he tells people: "If you really don't want to know the answer to the question, don't ask the question."

But I'm not sure that's fair. Because we don't even know we're asking it.

* * *

DNA testing companies don't always do the best job warning customers about the unexpected things they may find if they test. A few years back, researchers at the University of Leuven in Belgium looked at the English-language websites of forty-three direct-to-consumer DNA testing companies and found that few warned consumers about the possibility of discovering so-called non-paternity events. But even the warnings that are in place may not be up to the task of getting consumers to pause before spitting and consider what's truly an existential question. To understand why, it helps to consider 23andMe. Among the major companies, it is fairly good at making explicit that its customers may discover things they weren't planning to find out.

"Though uncommon," customers are told before they access 23andMe's "DNA Relatives" feature, "unexpected relationships may be identified that could affect you and your family." The company offers additional warnings on a before-you-buy page on its website and in its terms of service. Its repeated warnings may be due in part to the company's learning curve on health-related results and the transparency imposed on it by the FDA. The company had been selling its genetic testing service for health for several years when the FDA ordered it to stop in late 2013, calling the company's "Saliva Collection Kit and Personal Genome Service" a medical device that required approval. The

FDA said it worried, in part, about false positives for things like genetic risk of breast cancer leading to unnecessary surgery. After its health testing was shut down, 23andMe continued to offer ancestry information and worked with the FDA on how to better deliver its health risk results. In 2015, it gained permission to tell customers about genetic variants for certain diseases, like cystic fibrosis, that they might pass on to their children, and in subsequent years, it was given permission to add genetic health risk tests for diseases including late-onset Alzheimer's, Parkinson's, celiac disease, and three variants in the BRCA1 and BRCA2 genes that substantially increase the risk of breast and ovarian cancer in women. In addition to proving its accuracy, the company also had to demonstrate that it was helping its customers understand what their health results meant and what next steps they should take. That meant using language that "your everyday average consumer would be able to understand," says Stacey Detweiler, a medical affairs associate for 23andMe, plus a lot of repetition, and this consciousness about helping customers handle health risk news has to some extent carried over to how it frames the possibility of familial revelations.

Also, 23andMe had a brush with the bad publicity such revelations can cause early on. In 2014, *Vox* published a package of stories about the perils of at-home DNA testing, recounting the story of a certain "George Doe" whose parents split after a 23andMe revelation. Shortly afterward, 23andMe co-founder Anne Wojcicki reversed a decision to automatically opt existing users into seeing their close genetic relatives, and announced that she was hiring a chief privacy officer.

Still, warnings may be of limited utility given how we're engaging with DNA testing. It's hard to fathom life-changing results when you're testing for fun and entertainment. As Benjamin Berkman, a bioethicist at the National Institutes of Health who specializes in genetics and emerging technologies, points out, those undergoing recreational testing are not prepared to have their identities upended. "If you go get a medical test, you know to be worried," Berkman told me. "You don't necessarily know to be worried about the uncertainty of your family relationships." Or, as geneticist Joe Pickrell, CEO of a DNA testing company called Gencove, puts it, "The problem is, nobody thinks it's going to be them."

Over and over, I found this to be true in interviews: Most people think they already know what there is to know about their families. If they have an inkling that the vial they're about to spit into could reveal family surprises for some people, they don't consider this something they, in particular, need to worry about. Even Jennifer Utley, who spends her days at Ancestry steeped in the complications of other people's families—and whose company cautions that testers may "discover unexpected relatives, which could surprise you and your family"—did not think she'd be a statistic. Until the day she was stunned to discover, through her list of genetic relatives, that she had a previously unknown first cousin—a child her uncle and his girlfriend had given up for adoption decades before.

This can have implications beyond how we conceive of family, identity, and trust, though those can be profound enough. "I never ever expected this to happen—ever," geneticist Ricki Lewis said after discovering through an AncestryDNA kit that she had a half-sister, which led her to unravel that her late father was not biologically related to her and that she had been conceived by a sperm donor. In late 2017 Lewis, a genetic counselor and longtime science writer who authors a weekly blog called *DNA Science,* had written an entry headlined "DNA Testing Kits as Holiday Gifts Can Bring Surprises," never dreaming that less than a year later, the surprise would be hers. The revelation came at a crucial time, just as Lewis was deciding how to manage a precancerous condition after previous brushes with the disease. When the half-sister told Lewis about her own past cancer, Lewis decided against a wait-and-see approach and chose to have a second mastectomy. She believes finding out this news when she did may have saved her life.

In the worlds of genetics research and clinical testing, participants are typically given the choice of whether or not to opt in to what are known as "incidental findings": genetic information that's potentially medically important but falls outside of the original purpose of the test. Over time, debate has grown within the medical community over whether to ask patients if they want to be informed of incidental findings—and whether to inform them of such findings when they've asked to opt out of such news. The problem of incidental findings is an important one, in the context of both scientific research and "recreational" testing: If you can't anticipate what you might find out, how can you make an educated decision about what news you'll want to hear?

Thomas H. Murray, president emeritus of a bioethics research institute called the Hastings Center, told me he first encountered these issues over thirty years ago, when he was recruited to create a bioethics center at Case Western Reserve University. He took a call from medical professionals faced with a dilemma: They had a male patient with failing kidneys, and had tested both of the patient's daughters to see if they'd be suitable candidates for a transplant organ. Testing had revealed that one of the daughters wasn't genetically related to her father. Were the doctors obliged to tell her the truth of her paternity?

After much thought, Murray told them, *No.* "All you need to say is that you're not a good antigen match." What the doctors did need, Murray told them, was a policy for dealing with these situations, under which would-be donors were informed that the information they'd receive would be limited to whether or not they were good donor candidates. The doctors would act as gatekeepers, protecting the families from information that was not germane to the matter at hand and could be deeply destructive.

But within the direct-to-consumer world, of course, there are no gatekeepers debating the value of an incidental finding that doesn't feel incidental at all to a woman who just discovered she's the product of an NPE. Is more information always better? Is it always empowering? Information carries its own burden, which is that once you know something, you have to figure out what to do with it. This is true for health-related genetic information: Knowing you may be at elevated risk of developing Alzheimer's is living in an uneasy truce with time. Do you tell your children, or will the information burden them? And this also holds true for the kinds of familial revelations that come from ancestry testing, which is making us all into lonely bioethicists, forced into difficult decisions without the benefit of task forces or review boards. What do you do when you find out you have a paternal half-brother, with all that this implies about your parents' marriage, your father's loyalties, your own identity? What do you do when faced with the prospect of informing your elderly mother that her genetic father was not the man she knew, but a sperm donor?

"We are empowering consumers, patients, family members, caregivers to obtain all sorts of information that we couldn't provide before, and we're just saying, 'Here are the keys to this and go for it,'" says genetic privacy scholar Mark Rothstein. "There's a principle that we teach our first-year medical students: Don't run a test unless you know what you're going to do with the results one way or the other," he says.

We are blundering into these discoveries, assuming that the results of testing will line up with what we already know about our families. Engaging in recreational genetic testing is framed as an act of self-discovery, but it's often assumed those discoveries will come with layers of historical and emotional distance. Consumers I spoke with were often motivated by the desire to discover something cool and innocuous. The fact that DNA kits are given at the holidays and for Father's Day sets the context for how they are viewed: as gifts, the cost of which can be measured in mere dollars.

The major DNA testing companies know this; they style marketing campaigns around the major holidays, with prices dropping as Thanksgiving approaches. "It's the perfect gift for every person," 74-percent-Italian Kelly Ripa promised on Ancestry's home page in November 2018, advertising DNA kits for $59.99. The ad copy promised that "your loved ones" will "have new stories to share—and new ways to grow closer over the holidays." A few days later I got an email from the company: The kit was now $49, its "lowest price ever." In 2016 and 2017, according to Kantar Media, Ancestry spent more than $100 million in advertising each year, dwarfing its competitors, though 23andMe has clearly been trying to catch up. It went from spending $12 million in 2016 to $40 million in 2017 to $50 million in the first half of 2018 alone.

But what about the times when this gift does not help a family "grow closer"? When I wrote about this phenomenon a few years ago for the *Washington Post,* I was struck by a line from novelist Margaret Atwood: All new technologies have a good side, a bad side, and a "stupid side you hadn't considered." Now, consumers are forced to grapple with the unforeseen implications of their curiosity, their generosity. Now they call the company that tested them, confused about why Ancestry is telling them some strange man is their father.

A spokesperson for Ancestry told me it has a team of "member services representatives" to help customers with questions about their DNA results, including "a small, dedicated group of highly experienced representatives who speak to customers with more sensitive queries." The specialized team at 23andMe undergoes months of training for phone calls that require not only scientific knowledge but a great deal of empathetic listening, and the company even launched a special page to help people navigate "unexpected relationships," complete with links for finding therapists. Greenspan told me people who contact his customer service representatives at FamilyTreeDNA with unexpected results often presume the lab is in error. His reps will tell customers that the company has a policy: It'll test the person's sample again for half the price, and if the lab is wrong, the company will refund that money.

And his reps will tell callers that the lab is almost never wrong.

* * *

Even if you didn't mean to ask the question, once it's asked, it will be answered. And once it's answered—well, for many people, there's something pretty compelling about knowing there's a mystery man out there who gave you half your genetic material. How do you not open that box? How do you not want to see your face in his, or to hear the timbre of his voice? How do you not wonder: Would he like you? Would he be glad you came into his life? This is how seekers are made: One question leads to another.

Laurie Pratt inherited the hobby of genealogy from her grand-mother; as a child she used to go on excursions with her to look at headstones. As an adult, she spent many years tracing her own family's history, following her mother's colonial American ancestry and her father's French Canadian roots, before the price of DNA testing dropped enough for her to test herself and her dad, whose tree she was working on.

When Laurie's father's results came in, he did not show up as a genetic relative. Laurie called Ancestry and got a customer service representative who tried to reassure her. "I'm sure he'll show up," she remembers the rep saying. But he never appeared among Laurie's relative matches. So Laurie contacted CeCe Moore, whose name appeared online for her just as it had for Alice—as a kind of lifeline for early adopters of this technology who could not make sense of what a company was telling them. Moore suggested Laurie try testing herself and her father again at 23andMe. Laurie made up a reason why she needed to test her dad again, and again, they did not match. Moore was gentle. She guided Laurie to 23andMe's chromosome browser, showing her how the company had come to this conclusion: Laurie and her father did not have any overlapping genetic segments.

Laurie is in her fifties. She lives in Orange County, California, and works as a ground operations supervisor for an airline. Both of her parents have since passed away, but her mother was "my best friend, hilarious, amazing, always owned her stuff," she says. She went to her mother, who at first said the DNA results were "impossible." But over time, her mom's accounting of whether she could have been with another man changed to "not that I remember."

"'Not that you remember?'" Laurie repeated, gently teasing her mom. "You sound like Reagan at Iran-Contra."

The first time I spoke with Laurie, she told me she believed her mom honestly didn't remember something that had happened half a century earlier. Later, as she thought back on tiny clues as so many seekers have done, she would conclude that her mother probably did remember; she simply didn't want to have this reckoning with her daughter. But eventually, Laurie's mom recalled some bare facts about a short-lived relationship during a period when she and her husband were briefly separated. Her mother had two requests: "Don't tell your siblings until after I'm gone," and "Don't tell your father." That was fine by Laurie, who said she never would have told her father anyway. "This is the guy who did the job," she says. "I can't imagine calling anybody else 'Dad.' "

Still, the genealogist in her was deeply curious. She threw herself into figuring out who her genetic father was, and it took years. Using genetic genealogy techniques, she traced the family histories of cousins whose DNA she matched online. She combed obituaries for men who were likely candidates. Sometimes she texted her mom photographs of the men—*Is this the one? What about this guy?*—and Laurie's mother pointed out that the fellow she'd slept with hadn't looked like some old guy from an obituary. He'd been a young man.

At one point, Laurie's search became a kind of dark comedy of misattributed paternity. She enlisted the help of a third cousin she'd found on her relative match list, who agreed to test her own parents in hopes of helping Laurie—and in the process this cousin stumbled across the fact that her *own* father was not her genetic father. Even as Laurie searched, her own parents—who had divorced when Laurie was an adult, though they remained close—died within a year of each other and were buried next to each other.

For cases of unknown parentage, much of the process of searching for relatives through DNA databases relies on the principle of triangulation. For years, the technique was premised on something called segment triangulation, which relies on specific segments of identical DNA shared among relatives; in more recent years, CeCe Moore and Laurie Pratt and others have come to rely on a technique called pedigree triangulation. The exact techniques vary depending on a number of factors, but these days, Laurie told me, the basic idea goes like this for a seeker—let's say an adoptee trying to figure out the identity of her birth mother—using the biggest database, AncestryDNA. The adoptee looks at her closest relative matches and examines the amount of DNA she shares with them, as well as cousins they share in common and any family trees that might be available, to figure out how she might be related to them, attempting to isolate her maternal matches from her paternal ones. She might turn to the Internet to figure out the full names and geographic locations of these relatives, if she can.

Then, using genealogical records and tools available online at Ancestry, FamilySearch, and elsewhere, including obituary records, city directories, school yearbooks, and places like Newspapers.com, and drawing on existing trees, social media accounts, ancestral hometowns, common surnames, and shared ethnicity, she builds family trees for her maternal DNA cousins and goes back through time looking for the ancestor she shares with those cousins. (In genetic genealogy forums, seekers refer to this as MRCA, which stands for "most recent common ancestor.") Then, the adoptee builds the trees forward through time—something Moore calls "reverse genealogy"—figuring out what marriages took place, and who are the offspring of those marriages, to isolate which branch likely contains her maternal grandparents. If she's right, the DNA and family trees of her other relative matches should line up with her theory. Once she knows her grandparents, she can very possibly divine her mother, or at least one of several sisters as a candidate for her mother. Then, it's a matter of figuring out what to do with that knowledge, which can be its own kind of journey.

These days, Laurie told me, she thinks she could solve the question of her own paternity within a couple of months, because the databases are so much bigger. But for years, her matches were too distant for her to make headway, and many of her paternal DNA cousins hailed from a small town in Maine with families that had intermarried, sharing last names and genetic material in such a way as to make the job extremely difficult. She became one of the volunteer "search angels" in Moore's large DNA Detectives Facebook group, and when her own search stalled, she solved other people's cases.

Eventually, in early 2017, after researching and creating 118 family trees for DNA cousins in hopes of understanding where she fit into them, and building back in time to the late 1700s, Laurie traced her way to the man who appeared to be her genetic father. She scoped him out online, learned about his religious affiliation, his charity work, and his family's backstory, and then sent him a letter. A few days later, by happenstance, the man showed up as her father at Ancestry, having just tested his saliva. Perhaps he'd been given the kit as a gift from his daughter for Christmas, Laurie thought, since the man's daughter was listed as the administrator on his DNA account. If so, it would prove to be an ambiguous gift.

The man called Laurie at the number she'd provided in her letter, and it appeared he hadn't seen his matches yet, because he suggested that perhaps his brother was Laurie's father. "Your Ancestry results came in today and it came back that you're my dad," Laurie told him. "I didn't test for that," the man replied—because, after all, he hadn't. He didn't know he was asking *that* question. "Did you test in a tube recently?" Laurie asked him, trying to thread the gap for the man, to explain that he had, in fact, tested for his genetic relatives, even if all he'd wanted was one of those cool pie charts that show you how Irish you are.

They talked for a little while. The man did not remember Laurie's mother. "That's OK," Laurie told him, "she didn't remember you either." He was a religious Catholic and said he felt terribly guilty about the idea that he'd had a daughter he didn't know about, though she'd been conceived before he was married. That was OK, Laurie said again; she just wanted to thank him for his role in contributing to her existence. She told him she wanted to meet him. He said he'd be willing, but he needed a little time to process this news and tell his wife and daughter.

For two days, Laurie was walking on air. Then she logged on and discovered the man's test had been deleted.

Laurie sent the man another letter. She included a screenshot proving their genetic relationship, in case he hadn't seen it. She was not trying to push her way into his family, she wrote, nor to make a claim on his money, and she was willing to sign a legal document to that effect. "I certainly hope you harbor no feelings of guilt or remorse," she wrote. "Everything turned out in both our lives exactly as God planned." What she sought were stories about her biological father and his family, to help her build a sense of where she came from. She would fly to him; she would treat him to lunch. One meeting, a few hours, was all she asked—which, depending on your perspective, might not be asking for much, or might be asking for a great deal indeed.

I first spoke with Laurie a few days after she sent that second letter, when it was becoming clear that her biological father was not likely to write her back. She told me all the things she'd ask the man if he'd talk to her: What books did he like to read? What had his childhood been like? Had he played sports back then? Had he been close to his parents? Did he have photographs of them so that she could trace her own face and those of her children in theirs? Could he tell her stories about his family? Because his family was hers now, and she longed to know her own family, if only a little bit.

She did not feel embarrassed about how she'd come into the world, she told me, and she did not regret her own discovery. She had been "devastated that I wasn't who I thought, that I was made of a stranger," but she was nevertheless grateful to know the truth. Still, she found herself imagining how this man, this stranger in all but genetic material, must be feeling. "We all tell ourselves stories about who we are, and we pick and choose what fits into that story," she told me. "I feel like I don't fit the narrative that he's chosen."

Who decides what story we get to tell? Countless American men have by now been contacted and told they helped make a child once, and that those children would now like to introduce themselves. Some knew this day might come; others never imagined it. Many more will face this reckoning in years to come. And then there are men who find out the children they raised are not their own. "Sometimes it's hard, 'cause she reminds me of what my ex did," one man told *The Atlantic*, referring to the girl he now knows is not his genetic daughter. The revelation led to his divorce. Genetic counselor Brianne Kirkpatrick, who has started her own secret Facebook groups for people who've experienced DNA surprises, had a client who discovered at the age of seventy-eight that neither of his children were genetically his. "He was *pissed*," Kirkpatrick says. He wondered if he could sue his ex-wife.

But just as you can't anticipate what you'll discover when you test, you can't predict how the person on the other side of that test will react. I emailed with a man named Jeff Lester in Lebanon, Missouri, who told me the discovery of a daughter he'd unknowingly conceived at the age of sixteen was a miracle. She'd matched him on AncestryDNA during a period when he'd put his genealogical work on hold, so he never saw her messages there; instead, she eventually messaged him on Facebook. Jeff told me he thought at first this was a scam. He was fifty years old, and he did not remember getting any girlfriends pregnant back in high school. But he logged onto Ancestry, and there she was, this girl—no, this woman, already thirty-two years old, already a mother herself—listed among his matches as his daughter.

Jeff told me he could not get over the unlikeliness of his daughter finding him. Outside of DNA testing, it almost certainly would not have happened. She had been given up for adoption when she was a baby and did not know the identity of either of her biological parents. When Jeff helped her uncover her mother's identity, she turned out to be someone with whom he'd had a brief relationship he didn't remember consummating. It "sounds like a bad after-school special," he wrote me.

The other piece of the miracle, Jeff said, was that he was not supposed to be alive. He'd been diagnosed in his twenties with ALS, also known as Lou Gehrig's disease, and had been on a ventilator for two decades. With this sword of Damocles hanging over him, he'd built a life for himself, marrying and having three wonderful daughters. He was astonished that he'd not only outlived the odds but lived long enough to discover that, in fact, he had *four* daughters, plus a son-in-law and grandchildren; he felt the hand of God at work. He told me all this over email and Facebook Messenger, typing by way of a wireless head-controlled mouse attached to his glasses, because the ALS severely limited his ability to speak.

Jeff told me he was determined to live as fully as he could while he was still on this earth. He had forged a relationship with his oldest daughter, and he and his wife even got to host the grandchildren for a sleepover. He did not feel guilty or ashamed about her existence;

those were wasteful emotions, and his disease had taught him that none of us have time to waste. Under the shadow of imminent mortality, "you focus more on what is truly important in life," he wrote. For Jeff, the unexpected outcome of recreational DNA testing simply meant more love.

6

ALICE AND THE DOUBLE HELIX

When Alice sent her saliva to 23andMe, she did not know exactly what would happen to the vial inside its cardboard box. It was like magic: The box went off in the mail, and the next time she'd know any more would be when she received an email telling her that results were ready. By now, the DNA of tens of millions of people has gone on this kind of journey, revealing its secrets to us. On a hot summer Thursday in Houston, Bennett Greenspan showed me what that journey looks like.

He took me down a flight of steps from his office. He showed off the sign for the women's bathroom, indicated by an XX, and the men's, a big X and a small Y. We walked past staff sitting at desks, and Greenspan handed me a white lab coat to put on and blue booties to stretch over my sandals. Then we went through a secure door and into a sprawling series of rooms filled with freezers and robotic equipment, the linoleum floor abruptly changing color to indicate where the space had been expanded as the company grew.

The air was filled with the whoosh and whirring of machines. FamilyTreeDNA analyzes the cells customers gently scrape from the insides of their cheeks with swabs that look like long Q-tips, and near the door, an auto-sorter with a robotic arm was reading the label attached to each vial of cheek cells in preserving liquid and arranging it according to the type of test the customer had ordered—whether autosomal, mitochondrial, or Y chromosome testing. Another set of machines transferred the liquid from each tube onto a plate designed for extracting DNA by means of a chemical process that "rips the cellular wall apart," Greenspan said. "Now we can get to the DNA in the center." In another spot, about twenty machines performed an enzymatic process called a polymerase chain reaction, allowing the lab to "make more needles than haystack" by "amplifying" the DNA to generate many, many more copies so that it can be analyzed.

We walked through to another room, where Greenspan showed me what looked like a small rectangular metallic bookmark, lined with two rows of smaller rectangles. The DNA of twenty-four customers could be placed on this specialized Illumina chip, one on each little rectangle, binding to it through a process called hybridization.

"There's seven hundred thousand data points on each of those rectangles," Greenspan said—seven hundred thousand data points that could say A, C, G, or T. Nearby, along a row of metal shelving, eight laser scanning machines made by Illumina were reading these chips. In a minute and fifteen seconds, a machine could read all seven hundred thousand data points on a single sample; for an entire bookmark with its twenty-four samples, it took about a half hour.

Greenspan showed me a new robotic storage machine that had taken months to install and had the capacity for about two million samples of extracted DNA, ready for any additional testing customers requested; an older freezer held another half-million tubes of extracted DNA. The company stores extracted DNA for several years and, in another place, keeps a second set of vials with their DNA not yet extracted; these do not degrade as quickly. "I've got millions of them downstairs," he said. He said the lab had the capacity to auto-sort, prep, and extract DNA from about ten thousand samples a day. I asked if he was ever operating at capacity.

Oh yes, he said. During Christmas.

* * *

What happens to your DNA after it gets extracted and amplified, hybridized and scanned? There are several ways to answer that question. For one thing, it is analyzed; each company has its own method for determining consumers' ancestral origins, which is why your results will vary somewhat across brands. We'll come back to that, because it's a big topic, wrapped up in questions of race, culture, science, and identity, not to mention how the legacy of eugenics is woven into our understanding of biological difference. Another thing that happens after DNA is processed in a lab is that these results are delivered to customers, who layer their own interpretations atop the proprietary analysis that companies like FamilyTreeDNA and 23andMe have performed. But Alice was still a few weeks away from being able to see her results, and from determining whether they jibed with her AncestryDNA test, or whether this whole foray into consumer genomics would turn out to be a waste of time.

So while we wait, it's worthwhile to consider a third way in which we might answer the question of what happens to your DNA. The answer, depending on where you test, may be that it becomes profitable in a very particular sense. And not just when you buy a kit for Christmas, but for a very long time after that.

The way that the value of those letters, those As, Cs, Gs, and Ts, increases once they enter the corporate arena, and the places those letters go—whether or not you realize it—has led some observers to wonder whether consumers really understand what they're giving up. At a fascinating conference on ancestral DNA testing in Toronto in 2018, organizer Julia Creet, a professor of English at Canada's York University, suggested that "the information that you're giving them"—the DNA testing companies, that is—"is much more valuable, I would argue, than the information that you are getting."

To understand the potential here, consider the example of 23andMe. One of FamilyTreeDNA's big promises to consumers is that the company will never sell genetic data to third parties. But 23andMe has a totally different model, and it helps to rewind to how that company came to be.

By the mid-2000s, a number of health-related genetic testing companies had popped up, offering tests for conditions like ovarian cancer and cystic fibrosis directly to consumers. As one molecular biology journal noted at the time, the United States was especially ripe for this "do-it-yourself diagnosis": Americans tend to embrace an ethos of individualism, and besides, they were already accustomed to "marketed medicine and paying out of pocket for healthcare services." When a former health-care analyst named Anne Wojcicki co-founded 23andMe in 2006, in the midst of this growing field, the company had two goals: to offer consumers access to their own genetic information so that they could better manage their health, and to create a vast database of genomic information for research. "Solve health," Wojcicki phrased its mission. 23andMe collaborates with outside researchers to study the data, and sells it to pharmaceutical companies. In some cases, it gives away spit kits for free to people who suffer from diseases it wants to study. The company has said that this huge repository of genetic data, along with information participating consumers have provided, allows the company to move much faster and more cheaply than such research usually requires.

It goes without saying that genetic health information can be incredibly empowering. When the actress Angelina Jolie revealed in 2013 that she'd undergone a preventative double mastectomy after discovering she was a carrier of the *BRCA1* breast cancer variant, the news caused consumers to rush to 23andMe to buy their own kits. But 23andMe's business model, which it has described as a "virtuous circle," has also been subject to criticism and suspicion. The year after Jolie's op-ed, *New York* magazine described the company's approach of charging people for access to their own genetic information as "a sleight of hand so quintessentially American that Tom Sawyer might have dreamed it up."

In 2017, after a big round of funding, *TechCrunch* estimated 23andMe's value at $1.75 billion. "The long game here is not to make money selling kits, although the kits are essential to get the base level data," Patrick Chung, a board member and venture capitalist, told *Fast Company* in 2013. "Once you have the data, [the company] does actually become the Google of personalized health care."

The vast majority of 23andMe's customers consent to have their de-identified data used for genetic research, allowing the company to boast that "23andMe's research platform is currently the world's largest consented, re-contactable database for genetic research." Customers can consent to different degrees of having their information shared, whether in aggregated or in individual form. They also fill out surveys, answering questions about topics as varied as arthritis, skin color, body mass index, psoriasis, cancer, multiple sclerosis, rosacea, and cholesterol. The addition of this information provided by the subjects of the genetic research themselves gives that data much-needed context for researchers.

"We specialize in capturing phenotypic data on people longitudinally—on average three hundred data points on each customer," Wojcicki reportedly told a room full of researchers at a company event in 2018. "That's the most valuable by far."

The company's consent document explains that it may develop intellectual property or commercialize products and services based on studies it conducts and "in such cases, you will not receive any compensation." But it frames the invitation to participate as "becoming part of something bigger," and indeed, for consumers, the impulse to share often stems from magnanimity: If something in my genetic material could help someone else, have at it. With all that crowd-sourced genetic data, Wojcicki has said, "I can partner with academic researchers, I can partner with biotech companies, pharma companies, to say how can we all—who have a common interest—come together and say, I want a solution?"

Along with research and funding partners including Genentech, Pfizer, NIH, Stanford, the National Institute on Aging, and the Michael J. Fox Foundation, the company has collected data and produced research on, among other things, Parkinson's disease (Wojcicki's former husband, Google co-founder Sergey Brin, found out he was a carrier after taking a 23andMe test), allergies, nearsightedness, breast cancer, motion sickness, rosacea, migraines, depression, inflammatory bowel disease, and health disparities among African Americans. In 2015, the company launched 23andMe Therapeutics, a research and drug development group. By 2016, *MIT Technology Review* reported, the company had already sold access to its vast treasure trove of consumers' genetic information to more than thirteen drug companies.

Ancestry has had research partnerships, too. For a time, it was collaborating with a Google-founded longevity research company, Calico, to "research the genetics of the human lifespan." It has also collaborated with the nonprofit University of Utah. Kristen V. Brown, the Bloomberg journalist, said it made sense to her that a company like Ancestry would be looking beyond genetic genealogy. "If you want to expand your market, you eventually want to expand your product offering," she said. "I think the health market is much bigger." Brown was right. The space was heating up. In 2019, both Ancestry and MyHeritage announced they were moving into the arena, teaming up with physicians to oversee their new health tests, which meant the companies didn't need to get FDA approval to sell directly to consumers the way that 23andMe has had to.

But there are unresolved tensions about fairness, about money, and about how much everyday consumers of these services understand. When 23andMe announced in 2018 that it had entered into a collaboration to give British pharmaceutical behemoth GlaxoSmithKline access to its customers' genetic data as part of a four-year deal to (in Wojcicki's words) "accelerate the development of breakthroughs," with GlaxoSmithKline purchasing a $300 million stake in the company, ethicists lined up to voice concerns. Was that genetic data still safe as it was changing hands? Did people who signed up to participate in research truly comprehend how asymmetrical their relationship with 23andMe was—how much it and other companies stood to profit off the information they and millions of others had paid for the privilege of providing?

"Making money off our genetic data?" one outraged consumer of many tweeted in response to the GlaxoSmithKline deal, adding that he was revoking permission for 23andMe to use his data for research. "This smells like a Henrietta Lacks situation," someone else replied.

* * *

What is consent?

As one lawyer and bioethics researcher pointed out in the clamor following the GlaxoSmithKline deal, 23andMe did appear to be disclosing how it could use consumer data in its consent documents—"the challenge is that people don't read it." Indeed, all the major DNA testing companies state in their policies that they don't sell or share genetic data for research purposes without customers' explicit consent. Companies with research programs generally make them optional. You can also change your mind about participating, though your information can't be removed from research already under way. 23andMe allows that, in addition to conducting research with consumers' consent, it does "use and share aggregate information with third parties" but says it goes to great lengths to ensure that it protects consumers' private information.

But the problem is, it can be difficult for consumers to understand exactly what they're signing up for—both how their data could be used and how it is being protected. I spent many hours poring over the companies' language about privacy, and then diving deeper into their various policies, all of them thousands of words long, trying to parse the definition of terms like "personal information," and thinking how much easier this would be if I'd gotten that law degree. Do such contracts make sense given the way we behave online? One 2008 study estimated that if American Internet users attempted to read the privacy policies they regularly encounter online word for word, they'd spend upward of two hundred hours a year doing so.

"We have all been conditioned and socialized into 'I agree,' right? Get me onto the next page: 'I agree,'" philosopher and bioethicist Françoise Baylis has said. And it's true. A few years back, professors conducting an experiment gave 543 college students the opportunity to sign up for a fictitious social networking service called NameDrop. The students largely bypassed the site's lengthy Terms of Service and Privacy Policy, missing the part where they were agreeing to hand over their firstborn children. Poor regulation and overly long and complex contracts become even more of a problem with sensitive genetic information. In a 2018 study looking at the policies of ninety American consumer genetics testing companies, Vanderbilt University privacy researcher James Hazel found varying and often vague information about how long companies would retain genetic information, who owned it, and whether and how it would be shared with third parties. In the event of a sale, merger, or bankruptcy, the information might "be treated as an asset that would be transferred to the acquiring entity." And the companies' language also typically indicated they were permitted to change their policies whenever they liked. Privacy issues within the industry are largely self-regulated. Most direct-to-consumer genetic testing companies are not governed by the Health Insurance Portability and Accountability Act, or HIPAA, which protects the privacy of genetic data in clinical and research environments. Instead, they are regulated "by a patchwork of federal and state laws," writes Hazel.

Even Paul B. Allen, one of the founders of Ancestry back when it was just a genealogical subscription service, told me that while he thought his old company and 23andMe were developing amazing features with DNA testing, "I'm worried about the control of the data, and I'm worried about corporations owning this data."

And what if the data at one of these companies—or at a third party with whom they share the data—were to be breached? In 2018, this happened to MyHeritage. The email addresses and encrypted passwords of ninety-two million users (though, notably, not their DNA data, family trees, or credit card information) were found sitting on a private server. There have been breaches of other big data companies, including Facebook, Equifax, and Google. To be fair, just how informative that data would be is unclear. Debbie Kennett, the genetic genealogist and researcher, told me she thinks the information people share on Facebook and through things like supermarket customer loyalty cards is more revealing.

"If I was to send you my raw DNA data sequence, on its own, it's not going to tell you a great deal," Kennett says, even if the data were uploaded to a third-party health site like Promethease or to a genealogical research site like GEDmatch. "I think people expect the DNA to tell them more than it actually can, and they expect it to be predictive of health issues, which it is not," she says.

Still, there are larger, almost existential questions about how we treat information that clearly has great value once it is parsed and put in context. It has the power to tell us who are relatives are and aren't, and to reveal what diseases we may be at elevated risk of, and—here in the United States, at least in theory—to make it difficult for us to obtain certain kinds of insurance. (I'll come back to that.) Some privacy scholars believe HIPAA's protections should be extended to cover companies like Ancestry and 23andMe. Bioethicist Thomas May wouldn't go that far, but he thinks more could be done to protect people's information. The direct-to-consumer world is operating like the wild west, he's written: "In the health care context, we would not be so lax about a sample's chain of custody, nor return test results without following clear procedures for establishing who has a right to such information."

In part, this is the cost of the empowerment that Anne Wojcicki envisioned. Consumers themselves are often handling one another's genetic information; if they submit a spit test to 23andMe, for instance, they are stipulating under its terms of service that they have "legal authority" to do so. The major companies make it possible for parents to consent on behalf of their children—a fact that profoundly troubles some bioethicists for reasons having to do with informed consent, with children's autonomy, and with the potential for years of anguish and anxiety on the part of parents *and* children should they learn something troubling. One blogger wrote about devising a method perfect for infants, the elderly, and anyone who couldn't or wouldn't generate the needed liquid for companies like AncestryDNA that require saliva samples. She made a video demonstrating her method for mixing a cheek swab with faked saliva on her infant nephew. Commenters beneath this thanked her, claiming her method had worked for them.

Elsewhere online, prominent genetic genealogists explain to seekers how they can swab their dead family members and send the samples to FamilyTreeDNA—and one blog chronicles how, in an instance when the cheek swabs failed because too much time had passed since death, hair follicle samples worked. In an attempt to create some guidelines, leading genetic genealogists have written a set of standards for their community, stipulating that testing DNA requires consent. And meanwhile, some companies have announced that they're offering, or plan to offer, what's known as "artifact testing" for old envelope flaps and stamps, perhaps from love letters sent between grandparents, raising profound questions about the ethics of testing those who can't consent because they're no longer alive. MyHeritage founder Gilad Japhet, who collects autographs, has spoken of planning to test the DNA on envelopes he owns sent by dead notables. "Their DNA is coming to MyHeritage very, very soon," Japhet announced at a conference in late 2018, "and I think it will be amazing for people to find if they're related to Winston Churchill or to Albert Einstein."

The wild west, indeed.

* * *

One day late in the summer of 2012, Alice got an email from 23andMe. Her results were back. She went to the website and logged in. She navigated to her list of DNA relatives. Her strongest connection, with about 12.5 percent shared genetic material, was to a man she hadn't known she'd find in the database: the son of her brother Bill.

Alice realized in an instant what this meant: Her nephew was her full genetic nephew and her brother was her full genetic brother. She was not adopted. She was not the product of an affair. As far-fetched as those ideas had seemed, she could not be certain till this moment. This meant Jim Collins, the kind, taciturn, hardworking man she'd loved, was indeed her biological father. She felt tremendous relief.

In retrospect, of course, it seemed impossible that Alice Collins Plebuch was *not* the daughter of Jim Collins. Not only does she share his eyes, his cheekbones, his wide feet, his stature, and his smile, but she and much of the Collins family inherited her father's facility for numbers, for systems, for keeping track of vast quantities of stuff. "My dad was pretty much a mathematical whiz," Jim's son Jim the younger told me. "When I was in college he had only gotten his high school equivalency, but he was able to do my calculus." Jim seemed to understand math on an intuitive level, though he couldn't necessarily explain his methods if you asked him to. He did it all in his head. Sometimes, for fun, his kids tested him. "You could have a row of numbers this wide and this long," Alice told me, making a square with her hands about the size of a dinner plate, and "he could add them up and get the result faster than you could read them or put them into a calculator. He would always be done before you, and he would be right."

This talent has trickled down: Jim the younger is a retired software and systems engineer who at one point worked on NASA supercomputers. Another brother, John, majored in electrical engineering and then worked in systems design at IBM. Brian was a materials manager who did purchasing, warehousing, and shipping. Bill was a facilities supervisor at a large research center. Ed managed warehouses, analyzing and overseeing the computer ordering and fulfillment systems. "It's some form of synthesis," Alice's best friend, Kathy Long, told me. "They take a lot of data and they synthesize it down to bite size, whereas the rest of us don't quite organize things in the same way." And then of course there's Alice, who during lunch one day with me and her niece, Cherie Collins, explained how, for purposes of genetic genealogy problem-solving, she used to keep a bar chart of a certain set of variables in her head. "You can see it's passed on—whatever he had," observed Cherie, who owns her own web development company and appears to have inherited Jim's particular mental prowess, too. (Gerry, a teacher-turned-homemaker, once joked that if any of the Collins kids was going to turn out to *not* be a Collins, it would be her; she was the only one of them "not on the cutting edge of technology.")

Jim the elder excelled at these kinds of things during his twenty years in the army, first at West Point and then all over the world: learning and teaching military engineering, and overseeing a system to purchase, store, inspect, repair, pack, and issue supplies, tracking it all on thousands upon thousands of stock record cards. The maintenance of all this—of records, of workflow—was not unlike what Alice did in her career with benefits, only Jim didn't have the opportunity to automate things using computers and the Internet. Outside the army, he was incredible at card games because he could keep track of everything that had been played and seemed to see through the cards in the other person's hands, a tactic that made him a rather unpopular opponent. Sometimes he had to lose on purpose so that people would continue to play with him. He also played poker and bridge and dice, and killed it at pool. After the orphanage and before the army, during the "misspent youth" he sometimes liked to talk about, he ran numbers for bookies managing illegal gambling rings, collecting money and helping keep their records.

But back in 2012, the results from 23andMe confirming that Jim Collins really was her father were nonetheless a big deal to Alice. Reassured by her connection to her nephew, she looked to see what the company could tell her about her ancestral background. Would it tell her AncestryDNA's ethnicity estimate was wrong? But instead of clarity, she found the results hard to parse—far more difficult in those early days of the industry than they are now. As best as she could tell, from staring at an image the website showed of her twenty-three chromosomes, with little flags showing the apparent geographic areas her genes hailed from, she had a lot of Northern European ancestry, as well as ethnicity from places in Europe including Lithuania, Belarus, Ukraine, and Russia. She could not tell just how much Eastern European ancestry might have been passed down to her—or how far back in time she'd need to look to solve her mystery.

She emailed her nephew, who informed her that his dad, Alice's brother Bill, had also tested through 23andMe; he just wasn't showing up as a relative because he hadn't turned on a feature to make himself findable. Bill Collins had tested, it turned out, at his wife's behest, but hadn't been interested in his ethnicity results, figuring they were junk science.

Alice sent an email to the professor in Illinois whose name she'd seen on the Yahoo group DNA-NEWBIE. "I'm so confused," she wrote. "For the last sixty-four years I've been 75 percent Irish and the remaining 25 percent split equally between British and Scottish." AncestryDNA had told her she was something else, and now 23andMe's results were proving difficult to interpret. "Would you please run it through your analysis?" she wrote. "My genome is attached." She had accessed her raw genetic data from 23andMe, downloaded it, and attached a zipped file, about 8.2 megabytes containing hundreds of thousands of As, Cs, Gs and Ts, possible clues to an answer if someone in possession of the right tools and scientific understanding could make sense of them.

She also downloaded her raw data file and uploaded it to GEDmatch, the free, publicly accessible genealogy database, along with her brother Bill's 23andMe results. Several years later, GEDmatch would become famous—infamous, some might say—when members of law enforcement and genetic genealogists began searching the genetic data of the million or so people on the site to solve particularly heinous cold crimes. But back then, it was a little-known destination for early adopters of this nerdy hobby, offering additional tools for people to analyze their genetic data, and a place for people who'd tested at different companies to compare their results.

These days, newcomers to their own DNA can fairly quickly get up to speed on how to understand their own ancestral origins and their relationships to genetic relatives. There are ethnicity pie charts with increasingly precise estimates of biogeographical origins. New tools at AncestryDNA and elsewhere make it easier to understand exactly how two people are related. The many Facebook groups geared to this interest are stocked with fellow citizen-scientists ready to share their tips. But in 2012, the knowledge was harder to gain and the process much slower. Alice had to understand the science behind all of this, so she studied up on SNPs, on the distinction between two people sharing genetic segments because they're related or by chance, on how to understand relatedness, and on the number of centimorgans—a centimorgan, or cM, is a unit for measuring genetic linkage—she could expect to share with certain relatives. As geneticist Ricki Lewis explains, genetic linkage reflects the way inheritance works, with DNA sequences that are near each other on the same chromosome tending to stay together as they're passed down. These likenesses are translated into patterns that help geneticists see evidence of biological relationships.

And Alice began to connect with her DNA relatives, asking to see more of their genetic material in hopes she could understand her own.

One DNA relative at GEDmatch, a man living in Canada, messaged her. They spoke by phone, and the man gave her an impromptu seminar, of sorts; he was the first of many genetic relatives moved by her mystery and eager to help. He walked her through how to compare the overlapping genetic data along her chromosomes with relatives at 23andMe, and he recommended she try testing a brother's Y-DNA at FamilyTreeDNA. Plus, he told Alice that, if the ethnic mix AncestryDNA was giving her was correct, she was likely descended from Ashkenazi Jews, a term for a group of Jewish people who trace back to central and eastern Europe.

Alice asked her Canadian cousin a question: *What was that word?* She'd never heard the term "Ashkenazi" before. It took her months to figure out how to pronounce it properly.

Judaism is, of course, not just a religion; it is also a culture and an ethnicity with a distinctive genetic signature, which explains why Alice could be considered ethnically Jewish even if she was not practicing the faith. "The Ashkenazi gene pool looks half European and half Middle Eastern, but it's unique in its combination thereof," says University of Arizona's Michael Hammer. Population genetics papers by Hammer and others trace the geographic journey of this community and document how its unique history, including a long-ago population "bottleneck," gave Ashkenazi Jews an unusually high prevalence of certain disease-associated genetic mutations. As Ashkenazi Jewish people traveled, and as many were chased eastward in Europe by persecution, they married among themselves, remaining "distinct from the local community," Hammer says. They suffered mass killings in the crusades, in pogroms, and, of course, in the Holocaust.

After several hours of genetic briefing by the man from Canada, Alice hung up the phone. In coming months, the enormity of the task in front of her would become clear. There were many, many genetic relatives like this man, and each one of them might be a clue. At heart, she had a data problem. First, she had to figure out exactly what her ancestry was, and—if she were indeed of Ashkenazi Jewish descent, as it appeared—which parent was passing down this history. And then, she might have thousands of genetic relatives, linked to her by count-less segments of shared material. She had to figure out how to organize it all, how to separate the useful information from the dross, how to compare the useful bits, and how to use all those bits to figure out, if she could, where her people came from and how her family had gotten its history so wrong.

Thank goodness for that analytical brain—Jim's legacy. She was prepared for this. During her career at the University of California, she had excelled at organizing the flow of information. In the late 1990s and early 2000s, when she was in charge of "employee direct systems," she was tasked with overseeing the interactive voice response system that people had to use to enroll for benefits (remember those?), with its pitiless choose-your-own-adventure logic: press 1 for this option, press 2 for that. She hated it, and fought to move enrollments to the web—and then, once everything was online, what a pleasure *that* was. By god, she had wrestled with the phone tree and won! She could do this. She loved numbers, and statistics, and logic, and she was retired. She was more than ready for this moment.

Alice Collins Plebuch was ready to build a database of her own DNA.

7

EUREKA IN THE CHROMOSOMES

Linda Minten's childhood was lonely. She describes her mother as self-involved and cold, her father as loving but beset by poor health and a painkiller addiction. Her parents split up when she was six. Linda lives now on a farm in rural northwestern Oregon, where she has a family, raises and sells a few beef cattle, grows timber nearby, and acts as a kind of keeper of lost creatures not unlike the child she once was. They include a blind chicken she takes for daily walks, and a goat she bottle-fed and raised in her house.

In her early fifties, Linda began experiencing symptoms of what her doctor believed might be ovarian cancer. She was bloated, bleeding, and in pain. An ultrasound was inconclusive, but testing found an elevated level of a tumor biomarker. The fact that Linda had recently learned she was likely about half Ashkenazi Jewish through her mother worried her, because it increased her likelihood of having a *BRCA1* or *BRCA2* variant, which would put her at higher risk of that kind of cancer. Wondering if her father might also have Ashkenazi Jewish ancestry, further increasing her risk of these variants, Linda decided to find out more about her ancestral heritage through AncestryDNA. She also ordered a clinical-grade test from a company called Color, to look for many mutations that increase cancer risk. Linda's father had passed away decades earlier, but Linda told her mother what was going on and consulted her on the family's health history, which included two maternal relatives with cancer and the fact that Linda's mom had had a hysterectomy. Linda brought all this information back to her doctor, who considered all the risk factors and recommended she have an emergency radical hysterectomy. "We couldn't wait for it to pan out," Linda told me.

The Color test came back the day before Linda's scheduled surgery. It didn't find evidence for the genetic mutations Linda was worried about, but her doctor felt the procedure was still warranted. So she had the surgery, and afterward, her doctor told her there was no evidence of cancer in the organs that had been removed. Linda was relieved and felt they'd made the right decision based on the information they had. But a week and a half after the surgery, as she was recovering, her AncestryDNA results came back, and that's when things took a turn. Her relative matches listed a man she didn't know as her father, as well as a half-sister and cousins she'd never heard of.

Understanding didn't come right away. Linda had to piece the clues together. She had to call her mom, who said she didn't know why the test had predicted some stranger was her father. She had to call Ancestry and speak with a kindly customer service representative, who assured Linda that the results were almost certainly accurate and that she was far from the first person to call with unexpected results. She had to log back onto Ancestry.com and look at her ethnicity, which suggested she had no Jewish ancestry at all. She had to examine the language on her birth certificate until she realized it had been amended and issued a year after her birth. She had to call the county in Arizona where she'd been born, and find out that her original birth certificate was sealed. And, she says, she had to call one of her brothers and wring the truth out of him.

This was how Linda found out she was adopted—a plain fact she says everyone, including neighbors, teachers, and her two older brothers, had known, but which she herself had never been told. Linda was fifty-one years old.

And she was dumbfounded. Dumbfounded that her mother had lied when asked, and hurt that her brothers hadn't intervened despite their mother's edict not to. "I just went through a hysterectomy," she told one of them. If she'd known the truth, she wouldn't have been so panicked by what she thought was an elevated risk. "I would have had more time. I could have retested." She might not have needed the surgery at all.

It was strange. Linda had never suspected she was adopted, even though as a child, she'd sometimes fantasized that she was. Now, the revelation of her adoption was freeing, making her mother's seeming dislike for her feel less personal. When she spoke with her mother again, Linda told her what she knew, and she says her mother told her, "We did the best we could." Mary Rubens, a close friend to Linda who is also the ex-wife of one of her older brothers, confirmed the basic timeline of Linda's story and told me that even *she* had known Linda was adopted since she'd been a teenager dating Linda's brother. Mary had long expected Linda's family to tell her, but she herself did not, feeling it was not her place, and feeling bound by a promise she'd made to Linda's father decades earlier that she never would. Once upon a time, some thought it best that adopted children not be told they were adopted. "I just think that he didn't ever want her to feel unloved," Mary told me.

And now what? At once distraught and relieved, and eager to understand where she'd come from, Linda set out to find her birth parents. She started with her birth father, who was already in the AncestryDNA database under his initials. She thought he might have tested his saliva in hopes of finding her. But the first few times they spoke on the phone, the man's reaction surprised her. "'I'll be darned,'" she says he remarked. He said he hadn't known Linda existed, nor did he remember her mother. Linda told him she was conceived in March 1965—did that ring a bell? Nope. Her mom might have been a redhead, like Linda—did that help? It did not. In their second conversation, he told Linda that these phone calls were getting him in trouble with his wife, and she apologized, feeling again like the child who was unwanted and out of place.

Still, she kept trying for connection. "Look, it's been a month," she told her biological father. "If you're not going to tell your kids, I will." One of her father's daughters was already matching her as a close relative in the database, and on Facebook she'd become friends with several paternal relatives. She figured it was only a matter of time before her father's three daughters discovered her existence. "I knew what being the last person to find out a secret felt like," she said. So she sent them a note.

All I could think when I read the Facebook Messenger exchange between Linda and her half-sisters was what a jarring and thoroughly modern way this was to meet one's genetic family. Linda's tone is effusive, vulnerable, alternating between being eager to know her siblings and allowing they might want nothing to do with her. She explains that she was conceived before their parents were married, but she's aware that her existence may nonetheless cause distress to them and their mother. Their replies are terse, stunned. Later, Linda would say she was so excited to have half-sisters that she didn't fully appreciate how her story posed a fundamental threat to *their* story.

Linda also set out to figure out her biological mother's identity, working with family trees she created based on maternal cousins among her relative matches, and homing in on a woman who was a classmate of her biological father's. When she figured out the identity of the woman she believes was her biological mother, she learned the woman had passed away six months earlier, leaving Linda to piece together a biography from clues: an obituary, which she read countless times, a photograph on Classmates.com, and conversations with maternal relatives. "You don't get to know the person, so you hold on to these little tiny scraps," she told me. The woman also left behind two grieving daughters who said they'd never known of Linda's existence. As with Jacqui's experience, the news Linda bore appears to have contributed to poisoning her overtures. She forwarded me an email sent by the husband of the deceased woman, calling Linda "deplorable."

Linda kept pursuing various members of both families, certain that if she could just reach the right person, she would be heard and, perhaps, welcomed, but for the most part, she's been shut out. She says she's heard that some in her father's family believe she is interested in money, an accusation that deeply wounded her. She forwarded me a "discontinuance of contact" letter sent by an attorney for her biological father and his wife, telling her that his clients believed her "assertions are based on unproven guesses and conjecture," and asking her to "slow down and research this in a more measured, analytical manner."

Negotiations over having Linda and her father submit to formal paternity testing broke down amid questions of who would pay for what, and mistrust on both sides.

"When you're fifty-one years old and this hits you, you feel like you're six," Linda told me one day, her voice breaking. "You just feel so vulnerable, so raw and vulnerable." She had a neediness I'd come to recognize from other seekers who felt they'd been denied their identities; entire conversations were conducted in a minor key. Linda told me she understood why her mother had given her up for adoption—"she was young, scared, and the love of her life just married someone else." But being kept hidden hurt so bad. Perhaps the way she'd come into the world had once been taboo; perhaps her mother's pregnancy had been a problem to be taken care of. But she herself did not want to be a taboo or a problem. She did not want to have to apologize for her own existence.

* * *

In this moment, we are learning just how painful and damaging it can be to discover that the truth of your genetic identity has been kept from you. One study out of the United Kingdom, looking closely at the experiences of sixty-five people conceived by sperm donation, found that nearly half hadn't been told how they were conceived till adulthood. For some, this created relationship problems with their families, and especially with their mothers, who they saw as having had a particular moral responsibility to tell the truth. They were, as the study put it in dry academese, "frequently more concerned about prior parental deception than by their parents' use of donor conception." However well intentioned their parents' decision was, these subjects experienced the information kept from them as a particularly "toxic" kind of omission.

Psychologist David Brodzinsky told me that in cases he's encountered over the years of patients who come in grappling with not having known until adulthood that they were adopted, "almost universally, it's the lie" that troubles them, rather than the fact of having been adopted.

"Relationships are built on trust. If you go through life thinking you are who you are, that your family is your family—that is the rock bottom foundation of who you are—and you suddenly realize, 'Wait, I have a different family, at least biologically; I've been lied to all these years,' it can be extraordinarily upsetting," he says.

There's a profound cognitive dissonance for the seeker whose revelation comes abruptly, accidentally, and well into adulthood. And on the other side of the secret there is sometimes a whole architecture built around the denial of the plain biological fact of someone's existence. Big secrets can be corrosive, and in the wake of their discovery, families can come apart.

"It's a secret, but at some point it becomes a lie, too," Linda Minten told me one of the first times we spoke. When I touched base with her again almost a year after our first conversations, she told me she hadn't spoken with her adoptive mother in well over a year.

* * *

Genetic revelations can be life-changing. This is what the early adopters knew before the rest of us had even heard of 23andMe. If questions of ethnicity and family relationships can be dramatic, revealing whole new categories of relationships—genetic siblings who deny their siblingship, for instance—health-related results are at least as profound. It's just that they're often also harder to understand.

Genetic counselors will repeat this till they're blue in the face: The results of health-related genetic testing can be very difficult to interpret. Bioethicist Thomas H. Murray points out that when people think of genetic risk for disease, they often conjure up examples like Huntington's disease, a rare inherited disorder in which the brain's nerve cells degenerate, typically causing uncontrolled movements, difficulty thinking, and psychiatric problems. Having a mutation in just one of the two copies of the relevant gene is enough to cause the disease. A person with the mutation will almost always develop the disorder.

But the vast majority of diseases in which scientists believe genetics plays a role are complex disorders, in which many mutations (or variants) in combination with each other and with the environment appear to play a role. The individual contributions of these variants tend to be subtle on their own. In other words, in most cases, health-related genetic testing is probabilistic, rather than deterministic; it can't tell you for certain that you have or are going to develop a disease. A positive result for particular mutations may simply indicate an increased risk of developing something—and how meaningful that increased risk is will vary. A negative result may not mean there's no risk for that disease. Other variants, as well as factors like a person's medical and family history, what they eat, whether they smoke, and what medications they take, will weigh into determinations of risk, which is why many doctors say health-related genetic testing should take place in the context of a medical consultation. It can be difficult to understand this kind of risk. "We have a hard time understanding probabilities—it's not at all an intuitive way of thinking," writes cultural psychologist Steven J. Heine.

23andMe's testing illustrates the complexities of understanding genetic risk. Beyond the three *BRCA1* and *BRCA2* gene variants that the company tests for, there are about a thousand other potential variants in those genes, any one of which could, on its own, elevate the risk of cancer. And beyond that, mutations in other genes can put people at elevated risk of hereditary breast cancer. " 'F.D.A.-approved' does not necessarily mean clinically useful," the *New York Times* editorial board wrote in a criticism of 23andMe's testing approach.

23andMe argues that more information is better, and says it has worked hard since its time in the FDA-imposed wilderness to make this information comprehensible, and to communicate to consumers the context and limitations of its testing. (For late-onset Alzheimer's and for Parkinson's disease—"serious conditions that do not currently have effective treatments or cures"—23andMe customers must explicitly opt in.) The company told me many customers who've tested positive for *BRCA1* and *BRCA2* gene variants would not have otherwise known they needed to worry; they were unaware of Ashkenazi Jewish heritage that put them at higher risk, or didn't know of close relatives with cancer. One customer, who was told by her doctor that she didn't need to test for these mutations, discovered through a 23andMe test that she had a *BRCA1* gene mutation that put her at high risk. She had a medically-recommended double mastectomy after confirming the finding with medical testing, and credits the company with saving her life.

"Our philosophy with genetics and science is anyone can be a scientist," 23andMe's Anne Wojcicki said in an episode of her company's podcast, *Spit*. "Here's your genome, you can learn about it." She has framed the scientific community as somewhat elitist, and has suggested those pushing back against the delivery of health care information directly to consumers are reminiscent of paternalistic fears in the 1960s and '70s that women couldn't handle at-home pregnancy tests. But scholars and medical professionals worry about the unintended consequences of consumer genetics. They warn about what one calls the "Walking Sick," who are inclined to skip screenings after consumer testing or forgo making lifestyle changes because of a false sense of security, and the "Worrying Well," who are driven to take more tests and perhaps, undergo unnecessary procedures. In a *New York Times* opinion piece, a genetic counselor wrote about the case of a young man who uploaded his raw data from 23andMe into Promethease, a third-party service that compares personal data against a database of medical literature. The disturbing results indicated a mutation associated with what the man understood to be a type of inevitable early-onset Alzheimer's. But after contacting a doctor, a geneticist, and a genetic counselor and finally getting a clinical test, he learned he didn't have the mutation after all.

23andMe told me that when it issues reports for the health risks that the FDA has given it permission for, like specific *BRCA1* and *BRCA2* gene mutations, it validates the results for accuracy in a number of ways. But the raw data are not validated in this way, and there can be errors. The company warns customers of the limitations of these "uninterpreted raw data," but it seems almost inevitable that more Worrying

Well will be created by an environment in which we're all supposed to be scientists. A small study found more of these errors, both within the raw data issued by consumer genetics companies and within the interpretation of those results by third-party services.

A number of health-related genetics testing companies have a hybrid model that requires physician involvement even as they market directly to consumers; some of them have begun to move toward working more with large health care institutions. Bloomberg's Kristen V. Brown believes that over time, there will be more convergence between the consumer genetics testing space and the medical establishment. "This kind of information about your health is never going to be not complicated," she says, "and you need an expert to help you interpret it."

* * *

There's another privacy issue inherent in genetic testing, given its power to reveal the risk of future disease. Both AncestryDNA and 23andMe say they don't share information with insurers or employers but warn that, in the event of a security breach, if genetic data could be linked to a person's identity, this could impact that person's ability to get certain kinds of insurance.

That's because in the United States, protections against genetic discrimination are imperfect. In 2008, Congress passed the Genetic Information Nondiscrimination Act, or GINA, a kind of preemptive civil rights law prohibiting discrimination based on genetic information in employment and health insurance. But there are loopholes in the federal law, as well as areas it doesn't even attempt to cover. For instance, GINA doesn't protect against discrimination in life, long-term care, or disability insurance, and, as *Fast Company* has reported, these types of insurers can ask "about health, family history of disease, or genetic information, and reject those that are deemed too risky." Only some states have laws that attempt to address the gaps in GINA.

"One of the situations I fear is, I apply for life insurance and they ask me, as they are wont to do, to sign HIPAA-compliant authorization releasing my medical records," says Mark Rothstein, a lawyer and bio-ethicist at the University of Louisville and author of *Genetic Secrets: Protecting Privacy and Confidentiality in the Genetic Era*. "Well, I haven't been tested by my doctor—but I've had a 23andMe test. And then they say, 'Have you undergone a genetic test in a research setting or direct-to-consumer?' And if I say no, I've committed fraud and they may deny paying me. And if I say yes, they say, 'Great, send it.'" Indeed, 23andMe warns of this very possibility, saying that failure to disclose health-related genetic information when asked by an insurer may be considered fraud. "I'm not aware of it happening, but it's just a matter of time," Rothstein says.

Advocates for genetic research worry that this threat may contribute to a chilling effect, discouraging people from participating in stud-ies and being proactive about their own health. And insurers fear that their model may collapse if lots of people who've undergone predictive genetic tests and know they're at high risk of developing certain dis-eases start taking out plans.

The idea of insurers using the possibility of future inherited dis-eases to deny us coverage is deeply discomfiting, dystopian even. Is genetic information sufficiently different from other kinds of medi-cal information—more precious, more private, more predictive of future health—as to warrant greater protections than other kinds of health-related information? This argument, known as "genetic excep-tionalism," is at the heart of GINA, and it's been the subject of long and heated debate. I sought perspective from Thomas Murray, the bioethi-cist who once wrestled with misattributed paternity in the case of the man with failing kidneys, and who many years ago chaired a Human Genome Project task force assigned to look at issues related to genetic information and insurance. Murray said that when he started on the task force, he was a proponent of genetic exceptionalism. The argu-ments in favor seemed compelling: Genetic information is a particu-larly personal and powerful kind of data. It is information that can serve as a kind of coded "diary" of a person's future health state; information that can reveal things not only about ourselves but about our relatives as well; information that, if it falls into the wrong hands, could give others the opportunity to discriminate against us. Insurers, sure, but maybe also totalitarian states.

And yet, as Murray turned the arguments over and over, he found himself unpersuaded that genetic information warranted special protection. After all, nongenetic information, like cholesterol levels and a penchant for skydiving, could also serve as risk factors in considering future health. And genes are not fate when it comes to disease; they are typically probabilistic. Murray worried that if he argued that genetic information was so different as to require its own class of protection, he'd be effectively arguing for a kind of genetic determinism, a mindset that genetics matters more than the countless other ways in which we define our own identities—an echo of what Nelkin and Lindee had warned about in The DNA Mystique. "Genetic information is special because we are inclined to treat it as mysterious, as having exceptional potency or significance," Murray concluded. He told me he was not opposed to GINA protecting genetic data, but that it didn't go far enough: The government should offer better protection for all medical information, not just the sort related to DNA.

The question of genetic exceptionalism is relevant to realms beyond the medical. Leading genetic genealogists argue against the idea that genetic information is so much more revealing and disruptive than what can be learned by traditional genealogy that it deserves to be treated differently. After all, CeCe Moore told me, genealogists could and did uncover family secrets back in the days when they worked with paper records alone. Looking at marriage and birth dates, they'd long noticed a number of children born after only seven months' gestation—implying either a surprising number of preemies or, more likely, a whole lot of shotgun weddings. And, genetic genealogist Blaine Bettinger has argued, when a state passes a law allowing adoptees access to their original birth certificates, it proves the point that "DNA does not have a monopoly on revealing family secrets." Besides, family members in the know have always had the ability to reveal secrets to one another, a fact of which adoptee Linda Minten is only too aware.

Still, as a practical matter, the amount of effort involved in making many such discoveries through old-school methods meant that in the pre-genomic era, genealogical revelations were limited to a smaller number of people—serious hobbyists who presumably became familiar with the risks inherent in their hobby as the years crawled by. You had to really want the truth back in the old days. Now the bar to entry is low, and many of the people testing are not familiar with the implications. The same factors that make autosomal DNA testing so powerful for helping researchers break through brick walls in their family trees also mean DNA can be much more jarring to people who otherwise would never have questioned their place in those trees.

Besides which, some revelations simply never could be made without DNA testing. Like Alice's.

* * *

For more than a week, Alice stared at her 23andMe results. She clicked through the features on the website, scrolling through her genetic cousins, and staring at the little flags in her chromosomes meant to tell her where she hailed from. Gerry's results hadn't come back yet, though they shortly would. Eventually, all the siblings would test, and they'd all turn out to be full siblings, with the same ancestral mix as Alice. But for now, Alice had only her brother Bill's results and her own, and by poking around, she found a feature that purported to show segments along their twenty-three pairs of chromosomes associated with people of Ashkenazi Jewish ancestry. Those portions were blue, and the rest of the chromosomes, represented as bars, were gray. She stared and stared, flipping back and forth between the image of her brother's chromosomes and her own.

Alice did not know what she was looking for, she'd tell me later, but she was stuck on these images. She was doing something she calls "thinking sideways," and over the course of her genetic genealogy research, she'd come to use this technique often. Thinking sideways,

Alice told me, means using intuition to make certain educated leaps, and it's also the trick to solving the crossword puzzles she does every morning. You can't be too literal with crosswords—usually there's a play on words going on, and you have to let your creative mind do some of the work.

So, Alice sat at her computer and clicked and stared, and after a couple of days, all of a sudden, she saw it. The truth was right there, plain as day, in the X chromosomes.

Alice's X chromosomes were represented by a single bar that was part blue and part gray, which meant part ancestry associated with Ashkenazi Jews and part ancestry not associated with Ashkenazi Jews. She knew she had two X's, one from each parent, so seeing all that blue told her only that she'd inherited some genetic material associated with Ashkenazi Jews; it did not tell her which parent was contributing this. But Bill had only one X, given to him by their mother, Alice the elder. He'd inherited his other sex chromosome, his Y, from their dad. And when Alice clicked over to Bill's results, she saw that the bar representing his X was all gray. This meant Alice the elder had contributed no Jewish genetic heritage to her son. It was through their father, Jim, that the mystery lay.

Alice dashed off an email to several siblings, including screengrabs of the images, and walked them through her reasoning. "It is dad's side of the family and I can prove it!" she wrote. "We'll probably never know if it's from our grandmother or grandfather," she added, "but at least the mystery is not quite as big."

But even then, she still didn't know exactly what she was looking at or how big the mystery would become.

* * *

In mid-September 2012, the professor in Illinois known for his skill parsing people's ancestry compositions got back to Alice. His email was filled with numbers—percentages and approximations, and references to "population sets and their fractions," and plus or minus this and that. He attached several images, including a scatterplot, or mathematical diagram, with tiny colored dots meant to indicate different ethnicities.

But the bottom line was this, he wrote: "What you are is 50 percent Jewish and the rest English/Irish/Scots or at least Western European. This is in fact as solid as DNA gets, which is in this case very solid indeed."

This was the moment when Alice knew for sure that the thing she'd discovered by sending in her saliva was way bigger than she'd imagined. The professor's email, combined with Alice's eureka moment in Bill's chromosomes, was clear to her now: It meant that her dad, Jim Collins, was not of Irish descent at all. He was not who he thought he was. The context for his life, and for Alice's, had been altered in a fundamental way.

Alice mulled over and over who her father was—who he'd been and who he'd become. He'd been dead for thirteen years, yet now his life was deconstructed, as if some hand had picked the stitches in one of Alice's beautiful patchwork quilts. He was Irish, *so Irish*, Alice had always thought. "He was raised in an orphanage—he didn't have anything else," she said. "He had his Irish identity." When Jim died, Alice told me, "he wanted an Irish wake. He didn't want people to be sad when he died, and so there were a lot of drunken grandkids." And Gerry, then a cantor at her Catholic church, had sung their father's favorite song, "Danny Boy."

The kids thought their dad looked Irish because he was Irish, and therefore this was what an Irish person looked like. "I grew up thinking Irish people were very little," Gerry told me. Their mom, who was half-Irish, was also short, after all. But Gerry remembered traveling to Ireland in 1990, with her husband and boys, and searching all over the country for her father's face in the crowds. It wasn't there.

"There was nobody that looked like my dad," Gerry said.

As a kid, Alice and Gerry had always thought their dad reminded them of the man Alice called "that famous Irish actor, Peter Falk," the guy who played the low-key, brilliant New York detective Columbo on the TV show. It was lots of things about Falk—the thick New York accent, his self-effacing nature, the social awkwardness coupled with that deeply analytical mind. Plus, Alice thought her dad *looked* like Columbo: the face and the build. It was only in adulthood, when Alice got her unexpected DNA results, that she looked up Peter Falk and learned he was Jewish, born to two people of Eastern European ancestry.

Alice sent the professor's conclusions to some of her siblings. Now she needed the genetic data from their two first cousins more than ever, and the wait was killing her. The kit for her maternal first cousin in Long Island had been sent to the wrong house and went missing for weeks before it was retrieved, while the saliva from her paternal cousin Pete's kit had to be processed twice by the 23andMe lab. Alice and Gerry no longer needed the cousins' results to tell them which side of the family had passed on their link to the Ashkenazi Jewish people—they knew now it was their father. But cousin Pete's DNA could help them understand why their father had never known about this heritage. Could their paternal grandparents, John Josef and Katie, have been Jews who assumed new identities when they came to the United States? Or perhaps the mystery was bigger than John Josef and Katie—perhaps their own ethnic backgrounds had been hidden even from them?

8

SEARCH ANGELS

When CeCe Moore started doing the work she'd eventually become known for, there weren't even words for the things she was doing. She and other citizen-scientists helped popularize the term "genetic genealogy" to describe their work, and Moore started using "unknown parentage" to describe situations in which a person doesn't know the identity of one or both biological parents.

The major genetic testing companies have made all sorts of innovations over the years to make it easier for people to find genetic relatives and to understand where their ancestors may have come from. But many of the techniques for deciphering family connections have come from citizen-scientists, a good number of whom volunteer their time to solve other people's genetic puzzles. Michael Hammer, the University of Arizona geneticist and FamilyTreeDNA scientific advisor, watched this unfold. "In the early days we'd go to these conferences and have workshops and train people on basic genetics," Hammer says. "Over time, because there were folks who were actually getting the data and working with the data, they started asking questions I couldn't answer." He remembers seeing retired physicists and engineers putting their genetic information on spreadsheets and writing their own code to analyze it. Many of these citizen-scientists told me of staying up late to try out new online tools, often getting lost for hours or even days in their research. Roberta Estes talks of these elaborate distractions as "chasing squirrels."

"I have a couple of collaborators I work with on genetic stuff," Estes says. "We'll be working on something and one of them will say, 'Hey, did you see this, did you try this?' and hours later it's three in the morning, we're all exhausted, and someone will say, 'Hey, whatever happened to the thing we were working on?' And someone will say, 'Squirrel. We saw a squirrel.'"

Estes told me that she spent so much time on surname projects traced through Y-DNA data at FamilyTreeDNA that years ago she was able to make a presentation on mutation rates on chromosome alleles among many generations of men born to a common ancestor born in 1495. The rates "differed dramatically from ones that had been published" to that point, she told me, and afterward, scientists attending the conference "came running down the aisles" to her. "This is so exciting!" one told her. "Can we have your data?"

Some genetic genealogists publish peer-reviewed papers on their findings alongside academics, lecture and teach workshops, offer their services to private clients, and act as consultants or unpaid, informal advisors to the DNA testing companies. They publish research in *American Ancestors* magazine explaining how they traced a three-hundred-year-old NPE using thirty-seven-marker Y-DNA tests to distinguish a man from haplogroup subclade E1b1b1 from a family of men with haplogroup subclade R1b1a2. Debbie Kennett publishes scholarly papers on the history of the ancestry testing industry and its social and privacy implications, and gives talks with names like "The Joy of Surnames" and "Mysteries of the *Titanic* Solved by DNA." Blaine Bettinger, an intellectual property attorney with a PhD in biochemistry and molecular biology, crowdsourced and compiled the shared centimorgan data for more than twenty-five thousand known genetic relationships, putting the ranges into a chart to help people decipher how they may be related to each other. He publishes books, runs a subscription help site called DNA-Central with courses, webinars, and a newsletter, and has a blog with posts like "Heteroplasmies and Poly-Cytosine Stretches—An mtDNA Case Study." The International Society of Genetic Genealogy, an influential volunteer-run organization founded in 2005, promotes genetic genealogy, tracks the industry, educates consumers on the science, and maintains an exhaustive wiki. There is no certification for becoming a genetic genealogist, though leading practitioners have collaborated on a set of standards intended to deal with ethical challenges and best practices.

If the genetic detectives can be obsessive, CeCe Moore leads the way, pulling all-nighters on behalf of other people's mysteries, pulling at threads until decades-old secrets unravel. In her work as a "search angel" for those trying to solve genetic mysteries, she's been known to work sixteen-hour days on her beige couch in the Orange County, California, home she shares with her partner and teenage son, staring into her MacBook. In the late 1990s, Moore was a struggling actress, artist, model, and singer, constantly hustling to make a living, when she decided to put a family tree together for her niece as a wedding gift. She considers this the moment when she was bitten by the genealogy bug. But before that, she'd long had an interest in the sciences and, as a young adult, had briefly thought about becoming a population geneticist before settling on music and the theater at the University of Southern California.

Moore started testing herself and family members, eventually getting forty relatives to test. She began answering questions at the International Society of Genetic Genealogy's Yahoo group DNA-NEWBIE, learning as she went along. In 2010, when far fewer than a million Americans had undergone commercial genetic testing, she started a blog series called "Known Relative Studies," to see how patterns of genetic inheritance played out in a real family—her own. "At that time, all we had was computer-simulated data," she says. Moore posted screenshots of various family members' shared genetic material, comparing three generations of her immediate family against a second cousin. It was a new way of approaching genealogy. Rather than looking solely at a paper trail, she could watch the biological trail of her great-grandparents—George, an immigrant from Australia, and Fredrikka, an immigrant from Norway—play out in their descendants, the shared segments of DNA getting chopped up and disappearing over the generations.

Moore told me that when she examined the Google analytics for her blog, *Your Genetic Genealogist*, she was struck by the fact that its readers had IP addresses from places like Harvard and Stanford. After spending so many years being ground down by a field focused on image, she found it gratifying to be recognized for her brain again. It drove home to her how wide-open this new field was—if academics were looking to a citizen-scientist like her for wisdom, she thought, there might truly be a place for her in it. She was in her forties by the time she found her calling, "the thing that I always should have been doing, but it didn't exist."

Eventually, all those years of struggling in the entertainment industry became precisely the background she needed to wind up in front of the cameras again, this time on behalf of science. Moore's talents lie not only in her ability to keep track of details, to multi-task, to follow her squirrelly obsessions up trees, but also in her ability to see patterns in what the DNA is telling her and to translate those patterns, and the science behind them, into explanations that others can understand. Moore has since become a regular fixture on television, the most recognizable face of genetic genealogy. She tells of her exploits solving unusual cases on 20/20, *Good Morning America*, and *The Dr. Oz Show*. There exists endless B-roll of her typing into her laptop, or scrawling family trees on whiteboards, or strolling with a thoughtful look on her face. Her paid media work has made possible the work she did for years for ordinary people, for which she didn't charge. She and her volunteer administrators, many of whom she helped train, run a Facebook group called DNA Detectives, which has well over 110,000 members, as well as several subsidiary DNA search groups, including one for foundlings, one for the donor-conceived, and one for discussing the emotional ramifications of genetic genealogy. In the past, to find relatives in databases, Moore often advised people to "fish in all four ponds," paying for testing at 23andMe and AncestryDNA, and then uploading their data at FamilyTreeDNA and MyHeritage, which accept transfers. She'd say that so long as she could find a predicted second cousin, she could usually solve a case, and she never knew which company, if any, might have that second cousin. These days, though, with the size of AncestryDNA's database, she simply tells people to upload there; combing all four sites is often not necessary. As recently as 2012, when Alice started her search, "getting a second cousin was winning the DNA lottery," Gaye Sherman Tannenbaum, a retired forensic accountant who's now a DNA search expert, told me. "It's more rare now *not* to get a second cousin match."

When possible, Moore and her fellow angels teach the seekers how to solve cases themselves—"teach a man to fish," the saying goes—but for years Moore pitched in on the most confounding cases herself, giving the seekers her long nights. She helped solve the case of Paul Fronczak, a complicated, years-long mystery involving a man who was abandoned as a baby, and then brought up by the wrong family after being mistaken for *their* baby, who had been kidnapped after birth. In 2015, she and her team cracked the true identity of an amnesiac known as Benjamin Kyle, who'd been discovered unconscious, naked, and injured behind a dumpster over a decade before. And, when DNA testing by Moore's own brother-in-law revealed previously unknown African ancestry, Moore began to dig and eventually traced his direct ancestry back to Madison Hemings, who most historians believe was the child of Thomas Jefferson and his slave, Sally Hemings.

Moore also helped a family uncover why their young adult daughter's DNA test revealed her not to be related to her father—it turned out she was the biological child of a now-dead convicted felon named Thomas Ray Lippert, who in the 1980s and '90s worked at a fertility clinic where they'd sought artificial insemination. Moore believes Lippert, who she says was a prolific sperm donor in the Salt Lake City area, intentionally swapped in his own sperm for that of the father in this and several other cases. The work she does, and what it reveals about human nature, can be disturbing. There is incest more often "than I think anybody could have imagined," she says. She offers resources to the adult children of incest and started a database to study the laws of inheritance in such cases, in order to make better predictions about how seekers' parents may have been related. She hears many stories about the mistreatment of unmarried pregnant women in past decades. "People thought they could get away with everything," she says—and indeed, they did, until genetic genealogy came along. In Moore's view, DNA is an equalizer, a revelatory force with the power to right past wrongs.

I got to know Moore a little through phone conversations over several years, though I met her only once. We were both speaking at a genealogical conference in Waltham, Massachusetts. She was there to give the keynote, as well as a basic primer on using DNA in genealogical research. She relayed so much information that by the end of the first hour, I had filled eleven pages of 8½-by-11-inch paper with cramped handwritten notes. She arrived just before her first talk and left right after her second, and in between, a stream of people waited to introduce themselves to her.

When we spoke—often when Moore was in the car traveling back and forth from various speaking engagements—she always sounded frenetic, her work life and her personal life bleeding together to give the impression that she was never really off. She had a particular kind of drive: Despite her accomplishments, she seemed perpetually dissatisfied. She worked seven days a week, eighty-hour weeks at minimum, for years splitting her time between working on *Finding Your Roots* and helping seekers pro bono. The only time I heard her talk about having free time was after she broke her shoulder on a trip abroad; then we talked for hours because she was having trouble using her computer.

* * *

One of the major forces of genetic genealogy, and a big chunk of the population that search angels like CeCe Moore have helped, is the adoptee community. Unlike Linda Minten, most set out with the knowledge that they're adopted, and use DNA to make an end run around state legislation that in many cases still limits adoptees' access to their original birth certificates. Many of the techniques pioneered to help adoptees find their birth families were created on behalf of, and by, adoptees. In their search for origins are lessons for many other seekers who never expected to find themselves on such quests.

Before the 1990s, most American adoptions were closed, says David Brodzinsky, a professor emeritus of clinical psychology at Rutgers University who spent his career researching psychological issues within the adoption and foster care systems. Now, the majority are open, which means many young adult adoptees grow up having some relationship with their birth parents, or at least knowing something about them. But among those born many decades ago, and among international adoptees to the United States, where openness is less common, gaining access to basic facts about themselves—their own birth names, ancestral backgrounds, the names of the women who bore them—can be incredibly challenging.

"The way we handle adoption in this country still is archaic," says Amy Winn, the president of the American Adoption Congress, which advocates for unrestricted access to original birth certificates, framing it as a basic civil right for all adopted people. Yet with DNA testing, Winn says, "we can go out and find ourselves." It's possible within ten years none of these laws will be barriers to adoptees in search of their origins.

Winn, a Sante Fe, New Mexico, therapist who works on adoption trauma issues, is in her early seventies, and she told me she began searching for her birth parents as a young woman. Like many others, she eventually resorted to hiring a search agency to help her. She was forty-five by the time she gained access to a short letter written by her adoption agency with basic facts about herself. Those three paragraphs with the most meager of biographical details—the fact that her mother was of English and French Canadian descent, and that her maternal grandfather suffered from bad eyesight—meant the world to her. When she spoke to her birth mother for the first time, eighteen months before the woman died, "I heard her voice and I literally felt myself go upright—I had roots to the earth for the first time," Winn said. Friends told her even her voice sounded different after this.

Winn told me the first time she met her half-siblings she felt more comfortable in their presence than she ever had in the home where she was raised. "We are connected to other human beings on this earth—genetically, historically," Winn says. "It's like magnets. I may not know where the other end of the magnet is, but I'm being pulled to it. How can we answer anything about ourselves if we don't know what our roots are, if we don't know who our people are?"

When I spoke with Winn, I was struck by how similar her words were to those of other adoptees. Genetic genealogist and adoptee Gaye

Sherman Tannenbaum told me her experience of ancestry testing was "grounding, it's validating. You've put your stake in the ground—and it's in Belarus or it's in French Canada or it's in New Orleans. 'Wow, my people came from here.'" In the middle of the twentieth century, psychologists coined a term for the sense of disorientation and disconnection that fuels adoptee searches. They called it "genealogical bewilderment."

The desire to know more about one's birth family is normal, says Brodzinsky, and not indicative of a bad relationship between a child and her adoptive parents. "It's part of the universal search for self. It's finding out who you are. It's trying to understand characteristics. We look at our parents and older relatives and we can project ourselves into the future," he says. Searching for family doesn't always mean that an adoptee wants to meet his or her biological parents. But it can mean an adoptee asking questions about where in Russia she comes from, or what her biological mother might have looked like, or why that biological mother couldn't keep her baby. Brodzinsky told me that when he speaks with adoptive parents, they often ask what the chances are that someday their children will search for more information about their biological family. He tells them, "One hundred percent."

In explaining why they search, many adoptees also cite their desire to find out more about their genetic family health histories. They speak of a feeling of frustration and alienation at not being able to answer the most basic questions when they are at the doctor's office. "I just found out heart disease runs in that family," an adoptee named Krista Driver told me months after discovering the man she'd thought of as her genetic father was not. Now, she had learned that four people in her genetic father's family had died of heart disease. "When I go to the doctor, I'm going to ask them to do a full workup," Driver said. "I would never have had that information."

But the interest in medical issues is broader than managing the risk of a discrete disease. Researchers Thomas May and Harold Grotevant have found that adoptees want their genetic family health history because that information informs how they imagine and plan out their lives. The process of seeking out genetic information may serve another purpose as well, May and Grotevant suggest—it may allow adult adoptees, who had no choice in whether to be adopted, or by whom, to exercise their autonomy in making meaning out of it.

* * *

But what about the adoptee's biological parents? How do they feel about being found? Brodzinsky told me social science research suggests adoptive parents, adoptees, and biological parents all tend to fare better, psychologically speaking, under open adoptions—and that, historically, research into closed adoptions in which children have managed to track down their biological mothers shows that most of those mothers are ultimately glad to be found. But this is not a guarantee, and the word "ultimately" is key—seekers are on their own timeline, and it may not be the same for those they seek. It is a different thing to find someone than to be the one who is found, and it can take time to come around to it. "We tell people if they get rejected . . . wait," says Karin Corbeil, a co-founder of DNAAdoption.org, a nonprofit that helped pioneer genetic genealogy methods to help adoptees. "Many times, the shock is so overwhelming to people." This can lead to situations in which a seeker's interests can seem to be in conflict with the person they seek at the most delicate time—at the very beginning of a relationship that has the potential to be intense and intimate.

"What we're seeing now is a more direct conflict between the privacy rights of genetic parents who have given their children up for adoption and adoptees," says bioethicist Thomas May, who is himself an adoptive father. Technology is "advancing to the point that it's forcing us to rethink the very manner in which we adjudicate balancing these interests." Should parents have an expectation of privacy if they relinquished their children decades ago when the culture was radically different, when psychologists said the best thing for a child was not to know his or her biological family?

I have heard of cases in which birth mothers did not want to be found because their children were conceived by rape or incest, or where such traumas were the explanation for misattributed paternity. Brianne Kirkpatrick, a genetic counselor who works with those affected by genetic surprises, told me about one seeker she worked with: "The woman said, 'I told my mom that the man she told me was my biological father wasn't my biological father. And she had this look come over her face, and she had to tell me her uncle had raped her.'"

Because of the shared nature of genetic information, it's not always clear whose desires should be honored when it comes to the question of relatedness. In 2017, AncestryDNA made it possible for consumers to opt out of seeing or being seen on relative match lists. It framed the decision as part of its commitment to customers' privacy and control, but the comments section underneath the company's blogged announcement was riddled with dismayed messages, and one man accused the company of "discriminating" against adoptees.

Concerns about the privacy rights of mothers and fathers are why Roberta Estes, who consults to people who want to better understand their genetic results, told me she doesn't take on cases in which people are searching for their genetic parents. Estes doesn't object to people knowing the identities of their genetic kin; indeed, she wanted badly to know her brother, Dave, and she did not see her letter to him as likely posing an unwelcome disruption to his life. But, she said, it was impossible to know whether birth parents would welcome being found by their genetic children—if only because they'd never told anyone about what was seen back then as the sin of having a child outside of marriage. Estes worried that some practices she'd heard about—tracking down birth mothers through social media and sending registered letters to siblings—could be experienced as invasive and meant that "whether the mother chooses to or not, she's outed because of the way the child chooses to find her." The only way to know if a mother welcomed her child's contact was by contacting her, and the act of contacting in and of itself changed the mother's life.

"I'm not saying I'm opposed to it," says Estes, who refers cases such as those to DNAAdoption. "I'm saying I have a concern and I don't do that work."

DNA testing poses questions that people in the adoptee community have thought about for far longer than such testing has been around. Will my birth mother/father/half-sister be happy to know me? And, more broadly, what is that person to me? What do we mean when we speak of "family"? How much does genetics get to tell us about who we are?

Some people I spoke with, who did not know that one or both of their parents were not genetically related to them till they tested, told me that in retrospect, they always sensed it. Dani Shapiro, author of the memoir *Inheritance*, about her experience discovering by DNA testing that she was conceived by donor sperm, has said she thinks she always wondered, obsessed as she was for years with writing about family secrets and the search for self—which is not to say that, in discovering the deeply meaningful truth about her biology, the man who raised her was any less her father. He "loved me into being," she writes.

Stories like that of Shapiro's sixth sense raise the question: Could there be a biological imperative to seeking out one's own genetic family? This is what evolutionary biologist Jerry Coyne, author of *Why Evolution Is True*, suggested to me. During most of human history, "kinship was of real importance to humans because we lived in small groups and we knew each other and it was important to know who our kids were," Coyne says. "During this period of time we developed—not consciously but evolutionarily—a way of sussing out who shares our genes and who doesn't." We give special preference to our own children, as do other species throughout the animal kingdom, he says. Natural selection may foster "mate guarding" and sexual jealousy among males, who want to know that their children are biologically related to them. "It becomes a matter of special interest to humans to know who shares their genes," Coyne says, and it may be that adoptees' urge to seek out kin is based, in part, on this evolutionary urge we all share.

But it can be difficult to tease out biological impulses from cultural influences. Other seekers who discovered biological surprises through testing have said they had no inkling those surprises were coming. And I've also spoken with people who believed—hoped, even—that they were not related to their genetic families, yet testing confirmed that they were. Ours is a culture in which we make family many different ways—through traditional conception, through adoption, through donors, through stepparents—and in my interviews, I found people defined their families with nuance and deliberateness. Kimberly Leighton, a professor of philosophy at Georgetown University and herself an adoptee, is deeply critical of the term "genealogical bewilderment," which she says reflects an old-fashioned notion that heredity is key to self-understanding, and an assumption that the best kind of family is a genetic one. The real cause of distress among people who are adopted or donor-conceived and don't know their genetic families, Leighton argues, may be a response to a culture that values heredity as the true measure of family and identity. This doesn't mean that adoptees and donor-conceived people may not want to know more about their genetic parents; it simply means that wanting to know may have much to do "with how we value genetics," she says.

"I really think it's the quality of the relationship that matters, and genetics plays a very small part in that," says Thomas H. Murray, who has written extensively on adoption and genetic bonds. Parenthood is "propinquity. It's being with someone hour after hour, day after day, when they're crying as an infant, when they're sick and need your care." But if genetics pales in comparison to a lifetime of interactions, Murray says, it's also true that there's a long history of belief that inheritance matters, "partly because cultures cluster along with genetics, and culture matters—cultures are often what define alliances and enmities." The questions of what makes a family are as old as families themselves; only now, in this new era, many more families than before are grappling with them. In one of his books, Murray points to a heartbreaking declamation from a man in ancient Rome whose foster son had been reclaimed by his biological parents. When that child was away at war, the man asked the court, "Whose was he? Was it not I who hung from the city walls in suspense? . . . Who bandaged his wounds when he returned? Who washed away the blood?"

I asked Harvard Law School professor I. Glenn Cohen how US law defines family and whether that has been changing in recent decades.

He told me that, starting in the 1990s, courts have dealt with new kinds of families—many of them with children conceived by assisted reproductive technologies like donor insemination, egg donation, and surrogacy—by placing them into the context of an old-fashioned nuclear family unit. This has at once reinforced the traditional idea that the nuclear unit is the model for the American family, and allowed for a reinterpretation of the family to accommodate gay couples and single parents. In doing so, Cohen said, courts were also reinforcing a social definition of family over a genetic one. They did not want to encourage "strangers" stepping into the nuclear family—even if those strangers were genetic relations.

It remains to be seen, Cohen said, whether consumer genetics will push the balance in the other direction, toward a view that reinforces genetic essentialism; he does not think these tests will. And he added that discoveries of paternity as a result of genetic testing are unlikely to lead to major legal issues around inheritance or child support. Parents are under no obligation to provide for their adult children in their wills, so the adult child of an NPE situation would have a hard time making a claim on his or her genetic parent's estate. As for minors discovering their genetic parents via testing, "most states have a rule that paternity essentially becomes final after a period of time unless challenged," Cohen said, and besides which, sperm donors are generally considered not to have paternal obligations when they donate anonymously through sperm banks.

In recounting the story of finding his birth mother, adoptee Tim Monti-Wohlpart, the national legislative chair for the American Adoption Congress, told me that attempting to answer the question of nature versus nurture in definitions of family and how it shapes us is a fool's errand. So are assumptions that an adoptee could have room in his heart for only one mom. That represents a failure of the words we have for such relationships, not a failure of the heart, because both things matter—for adoptees and for parents. "People do a lot to have children that are genetically related to them," says Ellen Wright Clayton, who teaches law, pediatrics, and health policy at Vanderbilt University.

"But I believe down to the tips of my toes that family is about relationships, about the social part, however it's formed. And I think people get to choose."

* * *

There is something uniquely American about the growing mass of seekers that has arisen over the last decade. Genetic genealogist Debbie Kennett, who tracks the industry, estimates about 80 percent of the market for DNA testing is in the United States. The next biggest markets are in the United Kingdom, Ireland, Australia, New Zealand, Canada, Finland, and Scandinavia.

Europe's market is seen as several years behind the US market because of a complex tapestry of policy, pragmatism, and culture. In general, says David Nicholson of UK-based Living DNA, Europeans are more concerned than Americans with matters of privacy and security. The regulatory landscape for direct-to-consumer DNA testing abroad is fragmented, with some countries, including France and Germany, imposing many more restrictions on how tests are sold. Andelka Phillips, a legal academic, says the European Union's General Data Protection Regulation, which went into effect in 2018, as well as additional EU regulations, may force more transparency onto genetic testing companies and may even restrict how some American companies operate in Europe.

There are also cultural reasons for this reluctance. In some parts of Europe, there are concerns that using genetic data to search for ethnic differences is reminiscent of the rise of eugenics in Nazi Germany. "I have some German friends who think that the idea of testing to see what percentage of ancestral origins you are is the *opposite* of what you should be doing," CeCe Moore told me. Besides which, says Nicholson, "people in genealogy across Europe generally have their records going back thousands of years anyway, so how's DNA going to help them?"

There's long been an idea that the New World is where immigrants remake themselves free of the past, yet it's precisely because Americans are so divorced from our histories—because we lack the continuity and the ancestral knowledge that comes from generations living in the same place, in a largely homogeneous nation—that many of us wonder where we came from. We are intrigued by our roots, curious about all the things that took place before our ancestors traveled across the ocean, or across the American continent, and settled here. We want to know the specific bits and pieces that make us *us*.

A media industry has arisen around this cultural curiosity. The Harvard scholar Henry Louis Gates Jr. has been at the forefront of this, hosting *Finding Your Roots* as well as other PBS specials combining genetics and genealogy. Ancestry sponsors *Who Do You Think You Are?*, which has run for many seasons on NBC and TLC and also traces celebrity genealogies. A growing number of people are taking heritage trips to the countries they discover they hail from when they take a genetic test, and the DNA companies are getting in on the action. AncestryDNA teamed up with a travel outfit to offer guided ancestry tours to countries including Italy, Ireland, and Germany; and 23andMe got together with Airbnb to showcase "homes and heritage travel experiences based on a customer's DNA."

And on YouTube, the "ethnicity reveal" has become a thing. In thumbnails, the people who post these invariably have shocked looks on their faces and tend to employ a lot of exclamation marks. One woman—who titled her video "Reading My Ancestry DNA Results! I Been Lied to My Entire Life!"—filmed herself discovering that "my dad isn't really my dad." She pauses the video to call her dad, then gets back in front of the camera. "He told me that he always had an inkling," she says, exhaling slowly and trying not to cry. So far, it's garnered more than 1.5 million views.

Over the last few years, I've spoken or emailed with more than four hundred people about their DNA testing experiences. They told me about how they were converted into seekers, about the wormhole that is the search for the past and for the self through genetic genealogy. They told me about the many tools they used, about testing in multiple places, and about using all sorts of third-party inventions, including browser extensions, software, and tools at sites like GEDmatch and DNA Painter to sort, compare, and predict relationships, interpret DNA segment data, keep track of notes, and filter matches.

Some seekers subscribe to newsletters like biologist and genetic genealogist Leah Larkin's *The DNA Geek*, which chronicles techniques like the McGuire Method and the Leeds Method (to "help you depict and organize your DNA matches, respectively"). Online, seekers commiserate over their brick walls, and discuss their maternal and paternal haplogroups. Some talk about pulling all-nighters, about waking in the middle of the night to use the Ancestry app on their phones, and working it until their phones die. Their spouses call themselves genealogy widows and widowers. There's something terribly addictive about this hobby and the questions that it poses: How much do you know about yourself and your family? How much *can* you know? ("I have been called a crack dealer," FamilyTreeDNA's Janine Cloud told an audience at a genealogy conference, and everyone got it.) The seekers talk about checking for relative matches several times a day and about the kismet of noticing family surnames on street signs. One told me of a brick wall so befuddling and engrossing that at one point an entire wall down a hallway of her home was covered in the names of ancestors and relatives on sticky notes, which she spent her evenings and weekends researching and rearranging. Seekers are stalkers of the living and the dead.

In my experience, both seekers and search angels are more often female. This makes sense: Brodzinsky told me that among adoptees, it's the women who are more often seeking to connect with their biological families. Perhaps it's in part because of the way girls are socialized to express their emotions and "to actualize a search based upon those feelings," he says, and perhaps pregnancy, childbirth, and motherhood form a strong impetus for women to connect with their own biological mothers.

Laurie Pratt, who became a search angel after discovering she was the product of an NPE, told me that after working in genetic genealogy awhile, she started to develop an instinct for it. Piecing together people's family trees draws on all sorts of knowledge—knowledge of history, of psychology, of geography. For example, if a person wants to figure out the identity of a genetic relative who goes only by a nickname on Ancestry, and offers no family tree, she recommends searching that nickname with a variety of email address suffixes, like @gmail, @hotmail, @yahoo. "People are creatures of habit," she says, so if you're calling yourself, say, ShyBunnies123 at Ancestry, chances are you're using ShyBunnies123 someplace else. Pratt says reading the obituary of her genetic father's brother helped her get a feel for the family he came from, the way it saw itself and the things it valued. As a search angel, knowing the landscape of Orange County, California, helped her trace the identity of a baby abandoned in a supermarket, because she knew, based on where certain people lived, how easily they could have reached that supermarket by way of a certain freeway. And she could see, based on clues she was finding online, a pattern of troubles within a family that suggested certain people might be desperate enough to leave a child.

"As people, we're the sum of other people's choices: the kinds of jobs they take, where they choose to live, the stories they have," Laurie says. Whether through social media or hundred-year-old records, a genealogist can start to see patterns: Is a family upwardly or downwardly mobile? Have they lost a child? How does dysfunction beget dysfunction? These are just clues, not certainties, but they can help make the work of a laptop detective more efficient, telling her where to invest her efforts. Laurie told me her fastest solve involved a woman who didn't know the identity of her father, but had a big clue of a match: a close DNA relative, though exactly how close it was hard to tell. Laurie pieced together the DNA relative's tree using records online, figured out who his grandmother must be, deduced that the relative was actually a half-brother to her seeker, and from that discovered the identity of the seeker's father. It took about twenty-five minutes.

Pratt can use relative matching in her genetic detective work because it works so well. While the science of ethnicity predictions as carried out by DNA testing companies has been heavily criticized as not as precise or accurate as is implied by those pie charts of 3.1% *this* and 26.2% *that*, relative matching relies on a different approach.

"Matches are incredibly reliable," says geneticist Barry Starr, Ancestry's director of scientific communications. "What you're doing with relatives is you are looking for large chunks of DNA that are exactly the same between two people. Unless you're a close relative, you're not going to show identical DNA."

There are cases where relative matching isn't perfect. Predicting just *how* two people are related is a challenge; a company can tell you the amount of genetic material shared but, in many cases, can't tell you precisely what the relationship is. A few years back, MyHeritage DNA updated customers on improvements meant to address some cases of "false positives," in which they were overestimating the shared segments of very distant relatives, making them look closer than they were. And in one unusual case, genetic counselor Brianne Kirkpatrick told me a woman reached out to her after 23andMe predicted she had a "half-sister." In fact, the mystery "half-sister" turned out to be a first cousin who shared an exceptionally large amount of genetic material with Kirkpatrick's client, a fact she and the client were able to verify through additional testing of another family member. "She's still trying to recover her thinking around her father because for a month, she thought of him as a man who cheated on his wife," Kirkpatrick told me shortly afterward. "She said she's having a hard time shaking it."

Because of the way genetic material is shuffled and passed down randomly, with chunks reconfigured, trimmed, and lost in each generation, you may not share detectable DNA with, say, a fourth or fifth cousin or beyond, which means they won't show up for you on your list of matches. In some cases, this is even true for third cousins. As computational biologist Yaniv Erlich has pointed out, "Even if you are connected to a person who came over on the *Mayflower* in 1620 (about thirteen generations ago), chances are that you did not inherit any of this person's DNA. If we could do a genetic test, you and your *Mayflower* relative would look like two unrelated people." In other words, even though someone shows up on your genealogical tree, that doesn't necessarily mean they'll be in your genetic tree. (Science journalist Carl Zimmer points out that this throws cold water on our genetic essentialism, shattering whatever illusions we might have about inheriting some important ancestor's royal blood, or another ancestor's ignominy, via their DNA.) But the system and science of relative matching through autosomal DNA almost always works quite well for the research Laurie Pratt and other search angels do, relying on multiple cousins, ideally as close as possible, coupled with traditional genealogy techniques, to understand precisely how people are related.

Though, in truth, Laurie told me, by 2019 she was rarely doing search angel work anymore, because seekers were finding so many close relative matches that they could easily unravel their own cases in a day with just a bit of guidance. It turns out, seeking is increasingly the easy part. It's the finding that can be complicated.

* * *

What happens when the seeker finds? When an adoptee or another seeker has identified her long-lost kin, what does she do with that information? No chart or browser extension can answer this question, nor can algorithms predict how the people she's identified will react to being found. This is when some seekers prowl the Internet, looking for clues into whether they want to know these strangers in all but blood. It is, of course, breathtakingly easy to figure out your father's career on LinkedIn, his political leanings on Facebook. Social media makes it possible—once you've discovered *who* your father is—to learn *what* he is.

The question of what to do with genetic information once you have it has inspired its own class of advice. Not every seeker reaches out. "Who am I to upend your whole life with this knowledge?" one woman said about her decision not to tell her biological father of her existence. But many have a desire for contact. And then: How should a seeker contact a genetic relative, if she decides to, and in what fashion, and what should she do if she meets with silence? With outright rejection? Some people reach out through a DNA testing company's messaging system or by email, whereas others turn to old-fashioned snail-mail letters even as the rest of the world is moving away from them. (No matter how cutting-edge the technology for identifying a relative, there's nothing like the gravitas of a physical piece of paper for contacting him.) Some genetic genealogists advise first taking screenshots of family trees and other materials proving their match, in case a relative decides to delete his or her account after being contacted.

There are "two main schools of thought," says Stephanie Talley, who found her sperm donor father through genetic testing and, as an administrator for a Facebook group, sometimes counsels other donor-conceived people about how to write that first letter to a donor. The approach she suggests is "initially being very light and friendly and maybe not giving them everything right up front. Just saying, 'I've found this connection. Are you interested in getting in touch? This is who I am.'" The key is "making yourself seem like somebody who is friendly, not terribly dramatic . . . because it's a lot. It's heavy anyway."

The other school of thought, Talley says, is for the seeker to offer lots of information about herself because "what if this is the only chance I have, and they tell me 'Don't contact me again,' and I never got to say what I wanted to say?" Talley says she understands this instinct, but putting everything into a letter can make a seeker seem "a little stalkery" and produce "this really high-pressure letter for the recipient." Besides, holding back allows for more conversation fodder later. "Why give up everything that might be bait for them to meet you?"

Those just finding their parents in adulthood are often in a defensive crouch, scared of rejection, and this perspective can seep into their letters, Talley says. She understands it; she tells fellow seekers that if they don't throw up from the stress and anxiety of these early interactions with biological family, they're doing better than she did. "It feels like the most important job interview ever."

And if the seeker's parent does reject her, and she's laid her heart out on the screen, how will she feel then? Because, in the end, these may be genetic parents that seekers are reaching out to, but they are also strangers—strangers who can choose, in ways most parents can't or don't, whether to have relationships with their children. Talley does not want fellow seekers to feel like they made themselves vulnerable to a stranger who then breaks their heart. She does not want them to regret saying something—or, for that matter, *not* saying something. So, the guideline she gives seekers is this: Imagine the worst: the rejection, the silence, whatever it is. Then write a letter you can live with if you get that response.

Sometimes, in the search for family, a whole lot of clues can add up to the wrong answer. One woman, who discovered she was donor-conceived via 23andMe, told the magazine of the American Psychological Association that early on in her search she mistook the wrong man for her biological father, "resulting in some uncomfortable interactions." And sometimes, getting it right feels like a mixed blessing. An adoptee I'll call Lisa, who tracked down her mother as an adult, told me their relationship could be strange and awkward, akin to the early stage of dating, with the initial nervousness, the testing of the waters, the struggle to set expectations. When they first talked, her mother asked, "Do you hate me?" Lisa did not hate her mother for relinquishing custody; indeed, she was grateful her mom had done it. But now, "If I don't call . . . soon enough, often enough, she gets mad," Lisa wrote in one email. "I just can't give her as much as she wants."

Laurie Pratt never did hear back from her biological father. "I want to see him in person," she told me one summer day. By then, it had been about a year and a half since she'd first contacted him and he'd deleted his test. She respected his boundaries and did not contact him after that single follow-up letter. She told me she could understand how her existence complicated his life and why he might not want to know her. But she continued to get to know cousins from that side of the family; they were her cousins, too. And, after hearing from another seeker who did something similar, she was mulling the idea that she might one day travel to her father's area and look for an opportunity to discreetly view him. After all, he lived in the world. He went to public events. "I just think it's weird to say I've never seen my father in real life," she told me.

In 2018, a group of Canadian third cousins on Laurie's father's side contacted Laurie after reading her blog, to ask if she wanted to come to a family reunion. So she and her husband flew up to their rural farming community, and a cousin showed her the house where Laurie's great-great-grandmother had once lived. She came back feeling empowered by the gift they'd given her—the gift of knowing her roots. And she realized something: For years, she had walked around genealogical conferences with a name tag identifying herself by her maiden name, Laurie Pratt, which was how she was known in genealogical circles. She'd fielded questions about her last name—"Are you related to Enoch Pratt?" "What about the Pratts of Springfield?"—and felt like an imposter, because she wasn't actually a Pratt by biology. Now, she decided to legally change her name, hyphenating her husband's last name with her biological father's last name, in order to claim her identity as someone descended from his genetic line.

"I'm not out to anger him, and I certainly don't think he'd be like, 'Hooray!' but this is reality," Laurie Pratt McBriarty-Sisk told me. "I just want to deal with reality."

9

27 PERCENT ASIA CENTRAL

Among the theories that Alice and Gerry entertained in the early
months of their investigations were several that were historically
intriguing. Alice wondered whether her grandparents John Josef and
Katie might have been Jews who somehow claimed Irish identities
when they emigrated to America. Or perhaps the explanation was further back—maybe her grandparents were Irish Jews who'd hidden their
religion (or had it hidden *from* them) all their lives.

These ideas seemed far-fetched, but they were ways of explaining
something that seemed unexplainable. Besides, history is far-fetched.
There are groups of people who've survived by passing. These include
the converso Jews in Spain and Portugal who were forced to convert to
Catholicism, some of whom continued to practice their faith in secret;
and light-skinned African Americans who passed as white in America, escaping rampant discrimination at the steep cost of leaving their
families and communities. These secrets were in the headlines long
before consumer genomics arrived. Madeleine Albright did not know
her parents had been Jewish till a *Washington Post* reporter uncovered
it in 1997, when the secretary of state was fifty-nine years old. John
Kerry did not know his paternal grandparents were born Jewish until
adulthood. Genetic testing has revealed many cases like that of CeCe
Moore's brother-in-law, who did not know about his African ancestry
until he spit into a tube.

Recreational genetic testing can suggest ancestral homelands
and migration patterns that point to a family's true history, hidden for
decades or centuries, prompting testers to reexamine not only their own
families but their beliefs about race and ethnicity, too. Some seekers are
caught entirely unaware when they make this discovery through DNA,
whereas others seek out testing because they've long had questions. In

the face of those findings, they must come up with their own definitions of ethnic identity, informed by what they were told as children as well as by things they were pointedly never told. It forces a reckoning with history, with the things their ancestors did in order to be able to pass on their genes to them.

* * *

Rosario Castronovo grew up without much in the way of cultural identity, but he clung to his mother's story that she had Sicilian heritage. At twenty-one, when he decided to legally change his name to distance himself from a father he describes as abusive, he chose Italian first and last names in a nod to his mother's culture. He'd heard "Castronovo" meant "new castle," and that's what he believed he was building—a new life, a new identity. Nobody in his family stopped him, though he'd later learn that many of them knew this was a fable.

Come to think of it, a lot of the decisions Rosario made as a younger man were nods to a heritage he'd later learn he could not claim. He joined the Catholic church as an adult and was baptized. He studied and began singing opera. He proposed to his Italian American girlfriend after flying her to Italy for the millennium. He was "trying to define who I was." Many of us have family stories, memories, holidays, habits, and language to assist in the constructions of our ethnic identities. Rosario filled in the blank parts with what he thought it meant to be Italian. He was "searching," he'd say later. "What was going to make me *me*?"

It was when he was about to be married, in the early 2000s, that he decided to find out more about the background of his mother, an orphan raised in foster care. Rosario wanted children, and he imagined that one day they'd ask where they came from, just as he'd asked his own parents. He wanted to be able to give them an answer.

In the town hall in the small town in Vermont where his mother had grown up, Rosario found not one but three birth certificates for her, and they were rife with redactions. Strange. One listed the race of his mother's father as "negro." Rosario was mystified. Perhaps his grandfather had been a dark-skinned Sicilian mistaken for a black man. But more likely, he thought, putting the information together with old census reports that sometimes named his mom's paternal side as black and sometimes as white, the birth certificate was right, and there were important, fundamental facts he did not know about himself and his mother. He did not go to his mother yet because he wanted more evidence before he rocked her world. *Maybe she doesn't know,* he thought.

Over the next few years, Rosario used genealogy skills to research his mom's side, but there were details he could not fill in from the paper trail alone. So, he embarked on genetic testing. He would eventually learn that he is about 18 percent sub-Saharan African, as well as smaller amounts Native American and Asian, all through his mother's side. His elderly mother was uneasy when he went to her with the results of an early autosomal DNA test he'd persuaded her to take.

"I said, 'Did you know?' and she said, 'Yes,'" Rosario says. She said, "I didn't tell you because I didn't want you to go through what I went through." She told him about her difficult life in foster homes, about her foster father directing a racial slur at her and her little brother, about her brother being sent away to a boys' home because, she understood, "he had darker skin."

And Rosario began to learn about the history his mother had wanted to protect him from. He researched the small town in Vermont his mom came from, traced how the slate quarries and mills attracted black Americans, traced how the black families in the town peaked toward the turn of the twentieth century and then began to disappear. He learned of the death of his thirty-year-old great-uncle in what a 1930 newspaper article called a "fire of mysterious origin," when many black families had left the town, and wondered about the circumstances of that fire.

And eventually, he came to understand some of the context for why his mother had lived out her childhood as an orphan, even as her parents were both living nearby. He learned that his grandmother was white and married to someone else when she had two children with Rosario's black grandfather. Looking in archives, he discovered his grandparents had both been sentenced for adultery in the 1940s, though only his grandfather served time, going in when the two children they'd had together were small. Rosario did historical research and wondered if his grandfather's race and his relationship with a white woman were the real reasons he sat in jail. Part of the judge's ruling in Rosario's grandmother's case was that she no longer "make her residence" with his grandfather. She eventually divorced her white husband and married someone else, and Rosario's mother had only the haziest memories of seeing her biological parents as a young child. It was as if her family had never existed.

* * *

You can't tackle questions of ethnic identity without first considering what the sciences can teach us about the concept of race. In recent years, as research into human genetic difference has advanced, discussions about how to define race and ethnicity have become a kind of meme of American culture, hashed out in the pages of the *New York Times* and *The Atlantic* and on Twitter before eventually deteriorating into scientific racism in the Internet's darkest places.

Those of us who don't study population genetics or anthropology for a living often come into contact with this debate, implicitly at least, via at-home genetic testing, and those little pie charts that purport to be able to tell consumers whether or not to root for Germany in the World Cup. Around the 2018 Winter Olympics, AncestryDNA aired one ad that implied a creepy kind of genetic essentialism. It showed a figure skater majestically crossing the ice, while a voice-over mused on what made her so good, conflating certain cultural stereotypes with biology. An "ethnicity" pie chart popped up in a corner of the screen, suggesting it was, perhaps, her 48 percent "Scandinavia" that accounted for her "precision," her 27 percent "Asia Central" that accounted for her "grace," and her 21 percent "Great Britain" that explained her "drive."

"One of the biggest criticisms of this whole industry . . . particularly in the social sciences, is that the existence of these tests is going to make people think that race is buried in your genes, and reinforce this essentialist view," says sociologist Wendy Roth, who studies race and genomics. As the early history of genetics demonstrates, this is more than an academic concern. As Roth put it in a lecture some years back: "The view that certain groups are essentially or genetically different can quickly morph into the view that they are essentially inferior, and that has been used to justify oppression, forced sterilization, eugenics, and even genocide."

What does the science show? "If you do genetic analysis, you actually see that we are a mosaic of heritage, and we come from different places in the world, and that racism has no scientific basis," computational biologist and MyHeritage chief science officer Yaniv Erlich told me. Erlich said he sometimes wonders how history would have played out if modern genetic tools had existed a hundred years ago. On the one hand, the power to determine someone's ancestry would have been incredibly dangerous in the age of eugenics. On the other hand, he says, perhaps greater scientific understanding would have allowed people to see that eugenics "is bullshit."

The science shows that for most human traits shaped by genetics, there are more differences within population groups than between them. And, at the risk of crushing the hopes and dreams of neo-Nazis, the notion of racial "purity" is bunk, given the genetic overlap between different populations and the fact that most genetic variation is "distributed as a gradient," as the American Society of Human Genetics puts it, so there's no clear cutoff between one population group and another. Over the course of history, "human populations appear to have repeatedly split, merged, and interbred," University of California, Berkeley, bioengineering professor Ian Holmes writes. As a 2018 piece in *National Geographic* pointed out, there's more genetic diversity among people in Africa than on all the other continents combined because Africa is where we all began and where human beings have lived the longest—underscoring the point one geneticist made that "there is no homogeneous African race."

The concept of race as distilled into popular culture is the product not only of scientific research but of history, culture, and politics as well—the product, in other words, of flawed human beings. Columbia University sociologist Alondra Nelson has described the concept of race as "a way to sort human communities in such a way as to justify social inequality; this sorting is neither natural nor inevitable."

"Deciding which characteristics matter—whether it's skin color or eye color or height or parentage—that is a social process and it's something that different societies do differently," Roth observes, giving as an example a poet who came from the Dominican Republic to America in the 1960s. Back home, the woman had been considered "indio" or "trigueño," "intermediate categories that are part of a wide continuum" of racial terms. In the United States, where culture is influenced by racial purity laws like the one-drop rule that historically defined someone's race on the basis of a single sub-Saharan African ancestor, the woman was "black." The woman's biology was the same, but her racial definition, and all that came with it, shifted as her culture did.

But if race is, as Harvard geneticist David Reich puts it, "fundamentally a social category" rather than a biological one, it is nonetheless quite real as a lived-and-felt experience for Americans. And so is racism. Among social scientists, there's a good deal of concern over how to talk about the research that scientists are doing into genetic differences among populations. They are careful about the language because they know the power of words like "race," and the assumptions embedded in it, and because they know full well how the history of genetics is intertwined with racism and anti-Semitism. As if to underscore this deeply troubling entanglement, James Watson, the co-discoverer of the structure of DNA, has repeatedly asserted that black people are innately less intelligent than white people, and has linked skin color with sex drive. Watson "fell prey to the same temptation to which the old Progressive eugenicists had succumbed: he let genetic determinism amplify his prejudices and biases," writes historian Nathaniel Comfort. "It is, perhaps, an occupational hazard of those who think about genes too much."

So, how should scientists talk about the real differences among populations that they see? Because while the concept of race morphs from society to society, privileging some groups at the expense of others, Reich writes that there are differences, on average, among populations for a number of traits. "We now know that genetic factors help explain why northern Europeans are taller on average than southern Europeans, why multiple sclerosis is more common in European Americans than in African Americans, and why the reverse is true for end-stage kidney disease," Reich wrote in a 2018 *New York Times* op-ed tied to his book, *Who We Are and How We Got Here*, on new findings from the study of ancient DNA, many of which have come from his lab.

And understanding these differences is important not only for our understanding of disease, evolutionary biologist Jerry Coyne told me. "If there were no genetic differences between groups, it would not be possible to retrace this history . . . to know how we spread across the planet and where we came from," Coyne says.

That discussions of population differences have been misconstrued to justify bigotry is deeply problematic, Reich writes. But also problematic, he argues, is the fact that some scientists are suggesting there are no population differences for fear of having their findings misunderstood. "Arguing that no substantial differences among human populations are possible will only invite the racist misuse of genetics that we wish to avoid," he argues, by causing laypeople to feel misled and patronized. Reich believes the gap created by the reticence of the scientific community is being filled by people like Watson and science journalist Nicholas Wade, whose 2014 book *A Troublesome Inheritance* was denounced by more than a hundred population geneticists and evolutionary biologists, including Reich himself, as filled with racist stereotypes. Also filling the gap: "race realists" (modern-day proponents of old-fashioned scientific racism) and other white supremacists.

"While race may be a social construct, differences in genetic ancestry that happen to correlate to many of today's racial constructs are real," Reich wrote in his 2018 op-ed, provoking fierce debate and underscoring the difficulty scholars are having agreeing on the very language for this discussion. In *The Atlantic*, UC Berkeley's Ian Holmes wrote that while Reich had done a good job of describing "race's complex relationship to ancestry," his op-ed fell down when it used the word "races," albeit in quotes, calling it "a loaded term for scientific outreach." In *BuzzFeed*, more than sixty scholars, mainly social scientists, published an open letter denouncing Reich's op-ed, and elsewhere Joan Donovan, a sociologist who has studied white supremacists' use of DNA ancestry testing, described his words as "biological reductionism." Reich stressed to me that this was the very opposite of what he meant.

At a moment when so many discoveries are being made into human genetic diversity, some scholars are deeply concerned about what happens to genetics research when it gets distilled and simplified by the culture at large, not to mention how assumptions shape the science itself. They know that now, as in the eugenics era, human beings looking for patterns will see what they expect to see. They know history, and they fear its repetition.

* * *

The criticisms of ethnicity estimates are manifold. For one, there's the use of the word "ethnicity," which is not quite right, since there's a distinction between one's ethnicity and where one's people came from historically. Ethnicity is in good part about "culture and tradition and self-identity," CeCe Moore told me—things that you cannot measure with DNA. "Someone can have a cultural ethnicity without the corresponding scientific basis for it in their genome."

Anita Foeman, who studies perceptions of race and ethnicity in genetic testing at West Chester University of Pennsylvania, told me about one of her subjects, a young woman of Chinese descent raised by a German American family. When her boyfriend's parents inquired into her ethnic background, she told them she was German. They laughed, thinking she was joking. How strange, the young woman recounted, to have people look at you and tell you you're something different from what you are. "What does a DNA test tell you when it's 100 percent not how you see yourself?" Foeman asks. "What makes you German? And can she own 'Chinese' even though she knows nothing about that culture?"

Yet the largest of the consumer genomics companies, AncestryDNA, has popularized the term "ethnicity," and it has taken off in part perhaps because of what cultural psychologist Steven J. Heine calls our "essentialist thinking," which encourages us to conflate genes with identity. Besides which, other, more precise terms aren't consumer-friendly. "Biogeographical ancestry is such a mouthful," says genetic genealogist Debbie Kennett.

The other major criticism of ethnicity estimates is that, while they are billed as estimates, they are presented as offering a degree of precision that companies cannot actually deliver on, for reasons having to do with their imperfect methods for interpreting the data in our genes, as well as the complications of history. They are "a young and developing science," a MyHeritage DNA speaker told a genealogy conference, reiterating what anyone who's had their results "updated" by a company knows. A number of genetic genealogists have blogged and spoken about the differing ancestry estimates they've gotten from various companies. "I had almost 50 percent Scandinavian when I took Ancestry's first test," Roberta Estes told me. "You know how much Scandinavian I have now, several iterations of the test later? Guess what: I don't have any Scandinavian."

"It's dead easy to tell continents apart," Barry Starr, Ancestry's director of scientific communications, told me. The DNA of someone whose ancestry is from Asia looks quite different from that of someone from Europe or Africa. The ancestral estimates of different companies will generally agree when it comes to such broad geographic areas. But because continent-level predictions are usually unsatisfying to people interested in tracing their family histories, companies like Starr's have been working to give more granular results, and this is where things can get complicated.

So, how do these estimates work? Two major factors account for why they often vary between companies: reference populations and algorithms. Ethnicity estimates are by necessity all about patterns and probabilities; they are a process of comparing a person's DNA against that of known populations and looking for the best possible match. Reference populations are sets of people, typically culled from public datasets and from a company's own customers, who are meant to stand in for the world's historical populations anywhere from several hundred to a thousand years ago, because they have verifiably deep roots in those places. But the actual people chosen to represent those populations will vary from company to company, as will the number of them. The idea is to match the hundreds of thousands of DNA markers, or SNPs, in each customer's DNA up against these reference populations, looking for what Bennett Greenspan of FamilyTreeDNA calls a "best fit." But here the second factor comes into play: Each company has its own method of crunching the data. "Each company thinks, 'Our algorithm is the best,' but each algorithm has its pros and cons," says Yoav Naveh, who heads MyHeritage's DNA team.

Over time, companies have honed their algorithms and increased their reference populations with an eye toward making their results more accurate, and indeed, they're much better than they were a few years ago. In late 2018, for instance, AncestryDNA unveiled a new ethnicity prediction system. Previously, the company had been working with a reference population of three thousand people; now its reference population was more than sixteen thousand, augmented in great part by customers from its own databases whose deep roots in particular geographic areas had been robustly vetted and verified. In a flash, AncestryDNA's Benin and Togo reference population went from 60 to 224. In the old reference panel, there were 394 samples from the vaguely named "Asia East," but nothing was specifically marked as "Japan" or "China"; in the new panel were 592 Japanese samples and 620 Chinese ones, making it so that the company could better distinguish between these two groups of people.

The other thing that changed was the method by which the company looked at the DNA. Instead of looking at hundreds of thousands of individual markers along the chromosomes, Ancestry had begun splitting the genome into segments—1,001 short windows of between three and ten centimorgans in length—so that they could examine these markers in context.

"There is no French SNP or German SNP or Korean SNP," Starr told me in 2018. "Each SNP is just more or less likely to be from a different region. So when you look at individual bases"—that is, the As, Cs, Gs, and Ts that make up a person's genome sequence—"you're just looking at *most likely* ethnicity for that SNP. But when you look at a longer segment, you can smooth out the results." He went on: "Imagine you have a SNP and it's most likely Iberian—so, from Spain or Portugal. But when you look at it in the context of a longer stretch, what you see is that on each side, it's surrounded by SNPs from the UK, and you have this one Iberian one in the middle. What the algorithm can do now is correctly interpret the Iberian SNP"—it's actually only an *apparent* Iberian SNP—"as a UK SNP. Since the DNA of Europeans is all pretty similar, that Iberian one is probably found pretty often in the UK, as well."

Before AncestryDNA updated its approach, a consumer might have wound up with some small, mystifying percentage of Iberian ancestry, and wondered about family stories left untold. Now, when she looked at her results, the Iberian was simply gone, smoothed out by the new algorithm. Combined with another feature that looked for closer genetic ties within a larger ethnic group and combed users' family trees to identify common ancestral birthplaces and migration routes, the company was now promising to pinpoint more than 350 global regions and subregions for its customers. Since then, that number has grown even more.

Even so, the science was a matter of probability, which AncestryDNA tried to indicate by giving ranges to its estimated ethnicities. Just because the algorithm makes a prediction of a certain ethnicity "doesn't mean that that's the only answer," Starr said. "It's just the most likely answer." Likewise, 23andMe offers a "confidence slider," allowing customers to see what their estimates look like "at different probability cutoffs."

A few months after AncestryDNA's update, 23andMe updated its own results, telling customers it could now evaluate customers based on more than one thousand regions. Nothing in my DNA was any different, of course, but my results changed: The "Scandinavian" grew and the "French & German" fell. (If you want to get really technical, this apparent "French & German" ancestry likely did not, in my case, include any actual French ancestors; this was a kind of labeling imperfection, emerging from the fact that the two populations are geographically close and can be genetically similar.) That tiny bit of Korean that the company detected a year before had dropped down to 0.1 percent "South Asian" in a previous update and now disappeared altogether. So much for that lost Korean ancestor I'd been wondering about. Now, 23andMe informed me with astonishing specificity that the area in Sweden that the Scandinavian portion of my DNA looked most similar to was called Skåne County. I asked my dad and he told me that yes, this was where his grandmother had been born.

To attempt to read the past through the genes, you need more than knowledge of science, statistics, and algorithms. You need to understand history, and history is profoundly messy. MyHeritage's Naveh told me the knowledge of genealogists and population geneticists on staff is crucial because it is in the movement and mingling of populations that DNA remakes itself. Populations that in the past were geographically isolated (Naveh pointed to the Finns and the Japanese) or married among themselves (like Ashkenazi Jews) are relatively easy to pick out. But for other groups, it's harder. Wars, migrations, porous borders, shifting nation-states—all these things make it difficult to distinguish neighboring populations from one another, or to define them by modern standards. An ancestry prediction may say your people came from "Germany," but what we know as Germany only dates to

1871—your distant relative may have actually called Saxony or Prussia home. Different companies also define areas differently. Ancestry grouped England, Wales, and northwestern Europe into one category, Starr said, because "the English Channel isn't that big, and there was a lot of traffic between England and northwestern Europe, and so genetically they're more similar than you might have guessed." Perhaps in the future, with an even more robust reference panel from those parts of the world, the company's experts might be able to break down these populations even further.

Genetic genealogist T.L. Dixon has likened ethnicity estimates to cocktails, to be enjoyed responsibly. "My goal is to get beginner and intermediate genetic genealogists to think more like population geneticists ('5 percent of my DNA is similar to British Isles populations . . .') rather than genealogy astrologers ('I have 5 percent Irish in me . . .')," he writes on his blog, *Roots & Recombinant DNA*. Just because "a portion of your genes are similar to [those in] Ireland does not necessarily mean your ancestors are from Ireland," unless you can confirm it through the paper trail. It's not the customers' fault that they engage in such simplistic thinking, he adds: "It's the fault of the DNA companies, who falsely advertise that you can take a DNA test and it'll tell you what's *in* you."

CeCe Moore says she worries about the fuzziness of ancestry predictions. "I've had people contacting me and saying, 'My father must not be my father because I got British and he's German!'" In the absence of any other evidence, you can't make that leap based on an ethnicity estimate alone, she says. And she also worries that because ethnicity estimates get "a bad rap, and much of it deservedly, it has led to misconceptions about the reliability of [relative matching], which is solid science."

Another problem has to do with the "undersampling" of non-European populations, including those of African and Latin American ancestry, as well as of indigenous peoples. "No company actually has adequate African samples," says Dixon, who specializes in tracing African ancestry. Joanna Mountain, 23andMe's senior director of research, says undersampling goes back a century, to the early days of genetics research. "Most of the researchers have been in the US and Europe and they have studied people who have been in the clinics nearby, or in the military," Mountain says. "In early days, that tended to be people who were white." Razib Khan, a population geneticist and science writer who previously consulted for FamilyTreeDNA, told me he believes undersampling has persisted in the consumer genetics world in part because the majority of people who are seeking these tests are of European ancestry. "It's customer demand," he says. "It's a business decision."

In *MIT Technology Review*, journalist Antonio Regalado has written about how undersampling is a problem when scientists develop polygenic risk scores, which survey many thousands of locations across the genome to come up with a composite risk score for conditions like diabetes, some cancers, coronary artery disease, and irregular heartbeat, which are subject to complex genetic factors. "Because the algorithms were developed by using large databases of health and DNA data mostly from people of recent European ancestry, the tests aren't as accurate in other groups, like black Americans, Hispanics, and Asians," Regalado observes. Many scientists have been trying to rectify this undersampling, concerned that it will magnify existing health disparities. As bioethicist Thomas May pointed out to me, the current state of things is not only fundamentally unfair to the groups that are undersampled, though it is certainly that; it is also unfair to everyone. "If we don't have diversity in the people being tested, our data will be less rich and less useful," he says.

In the context of the ancestry results offered by a company like 23andMe, undersampling generally means that a person's estimates wind up being more regionally broad, Mountain says—instead of predicting that a person's ancestry is, say, Coastal West African, or Southern East African, their estimate might just say "Sub-Saharan Africa." "We did start thinking about rectifying the undersampling and lack of research on non-European populations pretty early in 23andMe's history," said Mountain, who heads up the company's research diversity efforts. Mountain's PhD in genetics focused on African populations, and as a Stanford professor, she studied genetics and linguistics among the peoples of Tanzania. She recalled a company phone call years ago between herself, Anne Wojcicki, and Henry Louis Gates Jr. during which the subject of 23andMe's genetic analysis for African Americans came up. She told them she was concerned about gaps in the company's service "because the research hasn't been done to support the product for African Americans. So, I'm not comfortable marketing as aggressively,'" Mountain recalled. "Anne said, 'Let's do the research that's necessary to fill in that gap.'"

That was the beginning of a series of initiatives to increase the diversity of the company's samples, both to increase its understanding of health issues among different populations and to improve its predictions about customers' origins. (The company's first big initiative, for instance, looked at the genetics of diseases that affect African Americans.) In some cases this has meant providing free kits to customers who can trace their ancestry to certain populations; in others, it has meant collaborating with researchers doing genetic work in parts of the world where 23andMe would like to improve its data and where research on such populations will improve the diversity of genetics research more generally. The bulk of its efforts have focused on Africa and Southeast and Central Asia.

Mountain told me that she worries about how consumers understand and interpret concepts like "race" and "ethnicity." She avoids using those words herself, concerned that they may invite reductionist thinking about how different we are from each other, and 23andMe itself favors "ancestry composition" when it returns results on where a consumer's roots may come from. Mountain says the company has at times consulted social scientists for advice on how to frame its work in order to avoid what's known as "racial reification"—that is, reifying, or reinforcing, the idea of essential racial differences. But, she added, the desire for nuance and scientific precision was sometimes in conflict with consumers' needs for results that were comprehensible and intuitive.

"They like the pie chart, we have found," Mountain says.

* * *

At a 2018 Toronto conference on the ethical, social, and privacy implications of genetic testing, a light-skinned blonde woman stood up and told the audience that she was black. Françoise Baylis is a bioethicist and philosopher at Dalhousie University in Canada. She told the audience that she knew some people questioned her identity when they looked at her; they might wonder whether she'd taken a DNA test to back up her claim. She had not. She did not need to. In the 1960s, her mother was "the first successful case of racial discrimination in employment in Canada. In testimony, my mother had to stand in court and answer the question 'Are you a Negro?'" If sometimes Baylis's own blackness was suspect in the eyes of others, she said, other times, it was her whiteness that was suspect. She was accused of "passing" when she was simply walking around in her own body. From this position of in-betweenness, and through her work looking at assisted reproductive technologies, she had thought deeply about questions of genetics and identity.

Baylis said the crux of her work was a theory that identity resides not in the body alone, nor in the mind alone. Rather, she said, she'd come to believe identity was a thing we work out in constant negotiation with others—"a narrative, relational account of who I am," a thing that shifts over time, with a person's experiences, and that is in part dependent on how other people respond to us. Identity is a lived thing, she said, a series of stories about a complex experience. She worried about beliefs that the age of genetic testing was giving rise to. "Identity is dynamic, it's social, it's cultural, it's political, it isn't this . . ." She gestured to the screen beside her, which showed a screengrab from the AncestryDNA website. "It isn't percentages," she said.

These issues around differing cultural, political, and scientific definitions of identity burst into the mainstream American political discussion less than a month later. Shortly before launching her presidential exploratory committee, Massachusetts senator Elizabeth Warren attempted to put to rest the taunts of President Donald Trump, who had mocked her claims of Cherokee and Delaware heritage with the nickname "Pocahontas" and, at one point, challenged her to prove it. Warren released the results of DNA analysis by Stanford University population geneticist Carlos Bustamante, showing "strong evidence" of Native American ancestry six to ten generations back. But instead of ending the matter, Warren's move succeeded in provoking raucous debate about who decides Native American identity, about the implications of politicizing claims of genetic ancestry, and about the conflation of culture and genetics.

Leaving aside the other big criticism of Warren's decision—that she "caved" to the race-baiting of the president—her decision angered many Native Americans. The experiences of the Havasupai with University of Arizona researchers, as well as other experiences in which Native and indigenous peoples have felt exploited by scientists, or have seen their cultures appropriated by white people, seemed to distill into this one moment. In the 1800s, naturalist Samuel Morton, whom the journalist Elizabeth Kolbert has called "the father of scientific racism," collected and studied skulls in order to divide humanity into a hierarchy of five races, with Native Americans second to last and blacks (or "Ethiopians") at the very bottom. The University of Alberta's Kim TallBear, who researches ideas of race in genetic science, draws a straight line from history like that to "old-school notions of racial purity that continue to inhabit genetics even today." Indeed, she says, Native Americans are understudied by geneticists in part because of tribes' resistance to this research, based on a history of mistrust.

The Cherokee Nation secretary of state, Chuck Hoskin Jr., issued a statement that "Senator Warren is undermining tribal interests with her continued claims of tribal heritage," and did interviews explaining that his tribe had its own way of determining membership, primarily through a "rigorous process" of documentation tying a person to the

Dawes Rolls, "a turn-of-the-twentieth-century census of Cherokees living in what was to become northeast Oklahoma."

"We fought long and hard for this status," Hoskin told NPR. "It's a legal status that we use to fight for the rights of Native children, to protect our lands. And when someone comes in and boasts that they took a DNA test so they're Indian, we think that erodes what it means to be a Native American in this country, even if they don't claim a specific tribal nation citizenship."

Do we define ethnicity by way of cultural heritage or genetic assessment? And who gets to decide? TallBear, a member of Sisseton Wahpeton Oyate, a South Dakota tribe, says suggesting that genetics alone can define Native American identity is in direct conflict not only with Native Americans' political authority but also with how Native Americans themselves define kinship—as a web of relationships tied to living people and to ancestors kept alive through stories. She says that DNA-only definitions are shallow and that autosomal DNA testing may even exclude some legitimate tribal members, given the fact that we lose some of the genetic material of our ancestors with each passing generation. She has written that Warren "and much of the US American public privilege the voices of (mostly white) genome scientists and implicitly cede to them the power to define Indigenous identity." (Warren later apologized to the principal chief of the Cherokee Nation and told the *Washington Post* she regretted claiming Native American heritage.)

"The question is not whether her DNA analysis is accurate," Columbia University sociologist Alondra Nelson wrote in the wake of Warren's release of her DNA results. "It's whether it can tell us anything meaningful about identity. The truth is that sets of DNA markers cannot tell us who we really are because genetic data is technical and identity is social." The concern voiced by Nelson, TallBear, and others is that believing too fervently in the truth-telling ability of the genes, as interpreted by imperfect science and imperfect human beings, may flatten and oversimplify our understanding of who we are.

And yet, as Joanna Mountain observed, and as much of the advertising around genetic ancestry testing testifies, these messages are also appealing. Roberta Estes told me she is troubled by that AncestryDNA commercial in which a guy named Kyle trades in his lederhosen for a kilt, both because it ignores the limitations of the current science of biogeographical ancestry analysis and because it offers a facile answer to a complex and deeply personal question.

"Believe me, we all hate it," Estes says. "The reason Ancestry runs it is because it's so effective for sales. Because people are like, 'I can take this test and I can get The Answer.' . . . Well, you know what? There is no The Answer." When it comes to the question of *Who am I?*, "there are many answers and a lot of it depends on how you ask the question."

10

WHAT TO CLAIM

How does consumer ancestry testing affect how we think about our genetic differences? Joanna Mountain of 23andMe says she's long wondered about that. Her company didn't set out to help people explore their ethnic and racial identities, she says, but over time it became clear that this was exactly the work they were doing. In 2008, the company debuted something called Ancestry Painting, which attempted to show customers' ancestral mixes by dividing their chromosomes into crude geographic categories—Africa, Asia, and Europe—based on the technology available at the time. "It was very disturbing for me because it aligned closely with the big three racial categories," Mountain said. She worried such a simplistic breakdown might reinforce people's notions of essential biological difference along racial lines.

But in time, the company's results improved, becoming more refined, and Mountain started hearing more and more stories about consumers discovering unexpected ancestral stories, including ones that crossed the color line. And Mountain wondered what effect such results might be having. When consumers discovered this diversity within themselves, would that make them more flexible in how they thought about race? Meanwhile, 23andMe was studying this diversity, as well as the legacy of historical constructions of race. In 2015, 23andMe population geneticist Kasia Bryc, along with Mountain, David Reich, and others, published a study from the company's consumer data revealing that close to 4 percent of those who identified as white Americans had at least 1 percent African ancestry, consistent with an African ancestor within the last eleven generations or so. (In southern states like South Carolina and Louisiana, about 12 percent of whites had 1 percent African ancestry or more.) Henry Louis Gates Jr. observed of these findings: "What hasn't been confirmed until now is how many self-identified 'white' women and men are walking around today with recent 'hidden' African ancestry in their families—you know, white people, who, at least according to the old, notorious 'one-drop rule' of the Jim Crow era, would have been considered legally 'black.'"

Research is demonstrating the importance of how we frame discussions of genetic difference. In *DNA Is Not Destiny*, cultural psychologist Steven J. Heine writes about his work showing that American subjects who tend to believe that genes determine life outcomes also tend to be more bigoted toward African Americans. "Thinking about genes as underlying essences and having racist thoughts seem to go hand in hand," he writes. But if the way we approach such questions can doom us, it can also save us. Heine points to more research showing that when we focus on our sameness, we do better, as in his study showing that adults prompted to consider genetic commonalities are more likely to offer environmental explanations for racial stereotypes.

Mountain hasn't yet been able to answer the question of what impact companies like hers are having on attitudes about genetic difference. She helped design a study conducted by Northwestern University that found a third of Americans believed that genetics "totally" determines racial identity, and she hopes future collaborations can gauge whether DNA testing causes this belief to grow or shrink. Beyond this, research on the topic is in the early stages. A 2014 study out of Columbia University found that having subjects read a newspaper article about genetic ancestry tests appeared to "magnify the degree, generality, profundity, and essentialness of the racial differences people perceive to exist," but additional work suggests that the effects are not so clear-cut. Professor Anita Foeman of Pennsylvania's West Chester University, who studies people's experiences of ancestry testing, told me she thinks the technology is opening up much more complex and sophisticated conversations about race. In one talk I watched, she put up pictures of a diverse range of people and invited the audience to speculate on who matched a particular pie chart. The blonde woman was part Asian, according to testing, while the child of Indian immigrants had European ancestry. It was impossible to predict, which was exactly Foeman's point—the genomics age can't help but blow up our neat categories.

Sociologist Wendy Roth, now at the University of Pennsylvania, has found evidence for both theories: that ancestry testing reinforces our sense of immutable racial categories, and that it breaks them down. The difference depends on the test-takers themselves. In a recent paper, she found that a tester's knowledge of genetics made the biggest difference in how he or she understood the implications of a test. Among more than eight hundred subjects, Roth found that for those with high knowledge of genetics, taking a consumer DNA test generally reduced their belief in genetic essentialism. For those who went in with very low knowledge of genetics, "there's some indication that it may actually increase belief in genetic essentialism," she says. In previous research, Roth had observed that people were also motivated toward racial essentialism—or away from it—based on what they wanted to believe. One subject pleased to discover apparent Basque origins said it reinforced her sense that race was something revealed by DNA, while another subject, who wasn't eager to embrace the Middle Eastern origins suggested by his test, preferred to see race as a social and economic construct.

"The way people interpret scientific evidence is not neutral," Roth told one audience, likening the way people interpret ancestry tests to what they see when they look at Rorschach tests. Which gets us to another point: How people determine what portions of their pie charts to claim and what portions to ignore is a complex process indeed.

* * *

People filter their ethnicity estimates through a complex web of cultural and personal realities. They bring ideas about truth and authenticity, fantasies about different cultures, and notions of the past. They bring their loyalties and longings and resentments, not to mention their genealogical knowledge for times when the paper trail clashes with the genetic one. In other words, they don't accept their results uncritically.

But sometimes, those pie charts can offer them just what they've been looking for. In 23andMe's podcast, *Spit*, company co-founder Anne Wojcicki tells the story of an adoptee who wrote to say how meaningful the company's ancestry composition report had been in her life. Upon discovering her true genetic heritage, she had quit her Bay Area job and gone to live with her people, an indigenous population near the Arctic Circle. She told Wojcicki she felt at home for the first time. "And I take this story," Wojcicki recounted, "and I walk over to my research team and I go, '*You better be right.*'"

Indeed, they'd better. Because consumer genomics is changing how some of us see ourselves. Wendy Roth has found that most testers tend to stick with the ethnic identity they had before the test, even when results reveal something unexpected. But for the sizable minority motivated to consider another identity, they pick and choose what to claim, often weaving the new knowledge in with an existing identity. What they choose is based on their aspirations and on how others respond to their claims to a newfound ancestry. In that sense, their ethnic identities are indeed—as philosopher Françoise Baylis describes it—a kind of ongoing social negotiation.

Interviewing one hundred early adopters of DNA testing, Roth and colleagues found that white testers seemed more driven than other groups to seek out new ethnic and racial affiliations. These seekers offered a host of explanations. They spoke of a desire to gain a sense of belonging and identity, and to distinguish themselves from what they saw as the blandness of being "just white." Roth told me she was struck in her interviews by how many European Americans had lost connection to their ancestral roots.

But Roth also noticed that white testers were drawn to the "cultural cachet" of being multiracial. "For our White respondents, it is not being non-White that is most desirable, but having a bit of something non-White," Roth observed. One subject, an adoptee who initially thought of herself as white, took a mitochondrial DNA test that suggested an African haplogroup. She incorporated that into her identity and began telling those close to her that she was "mixed," even as she allowed that she did not have to face the social consequences of blackness "because I don't have Black skin."

Roth found that testers who identified as black or African American were far less inclined to incorporate new ancestral knowledge into their identities. In part, that's because they tended to identify strongly and positively with their existing identities; unlike white respondents, they did not describe their race as boring and plain. As well, Roth says, the routine rape of black women by white men under slavery, and a historical construction of race based on notions of white purity, meant the African Americans she interviewed had always assumed they were more than one thing. Finding evidence of other ancestries "doesn't change the label that society applies to them and it doesn't change the way that they think about what it means to be African American," Roth says.

In her 2015 paper, 23andMe's Kasia Bryc found that African Americans in her study tended to have about 73 percent African ancestry and the rest mainly European. Bryc's study found that more than half of the contribution of European ancestry to African Americans occurred before 1860—in other words, during the time of slavery. History is ugly, Anita Foeman says, and sometimes, the ugliest actions of our ancestors are what carried our genes into the future—leaving African Americans who discover their great-great-something owned slaves and committed rape to grapple with the fact that "genetically, you're related to those people. It's not *they*; it's *you*," Foeman says. "That messes you up." This ugly reality is why so many people were outraged by a 2019 AncestryDNA ad that they argued whitewashed the systematic sexual violence, coercion, and exploitation under slavery, offering a romanticized explanation for why many African Americans have white ancestry. "Abigail, we can escape to the North," an 1800s-era white man tells a black woman in the ad, holding out a wedding ring. "There is a place we can be together, across the border." For black Americans, "DNA and documentary ancestry information is as painful and traumatic as it is illuminating," journalist Kimberly Atkins pointed out in an incredulous tweet. "These are not love stories." Ancestry apologized and said it was pulling the ad.

The DNA age forces reconciliation with the past, and with the countless ways in which it lingers and shapes our perspectives now. It forces reconciliation with what our culture, politics, and policies have told us about what race means. One of Roth's African American subjects described a sociopolitical motivation for not embracing the apparent European ancestry revealed by a DNA test: "African Americans number only about maybe, what, 36 or 37 million? . . . I don't want to dilute or make that count fall down, because my whole experience has been African American." Besides, he said, he did not want to run away from the historical traumas of his people. A subject in a different study conducted by Foeman put it another way: "My momma used to tell me, 'If the cops stop you, they're not going to ask if your momma is from Ireland,'" said a dreadlocked university professor who identifies as black, after getting DNA results suggesting he was primarily of European descent. "Even though I know that race is a social construction, it is as real as oxygen," he said.

Like Roth, Foeman has spent years looking at race through the lens of genetic testing. She's found that even as people of color expect more diversity in their ancestral backgrounds, African American testers routinely underestimate just how much European ancestry they have. Foeman, who founded West Chester University's DNA Discussion Project, which since 2006 has tested more than three thousand people, has found women are generally more flexible about their racial identities, as are younger people compared to older ones. She says white and black subjects both tend to overestimate how much Native American ancestry they'll find—a belief she and colleagues believe is partly connected to a romanticization of what it means to be Native American and to a desire for an "indisputable" claim on this country. What we want to be becomes what we think we are. Identity is aspirational.

On the one hand, Wendy Roth finds her work demonstrating people's selectivity heartening. It suggests that scientific knowledge is just part of how people construct their identities. On the other hand, who picks and chooses what new information to incorporate into their identity is telling. The ability of white people to claim multiraciality or blackness as something symbolic and "exotic," to be revealed on a case-by-case basis when it suits them, means genetic ancestry testing may be reinforcing the privileges of those who already experience them.

It's clear that law, culture, and practice have not caught up to an era in which we can easily and cheaply test for genetic heritage. There have been a few high-profile cases of people using genetic testing to claim benefits or other advantages associated with minority status, although these may be outliers; Roth says she did not encounter this as an incentive for testing among her subjects. An investigation by the Canadian Broadcasting Corporation found some Canadians were claiming indigenous status—possibly as a means to getting tax exemptions—after sending their cheek swabs to a company that appeared to reliably return reports of sizable Native American DNA to everyone. (The company even declared one man's chihuahua to be 12 percent Abenaki and 8 percent Mohawk.) And a Washington State business owner named Ralph Taylor applied to have his company certified through the federal Disadvantaged Business Enterprise program on the strength of a DNA test; in 2018, after Taylor was approved as a minority for a state program and denied by the federal program, he sued. The case opened up questions about the reliability of ancestry estimates, how the government determines minority status, and whether someone like Taylor—who, according to the *Seattle Times*, "lived most of his life as a white man," but claims he is 4 percent sub-Saharan African and 6 percent indigenous American—should have to prove that he's suffered discrimination. Sociologist Alondra Nelson points out that seeing black or white identity as embedded in genetic markers alone misses the point that race is a social construction. "You have two facets of identity: who you think you are and what other people say you are," she told the *Seattle Times*. "Generations have been disadvantaged based on what they look like, how they talk, or where they come from."

And what happens when a person is *not* motivated to claim something suggested by an ancestry test? Foeman thinks of our family history narratives as akin to lullabies, with the power to lull us into a sense "of security and predictability"; DNA tests can disrupt these stories. A woman in Foeman's project who hailed from a family proud of its Italian heritage reported that her father hung up when she informed him of test results saying she was only 19 percent Italian. Another woman, who identifies as Iraqi, was "outraged" by the results of a test telling her she was largely European.

In an extreme example, members of the racist far-right have used this consumer technology to attempt to prove their beliefs about racial purity—ones that nineteenth-century naturalist Samuel Morton, with his pseudoscientific ranking of skulls, would have recognized. Sociologists Aaron Panofsky and Joan Donovan decided to study what happens when genetic testing proposes a profound cognitive disconnect: when you learn you *are* the people you hate. Analyzing years of posts on the white nationalist website Stormfront, Panofsky and Donovan found that, in some cases, posters who confessed to something other than wholly European ancestry were shunned and shamed. But far more often, the community offered "emotionally supportive" responses (as in "I wouldn't worry about it. When you look in the mirror, do you see a Jew?") or attempted to reinterpret a tester's results in a better light. Sometimes they pointed to the very real imperfections of ancestry estimates as proof that unwanted findings were "statistical error," or theorized that the genetic testing industry is a Jewish-controlled conspiracy intended to confuse whites and spread multiculturalism.

In other words, this is an old story; selectivity is not a feature of the genomics age alone. As historian François Weil points out, genealogy has long been about not just who we think we are but who we want to be, going back to the days when Americans purchased suspect pedigrees in hopes of borrowing a little aristocratic luster. In *She Has Her Mother's Laugh*, his book on the science and culture of heredity, science journalist Carl Zimmer writes of an early-1800s Massachusetts genealogist who, in drawing her family tree, simply left off those branches descended from a couple considered "near idiocy." I've noticed people often explain surprising ancestry findings by looking for confirmation in preexisting preferences—as if they somehow always sensed their own essential Korean-ness or Italian-ness. This may have to do with the way human beings look for patterns to make sense of our lives, and the fact that—as professional fortune-tellers know—we're willing to do a lot of interpretation and filling-in-of-the-blanks when we're motivated to believe something. "Now I know why I like tulips so much, why I love those Belgian chocolates, and why I eat so many French fries," writes one man upon discovering Western European ancestry. "Now I know why I have always liked the Beatles!" writes one woman about her apparent British ancestry.

But if such explanations can seem facile, DNA testing can also be the means by which profound understandings of identity are forged. Many seekers make nuanced decisions about how much of their genetic backgrounds to own. They wrestle with what those pie charts mean. Is it ever too late to identify with one's roots, if you grew up as something else? And how do shifting loves and loyalties influence this? One woman I'll call Nancy told me that when she finally reconnected with her father in adulthood, and he told her how he'd tried to remain in her life as her mother refused him contact and moved from town to town, she also found herself motivated to claim the African ancestry he'd passed down to her. Nancy told me her upbringing as a white girl had left her feeling confused about being part black—"having this identity but not being able to truly embrace it." She straightens her curly hair; people ask if she's Italian or Hispanic. But after reconnecting with her father, she went to a doctor's appointment and for the first time checked white *and* black on the form. She did this out of respect for her father, she told me. If he wanted to know her, she wanted to claim him.

We may be selective in how we identify in the wake of genetic testing, but this doesn't mean we're wrong to be selective. We are the products of culture, memory, family. DNA is part of the picture, but it isn't everything. As Roberta Estes puts it, "It's not love."

* * *

T. L. Dixon wanted his own *Roots* story. He wanted to know his people.

Dixon was adopted, and traditional genealogy had helped him track down his biological sister in adulthood and to learn that his birth mother had passed away. He wanted to know more about his genetic family and its past, but further back the paper trail became difficult to follow. Dixon did not know his ancestors' names, did not have records of them the way so many other Americans do. He did not know when they were taken from their country, or even what their country was. He did not know what ship they were put on, what plantation they were taken to, what children they bore, how their lives began and how they ended. The profound losses of slavery were echoed down to Dixon in yet one more loss: the paucity of genealogical records about his enslaved ancestors. As the science journalist Christine Kenneally points out in her wide-ranging book, *The Invisible History of the Human Race*, "the thing about records is that they tend to proliferate as a matter of course around people with power." Those without power are silenced by the lack of records.

Dixon, a social worker who was previously a journalist, has written on his blog, *Roots & Recombinant DNA*, about how the 2018 movie *Black Panther* sparked a kind of *Roots* 2.0 for African Americans who want to find their own "Wakanda"—the fictional African country that is home to the superhero. And yet, where would they start? How could they find out more? "For the majority of African descendants in the Americas, our 'Africa' might as well be in Wakanda because our actual ethnic origins felt like a sort of mythical—and stereotypical—place," he observes.

The difficulties of African Americans trying to trace their ancestries are manifold. Many hit a brick wall trying to trace their ancestors before 1870 because that federal census, the first done after the Civil War, was the first to name former slaves. Before that, FamilySearch's spokesman Paul Nauta has said, enslaved persons "showed up as ticks or hash marks on paper." And another legacy of being treated as chattel means that people in a person's family tree may be spread out all over the country, the result of "enslaved Africans being moved around from place to place," Dixon says.

The longing, and the limitations of genealogical records, are what made Dixon turn to saliva back in 2012, and what made him a genetic genealogist. Through the bigger databases, he was able to find and make contact with several distant cousins who are of recent African descent. The genetic heritage they have in common with Dixon traces back to a time before his ancestors were imprisoned and shipped abroad, and they provide a kind of window into what their lives might have been like before the profound violence and disruption of the transatlantic slave trade.

On his blog, Dixon documents how he used these African cousins to trace his likely ancestry not just to broad areas within the continent but to particular regions of countries, including Cameroon, Guinea, Nigeria, and Ghana—and to specific ethnic groups within those regions. On 23andMe, for instance, Dixon was able to see he shared twelve centimorgans of genetic material with one distant cousin along their seventh chromosomes. He confirmed and contextualized that little piece of DNA by triangulating it with his American relatives, and reached out to his African match, who told Dixon about his people, the Fulani, and gave Dixon family surnames. Drawing on books, online articles, and old newspaper clippings, Dixon was able to understand the historical forces that likely shaped the life of his Fulani ancestor.

Dixon was still trying to find more about his specific ancestors, he told me when we spoke during the summer of 2018. He'd recently heard about a new collection of slave records he hoped to consult. But what he'd been able to achieve was very meaningful to him—and would not have been possible without DNA. "It was actually very fulfilling and amazing that I was able to find my actual African roots and not just say, 'Oh, I'm African,'" he said.

In her book *The Social Life of DNA*, Columbia University sociologist Alondra Nelson writes about how African American early adopters of genetic testing have used this technology to access their own histories. Nelson was struck by this, given a history that has left many African Americans mistrustful of medical research: eugenics and forced sterilization; scientific exploitation like the horrifying Tuskegee syphilis experiment, conducted by the US Public Health Service for decades on

African American men; a long history of segregated health care. Despite "black communities' accumulated mistrust of science," despite awareness about how intertwined the history of genetics is with scientific racism, these early adopters saw something special in the opportunity posed by DNA testing. Nelson argues that these early adopters were engaged in their own small-scale journeys toward reconciliation with the pain of the past, "in which DNA test results are used as a form of testimony."

Recreational DNA testing can be an unflinching history lesson. Marissa Kinsey, a graduate student in library sciences in Queens, told me about wrestling with the implications of this history up close after an undergraduate assignment to create a family tree sparked an interest in genealogy and DNA. (Kinsey is transgender and asked me to use gender-neutral pronouns.) Through genetic genealogy, Kinsey was able to confirm the outlines of their family's oral history: An enslaved great-great-great-grandmother named Isariah bore a baby girl named Fannie by her master. From probate documents, Kinsey was able to deduce that Isariah was likely about fifteen when she had Fannie in the early 1850s, and not long after that, mother and child were separated. The little girl, valued at one hundred dollars, was placed in "Lot 1" and parceled out to the white slave owner who was very likely her own father. The mother, valued at five hundred dollars, was in the next lot and went to a different family member. Kinsey was able to figure out the likely identity of this slavemaster and to connect with a white cousin descended from this same man. Kinsey told me of spending time in an archive in North Carolina with their mother, sorting through wills and slaves' bills of sale, the two of them by turns distraught, angry, and driven, wondering about the lives of these enslaved ancestors. The second time I spoke with Kinsey, almost a year after we first spoke, they were in the process of legally changing their name to Riah Lee Kinsey, both to reflect a gender-neutral identity and to honor the ancestor who had so much taken from her.

Kinsey told me they struggled with how to view the white slave owner they descended from. "What kind of man could sell a woman away from her child, but then take that child in and raise her?" Sometimes, Kinsey thought that perhaps "there was a crumb of human kindness in him that he actually was able to raise this child." And then, Kinsey wondered about this desire to find something good in him—about the possibility that perhaps they were driven to look for this good in this man because they were descended from him, about the fact that some things are just too terrible to look upon, and so we refuse to see the whole picture.

The rise of genealogy and DNA testing may be making it increasingly difficult for white Americans to engage in historical amnesia, to fantasize about their own family histories without ever considering how slavery might have figured into them. In 2012, when the *New York Times* reported that new research showed that Michelle Obama was descended from a slave owner's son and one of the family's slaves, descendants of the white line had varying reactions to whether they'd like to meet their extended black family or, indeed, be quoted in the newspaper about the news. One struggled with how to understand the relationship between her great-grandfather and Obama's great-great-great-grandmother. "The idea that one of our ancestors raped a slave . . ." she said, trailing off. Kinsey told me they have spoken with their white cousin, to whom they are related through that ancestral slavemaster. "I feel like it's a little bit more awkward for her because she hadn't known about us," Kinsey told me. "We kind of just appeared out of nowhere, and I'm pretty sure that's not what she was banking on."

* * *

Rosario Castronovo is still seeking. Over the years, the would-be Sicilian—who is, he knows now, the son of a biracial woman who hid her ethnic identity—has consulted spiritual mediums and woken in the night believing he'd heard his grandfather's voice. The last time I talked to him, he was planning to change his name again, this time to Jerome Lafayette Narramore, to honor the ancestral discoveries he'd made through genetic genealogy. "Some people say what's in a name?"

says Rosario, a restaurant server and union officer who lives in Manhattan. "I think everything is in a name."

Rosario struggles with how to think about himself and how to present himself to the world. "It's kind of weird for a guy who presents as white to say, 'I'm black,'" Rosario said. "I'm still getting used to it." Could he even claim blackness, he wondered, without that lived experience and without any of the implications of what it means to be perceived as a black man in contemporary America? "I'll never know what it's like to be pulled over by a police officer and fear for my life," he said. "Can I call myself black and never have to experience it? It almost seems unfair, or like I'm pandering." Yet his mother's elderly cousin, who is more fair-skinned than he—and, according to her DNA results, has less African ancestry than his mother—is quite clear in her identity as a black woman. "It's a social construct," Rosario told me. "I raised myself as an Italian man, I immersed myself in Italian culture. What do I do now?"

We were talking on Independence Day, and Rosario told me the more history he learned about his black family, about the lives of black people in Vermont and in the rest of America, the more he wondered what he should be celebrating on this day, and if he should be celebrating at all. "I have learned more about African American history in this country, and the more granular you get, the more bitter you become," he said. "I never got to know my people."

And yet genetic genealogy had also given him an opportunity to know about what had been hidden, to wrestle with his mother's pain, with his grandparents' sacrifices, and with the wrongs done to his family. DNA testing had given back to him and to his mother a little of what was stolen by the past.

Rosario's mother has Alzheimer's now, but there are moments of lucidity. One day, Rosario took her to a quiet spot in a church not far from her assisted living home, and handed her a framed copy of the only photograph he had of her father: the man's mugshot, from when he had served time for the crime of adultery with a white woman—for the crime, really, of creating Rosario's mother.

The old woman sat in a pew and stared at the image. She wiped her eyes and held the photograph to her chest and kissed it. She couldn't get over how handsome her father was.

"This is so wonderful," she said. "I could just sit here and cry."

Rosario sat down in the pew and told his mother they could do that together.

11

THE MYSTERY OF JIM COLLINS

On September 25, 2012, Alice went to one of FamilySearch's family history centers, the satellite offices offering access to the LDS Church's treasure trove of genealogical materials, not far from her house in Vancouver, Washington. She was doing research on her paternal grandparents' last names, investigating her theory that one or both of her grandparents were Jews from eastern Europe who had somehow come to the United States via Ireland, Irish-izing their names along the way. It seemed a thin tendril, but Alice's policy was that any plausible possibility had to be explored.

So anyway, there she was, trying to find a record of her paternal grandfather's birth in Ireland, when she got an email on her phone saying her paternal first cousin Pete's DNA results had come in. She got out of there real fast and raced home. She logged on, navigated to Pete's account, which she was administering, and looked at his results.

And she discovered that Pete was not related to her. Not one tiny little bit.

Alice felt crushed. But they *were* cousins! She was Younger and he was Elder, separated by precisely fifteen years. Younger was so fond of Elder. She'd lived most of her life knowing she and Pete shared a birthday, and not knowing what had happened to him. "I wonder what happened to Pete," her father had said on his last day on earth. And then Pete and Alice had found each other late in life by way of crossed letters that seemed like kismet. Alice could not bear to lose the cousin she'd only recently found. She was, in her own word, "devastated. And I was afraid he was going to reject me because we were no longer biological cousins."

Now, Younger called Elder, and told him the news. She was relieved by his response. "He was sad, but he also told me I was the best cousin he ever had."

Alice talked to Gerry. They knew what this meant. The sisters drafted an email to their brothers and titled it "Important Family Information."

"Most of you know I took a DNA test a few months ago," Alice wrote. She walked them through everything that had happened since: the unexpected ethnicity results from AncestryDNA; her fear that she wasn't Jim's daughter; the 23andMe results, which eased that fear by proving they had a different mystery on their hands; the eureka moment along Bill's X chromosome; and now, the latest confounding finding. In typical Alice style, it was thick with facts and details, but the bottom line was this: "We share 0.0 percent of our genes with Pete!" she wrote. "Furthermore, his admixture indicates that his origins are entirely in the British Isles, exactly what I expected to find for us. Gerry and I interpret this to mean that Dad is not the biological son of Katie Kennedy and John Josef Collins, but our aunt Kitty *was*."

In other words, Jim Collins had a different set of parents than his beloved sister. Looking back, the sisters could see there were always puzzles. Gazing at old family photos on her wall, Gerry had long believed she could see bits of herself and her siblings in the faces on her mother's side of the family. There was the Nisbet nose up through the generations, a steady marker of Nisbet genes. But on her father's side, she could not see that same genetic legacy. In the photo she had of her paternal grandfather, John Josef, he had deep-set eyes, a long face, sharp cheekbones, and a cleft chin—he looked like no one in her immediate family. Plus, John Josef was said to have been tall. Gerry had gone so far as to blow up this picture of her purported grandfather, alongside a photo of her father, and take them to Thursday morning bowling. "Can you believe they are supposed to be father and son?" she'd asked her friends.

But photographs can be poor proxies; comparing the living was rather more dramatic. Gerry remembers meeting their uncle John, her father's brother, when he came to visit California, in 1975, and noticing how different he was from her father. Uncle John was about six foot one, and his face and build were entirely different, with the long, narrow face that John Josef had and that she'd recognize decades later in her first cousin Pete. Her father, Jim, was eight inches shorter, with a round face, dimples, and an entirely different build. Even the brothers' mannerisms seemed different, Gerry thought. Even so, it took a family outsider to point out the obvious.

"I don't understand how your dad can be that short," Roger Wiggins, Gerry's now-husband, told his fiancée.

"Well, my dad was in the orphanage," Gerry said, parroting the family line. "He was malnourished."

Even then, though, no one suspected that Jim and John might not be related. It was the sort of thing that seemed obvious only in retrospect, because sometimes, you don't see something "unless you're really looking for it," Jim Collins's son Jim the younger told me. "There are always people in families that look a little bit different."

After Alice and Gerry sent their family-wide email, what ensued was a lengthy exchange of emails among the seven siblings, spread out across the country from Washington State to California, Minnesota, Colorado, and North Carolina. They talked about what to do now, and they talked about the past and about what this all meant. They thanked Alice for her doggedness. They talked about the possibility that they might have biological family they'd never known—a boon for the children of a man who'd spent his childhood in an orphanage and didn't have much in the way of extended family. They talked about their identity as Irish Catholics, what had changed and what hadn't. They talked about how they were lucky to be a family, however it had happened. "Does this mean we get more holidays to celebrate?" one brother wrote, before expounding on the elusive nature of truth.

Alice wrote that she never could have imagined this possibility when she first sent away for a DNA test. It was a strange time, she would say later. "When I realized that my dad was not who he thought he was, I really lost all my identity," she said. "I felt adrift. I didn't know who I was—you know, who I really was."

It had been only a few months since the first DNA test, and it would be years before Alice and her siblings would finally unravel the truth about their father and themselves. But even so, they'd learned so much more in those few months than they ever could have anticipated—so much about science and family and the vagaries of life, and so much about the profound and disruptive power of DNA testing.

* * *

For months, Alice had been lurking on the Yahoo group DNA-NEWBIE, reading about the travails and techniques of other seekers. She knew she needed help now, so she posted for the first time. Her journey had begun on a lark, she wrote, and was now "extremely personal." She explained how she'd been certain the original AncestryDNA test was wrong, how she'd taken the 23andMe test in hopes of disproving it, and how she was now testing a brother's Y-DNA to try to find her biological grandfather. Was she doing the right things? Were there other tests she could be pursuing? Could anyone help?

CeCe Moore reached out to Alice, and they began to email and exchange information. The genetic genealogist had helped many adoptees trace their way to their biological parents, and she saw Alice's situation as similar. The late Jim Collins was in the position of many adoptees, with his biological parents unknown. Even back in 2012, when the databases of people who'd undergone genetic testing were much smaller than they are now, some adoptees were having luck by searching among their genetic cousins, using segment triangulation and family trees to deduce the identity of their biological parents.

But looking at some of Alice's autosomal DNA matches, which Alice gave her access to, Moore could see this was going to be a challenge. Ashkenazi Jews are considered an endogamous population, which means people marry within the group and, sometimes, within families, with cousins marrying cousins. There are a number of populations for whom this is an issue, including Cajuns, Polynesians, the Amish, French Canadians, and people descended from early colonial American populations, posing challenges to those among them who wish to use genetics in their genealogical work. Endogamy inflates the relationship predictions for Ashkenazi Jews, leading to the appearance of many seemingly close cousins who are, in fact, more distant cousins related to a tester in more than one way. And Alice's genetic match list along her father's side "was just a jumble of Ashkenazi Jewish endogamous spaghetti," Moore told me.

One by one, all of Alice's siblings were submitting their saliva to 23andMe, in hopes that having more Collins DNA and more genetic cousins to look at might help their search. As each one got results, Alice would upload the autosomal DNA data to FamilyTreeDNA, which accepts transfers from other companies, in hopes of maximizing the pool of relatives they might match to.

She signed up to take a class outside Seattle. Billed as a class for helping adoptees, it refined techniques Alice was already starting to use, which she'd learned about on DNA-NEWBIE and elsewhere. The teacher, a search angel named Diane Harman-Hoog, had been helping adoptees for free since before autosomal DNA testing was an option, and eventually became involved with co-founding the volunteer website DNAAdoption.org. She had previously retired from her work as a manager for Microsoft, and had a background in mathematics and an MBA in information technology and finance. Like Alice, Diane Harman-Hoog had spent years as the only woman in the room. And like Alice, she was not afraid of data.

Diane's class, held for two days at a local library in early 2013, boasted a table of contents that would have intimidated many, with such items as "Odds of Segment Persistence," "X Chromosome Inheritance Charts," and "Phylogenetic Trees." The heart of the class was a methodology for working with the FamilyTreeDNA database that Diane and others had only recently researched and formalized.

Diane sent me an early version of "The Methodology," as it is known, and I found the nineteen pages of instructions absolutely overwhelming. Suffice to say, a seeker would have to be pretty motivated to use it, particularly if she weren't tech-savvy. Adoptees using autosomal DNA back in the "early days"—which were, astonishingly, less than ten years ago—really had to work for their discoveries, which were based off distant cousin matches detected by shared genetic segments in chromosome browsers. To start with, the seeker needed to input and organize a ton of data on spreadsheets, which involved "several tedious tasks," Diane explained in her paper. But if you could complete the grunt work, the beauty of the spreadsheets was the ability to organize. The little groupings of relatives with shared genetic segments were placed together, along with their names, the length of the shared segment, and the locations of those segments along a particular chromosome. Then, starting with relatives with the longest overlapping segments—in other words, those likely to be closer relatives—the seeker reached out to ask for a family tree or any other information. This made the process a collaborative one, dependent on the time, interest, and accessibility of other family researchers in those early databases. Then, the seeker used genealogical research to expand on her cousins' trees, going back in time to look for common ancestors—perhaps as far back as 1600 if she was trying to understand her relationship to a fifth cousin. "Eventually," Diane wrote, "you will want to try to piece together as many of these as you can into one tree." With enough relatives and ancestral lines, the hope was to find an ancestor to the adoptee, and use that to determine at least one side of the adoptee's birth family.

The challenge of obtaining and synthesizing all this data was immense. And too often, Alice was learning, cousins wouldn't respond to requests to compare genomes or share family information. The trees of genetic cousins, when they were available, weren't always in great shape; Diane told me she sometimes worked with information written "on torn scraps of paper and paper plates." And then there was the problem of endogamy. At the top of Diane's "Methodology" was a caution that Alice read but which did not deter her, because Diane's approach was literally the only option for hacking her DNA. "Unfortunately," it said, this approach "probably will not work with Ashkenazi lines."

* * *

Alice and her siblings had eliminated a lot of theories by this point—from the most likely ones to the far-fetched. If the mystery of

Jim's ancestry didn't harken back to Europe, or to a New World affair, they needed to look to when Jim was young. His early years had been marked by turbulence and upheaval. It seemed possible something had happened then.

In the fall of 2012, Alice and her siblings decided to write again to the religious sisters located at the site of the old orphanage and ask for more specifics about the school's practices a hundred years earlier. They asked for details about how Jim Collins had been admitted and discharged from St. Agnes Home and School. They asked whether the Catholic institution might have accepted children of other faiths—one of whom, a Jewish child, could perhaps have been confused with another boy named Jim Collins and assumed his identity.

The reply, from the archives office of the Dominican Sisters of Sparkill, offered some important clues. The archivist elaborated on the school's policies and wrote that she was not aware of restrictions on who the orphanage took in. And she added that Jim and his siblings had been committed by the New York Society for the Prevention of Cruelty to Children, a step usually taken "if parents were deceased or unable to care for them."

This was news to Alice, Gerry, and their brothers, who had not known how the Collins children were committed to the orphanage. It opened up a new window into their father's early life.

Founded in 1875, the New York Society for the Prevention of Cruelty to Children was the first child protective agency in the world. (Notably, it was founded nearly ten years *after* the American Society for the Prevention of Cruelty to Animals.) Alice wrote to the organization and got back a lengthy summary of the files in its archives, which helped her piece together a rough idea of her father's first year of life. Jim and his siblings were very young when the society was first contacted in 1914 by a woman who said the children were "in need of everything." Society officers went to the top-floor apartment of an address on Southern Boulevard in the Bronx, and children answered to their knock. Their mother was sick, the children told the strangers, and had told them not to let anyone in. The officers left.

Not long after that, when baby Jim was nine months old, the children's mother died, and the rest of the file, consisting of the accounts of various "informants" and an official from the city's Department of Public Charities, is a confusing timeline of turbulence for the three Collins children. They were so little. The eldest, Kitty, was just four. At various times they were being cared for by three different women, at least two of whom were aunts. Various claims were thrown around and sometimes contradicted or walked back: that their father, John Josef, was "intemperate," that he wasn't, that the parents had separated. Money was tight, space was tight; the children were moved around and separated from one another. One relative, whose husband worked as a city street paver, explained that she was caring for the two older Collins siblings in addition to three of her own, without any financial assistance from their father. This could not continue. John Josef, then working as a "rigger's helper," earning twelve dollars a week, met with a society officer and said he intended to have his children "placed." At a court hearing, John Josef agreed to pay one dollar a week per child and have the children go into custody of the convent. They were separated, with the two boys going to the orphanage in Rockland County and their sister placed in a Catholic institution in Manhattan.

As the years went on, the society kept track of the children and of their father, who would never again be a family. Within a few years of relinquishing custody, John Josef was said to be in bad health, and an officer visiting his "poorly furnished but clean" apartment found him "in very delicate physical condition." He'd begun working as a longshoreman, John Josef said, but his health made it difficult; "this had resulted in his being in arrears in his support payments and therefore unable to visit" his kids. At one point, he arranged for his daughter to be released to the care of an aunt who appeared relatively better off, but the boys had to remain at the orphanage. At another point, John Josef visited one of his sisters to ask to borrow a few dollars.

Alice took note of several things in the file. First, she noted that her father's parents had apparently separated, which seemed unusual for Catholic immigrants at the time, particularly right around the birth of a baby. Why would they separate, and why at that moment? Looking at the dates, she was struck again by the fact that her father was given away exactly a week before he turned one. "His father didn't even keep him around until he had a birthday," she told me, touching her chest, where it hurt. The last thing she noticed was that, as far as she could tell, Jim and his siblings appear to have been transferred quickly from the care of their father to the Dominican sisters, lessening the possibility that Jim's identity had been mixed up by the organization that had taken him from his troubled home.

But that still left the theory of a mix-up at the orphanage. It seemed unlikely that Jim's older brother, John, wouldn't have noticed or said anything if his baby brother's identity were confused, but it was a possibility, and Alice was committed to exploring every single one. Here again, the Collins kids were butting up against the passage of time and the limitations of the paper records. They did not know a whole lot about their dad's time at St. Agnes, in part because Jim didn't talk about it much. "If only we had a time machine . . ." Gerry mused.

But one of the few stories Alice and Gerry had heard about their father's childhood was that once a week the orphans were brought to a classroom to meet with visitors. "Every Sunday, Dad had to dress up in his Sunday best and he would sit in a classroom with his hands folded on his desk and wait for a visitor," Gerry remembered her mother recounting. For nearly a decade, he never had one, at least not that he recalled. Then, one day when Jim was ten, he was told someone was there to see him. He was "terrified," Gerry said. "So afraid that he peed in his pants," Alice added. The man was very tall, "a six-footer," Jim would tell his children decades later, recounting the story with matter-of-fact delivery. "Do you know who that is?" Jim's older brother, John, asked. Jim did not.

"That's your father," John told him.

The tall man asked if there was anything Jim wanted. Jim asked for ice skates. Some time later, the tall man came back, carrying ice skates. And after that, John Josef did not come back, because he died, at Bellevue Hospital, of pneumonia and septicemia, in his early fifties. Jim had just turned twelve.

Alice had a no-stone-unturned mentality. How, she wondered, could she determine whether the baby placed at St. Agnes grew up into the man she and her siblings knew as their father? She hit on the idea that there might be a specialist who understood how faces change over time. She found a forensic artist and emailed the woman, recounting the story of Jim. "We have one photograph, purportedly taken of him, his father, and siblings immediately before the children were placed in orphanages," she wrote. "Is it even humanly possible to 'age' a photograph of an eleven-month-old baby enough to identify him as an adult?"

It was possible, the artist wrote back. It would cost fifty dollars.

Alice emailed her dad's only baby photograph, along with two pictures of him as an adult. The forensic artist wrote that she believed the baby and the man were one and the same. The two pictures showed the same domed forehead and earlobes, she wrote, the same mouth shape and expression, the same facial proportions. Looking at the portrait of baby Jim with his dad and siblings, she observed that Jim's big brother, John, "looks far more like his father than the baby does."

The orphanage theory had always seemed thin, and now Alice, the daughter of a skilled card player, concluded that few gamblers would give it good odds—especially not while there was another moment in time that seemed like a better bet.

* * *

Who was Jim Collins? How did he come to be?

The power of the DNA revelation is that it provokes questions about the past, which seems suddenly not like the past at all. It makes you question fundamental truths. It provokes an accounting—of a life in an orphanage, or a time when a couple was having trouble conceiving, or when a young unmarried woman went off for a mysterious vacation and returned many months later. It invites revisions to things we long ago analyzed and incorporated into our personal narratives. It suggests the past is never over, but a living thing that can be amended. How do you make sense of a life that starts with a mystery? How do you tell your story without a beginning?

There's a strange circularity to Jim's time at the orphanage. Admitted a week before his first birthday, he was discharged a week before his fifteenth. He had not finished high school. No one knows why he left when he did, but according to records, he went to live with his brother, John, who was only a little older than him. And this period of time, with two teenage boys attempting to live on their own, began the chapter Jim liked to call his "misspent youth."

He collected money for an illegal gambling ring. He made money playing pool and poker and craps. "He developed a math system—I couldn't explain how it worked—for taking bets for horse races," his son Bill Collins told me. "He told me he did so well with it that eventually the bookies in New York would stop taking his bets."

Adulthood and the army seem to have mellowed Jim. His kids knew he'd had a difficult childhood, but somehow, Gerry told me, it had not made him bitter. "He wasn't around a lot but he was soft and gentle and very easygoing," said Brian Collins, who could not explain how a man with such tough beginnings emerged with his sweetness intact. Jim valued family; without much of his own, he'd adopted his wife's family, calling Alice the elder's mother "mom" and coming to adore his wife's sisters. "My mother was close to her family," Alice told me, "but with Dad it was something almost transcendental."

And Jim worked hard. With seven kids, money was often tight. In his job as a correctional officer in a state prison in Soledad, California, after twenty years in the army, Jim worked double shifts and holidays. The few times he got angry were in arguments with his wife over money, one Collins brother told me. "It was always living paycheck to paycheck," another said. The Collins family lived in a three-bedroom, two-bath house in Salinas, California, all nine of them crammed into 1,250 square feet. The two eldest boys slept on a couch with a trundle bed that barely fit into a tiny connecting room intended as a den, and when Alice (no. 3 in the lineup) and Bill (no. 4) got into fights, Alice would run laps through the house, busting through the accordion door to that room in her attempts to get away. In the living room, their parents "had their chairs and they were sacrosanct and you did not sit in them," Gerry said one day when Alice and I were videoconferencing with her from Alice's sewing room. "You did not!" Alice echoed. "You got a seat and it was your seat unless you got up."

The kids had their chores and every Saturday they did a full housecleaning, top to bottom. They made their own lunches, and made their mother's breakfast every morning and served it to her in bed, and Alice told me she made dinner every night from the time she was ten. If the Collins kids were united in their affection for their father, they were also unified in describing their mother as difficult and overbearing. Jim Collins and Alice Nisbet were both older for their era when they married. He was thirty, and she was twenty-seven. What drew these two very different people together none of the Collins kids could quite say; their togetherness was simply a given. "He loved watching her," his granddaughter Cherie Collins says. "What it was that they had I don't know, but there was a spark." And later, when Alice the elder was sick for the last fifteen years of her life, partially blind and mostly bedridden after breaking her leg at the age of sixty, Jim was his wife's full-time nurse.

Jim Collins had qualities that often endeared him to people. At the prison, he was clear-eyed about the inmates, but he treated them with respect, and that seems to have gone far. Bill recalls visiting his dad's workplace for a college project, and his dad introducing him to some of the prisoners. "They were saying, 'Don't worry about your dad, we'll take care of him,'" Bill recalls.

Jim was a little socially awkward. He could ask weird questions, and mangle his words, and be slow and deliberative in his thinking. "My father's favorite thing to do when he came home from work was to sit in his chair and read the encyclopedia," Alice says. (It was this habit that led to one of her parents' bigger fights, when Jim discovered a hefty phone bill that his wife had hidden in one of the volumes.) He did the neighbors' taxes for free because he just loved doing taxes. Cherie talks

fondly of trips she took with her grandfather to the military commis-
sary, where Jim would pull out coupons and work his double points
to see how cheaply he could buy cartons of Jell-O. Cherie told me her
grandfather used to go dumpster-diving behind his apartment to find
old toys, which he'd clean up and give to his grandchildren. I stopped
her when she said this to make sure I was hearing right, since old toys
from dumpsters struck me as kind of gross, but Cherie insisted that "it
was very sweet." Her little sister got a favorite bunny that way. The thing
is, Jim never got toys as a kid. One year, he got an orange for Christmas.
"The way he was raised, you didn't throw away a toy," Cherie said.

When Jim got older, he began to talk more about his life. One day
in 1993, when he was visiting his daughter Gerry and her husband,
Roger, they pulled out a video camera and taped him at the kitchen
table. When I visited Alice, she showed me this grainy video, which
featured her dad telling stories about the time he and a colleague at the
prison were ambushed by a group of inmates hiding in a kitchen, and
the time in the army in Germany when he pretended he didn't know
how to play pinochle and managed to win a thousand dollars in an epic
night of gambling at the officers' club. Jim explained that he had learned
the rudiments of pinochle as a kid, and then, at great length and with
mounting excitement, he explained a series of complex maneuvers in
exquisite detail, playing out a lengthy game of double-deck pinochle in
his head, with Gerry as his partner. "All you have to control is two suits,"
he said. "That sounds strange but I'll tell you how you do it." He talked
of tricks and melding and round houses. He explained how he could
infer the cards in his partner's hand by the ones she had already played.
He switched perspective to describe how an opponent could counter his
moves. He went on and on, grinning at the salient points, connecting
the strands of this imaginary game as if he were following an elaborate
chart only he could see. "You add your six queens, you add your eight
kinds, you add each marriage, a king and queen, and you add the marriage, and trump is four . . ."

Jim Collins at age eighty had receding salt-and-pepper hair neatly
combed back from his forehead and a pair of thick-framed glasses. He

had the same round face, mischievous smile, and dimples he'd had in that photograph taken when he was a baby on his father's lap. He had an old New York accent: "oil" was "erl," "cards" was "cahds." Sometimes he put his finger to his nose, as if signaling his interlocuter to pay attention. Other times, he smiled and rubbed his hands together, with the air of imparting a secret, or spread his hands out wide to demonstrate the hopelessness of being outwitted by a cunning opponent. Morning sunlight streamed in through the window, making stripes on his forearm, and I let his words wash over me as if they were in another language. "Follow me?" he'd say, after explaining something, as if everyone else could see this complex imaginary game, too. "You got that?" "See?"

"No, I don't see that," said Alice, who was sitting next to me, watching her father on the screen. I turned to look at her and she shrugged, and I was relieved I wasn't the only one.

12

THE SIMPLEST EXPLANATION

Alice and Gerry talked and emailed constantly as 2012 rolled into 2013 and they chased down lead after lead. They were beginning to think only one explanation made sense.

If the mystery of Jim wasn't buried in the old country, and if it wasn't confusion around the identity of a baby who'd been relinquished to a child protection agency and then to an orphanage, could something have happened right around the time Jim was born, on September 23, 1913?

Early on, before Alice stumbled across the clue that changed the way they thought about everything, Gerry had come up with a scenario that could explain how a Jewish baby from eastern Europe could wind up with Irish Catholic parents. Perhaps their grandmother Katie Collins, giving birth in her apartment with the help of a midwife, had birthed a stillborn child. Elsewhere in the Bronx, Gerry imagined, a Jewish woman had had a baby but then she died—and somehow, the two families had found out about each other. Could a bereaved John Josef and Katie Collins have taken in that other woman's baby as their own?

But then, Alice found the clue, and everything changed. The funny thing was, she'd had it in her possession since she'd first started her genealogical research on her dad. Jim's birth certificate had arrived in the mail back in 1998, shortly before Jim had died, and Alice had looked at his birth day and year, and looked at his parents' names, but she'd failed to notice one key detail. This part of the story reminds me of that psychology experiment in which subjects asked to focus on a video of a basketball game fail to notice the person walking past in a gorilla suit—sometimes you don't see something because you're not looking for it. Alice finally saw the detail fourteen years later, when she was in the sewing room/office of her house in Vancouver, mired in the mystery of who her father was, and she studied his birth certificate again. The detail was right there, stamped on the bottom, and again, in looping script, on the line next to "place of birth."

It said: Fordham Hospital. In other words, Jim had not been born at home, as everyone in the family had assumed.

Fordham Hospital during the era when Jim Collins was born

She told Gerry, and they discussed the implications of this information. Gerry thought one explanation that might fit the facts was an update of her baby-dying-in-childbirth theory. Maybe Katie Collins's baby died in the hospital, and the woman in the next bed died, and voila, a bargain was struck. She knew it sounded "cockamamie," but "fact can be stranger than fiction," and it was true, the facts were already proving so strange.

But more likely, Gerry and Alice both thought as they digested the news, there was another, less complicated, explanation. "That's what we did at work," Alice says of her old job, when she needed the most direct approaches to wrangle phone trees or online enrollments. "You work on the simplest way of doing something." Of course, the theory the sisters came up with seemed simple and obvious only because the more prosaic explanations that people like CeCe Moore see much more often—an affair, an adoption—had already been eliminated. Even Moore would later say that the very first time she heard this theory, she thought it sounded too much like a soap opera to be true.

* * *

Alice did not know much about Fordham Hospital in the Bronx. It had closed in 1976 and eventually became a parking lot, and neither she nor Gerry—doing Internet research and calling around to places like the Bronx County Historical Society—could find any additional records pertaining to their dad's birth. What Alice would come to know later, when she began to research childbirth in America, was that 1913 was truly early days for maternity wards. As the childbirth historian Judith Walzer Leavitt told me, it was not until 1938 that even half of American births were taking place in hospitals.

Before the turn of the twentieth century, Leavitt says, "the ideal way to give birth was at home with family around, and a doctor coming in and helping, maybe a doctor and a midwife." Only a handful of women gave birth in hospitals—the poorest and most desperate urban women, sometimes unmarried and facing wholesale rejection from their families, or women with special medical conditions but without enough money to afford private doctors. But because notions around childbirth were starting to change in the early decades of the twentieth century, it's hard to say why Katie Collins chose to give birth at Fordham, a public hospital that had opened in 1892 and was part of a group of allied public hospitals run by a reform-minded board of trustees. Fordham and other public hospitals offered free care to the poor, and in 1913 they would have been serving lower-working-class and immigrant populations. Jim was born as the idea of hospital deliveries was becoming less unusual. As obstetricians advocated for their medical specialty, they promoted moving deliveries into the more controlled hospital setting, Leavitt says, even though hospital births weren't necessarily safer yet.

Before John Josef and Katie Collins started their family, Fordham had been located in a cramped and out-of-the-way location, but by 1907, it had a new facility "in the heart of the swarming population of the Bronx," one chaplain wrote. The new hospital, erected at a cost of about $600,000, was "thoroughly modern in its plan and equipment," a publication gushed, "ideally situated on Southern Boulevard, facing the beautiful Botanical and Zoological Parks." The huge wards were painted white and filled with natural light, and patients sometimes convalesced in rocking chairs. Doctors wore suits and ties, while nurses wore crisp white uniforms with voluminous shoulders and full skirts. Fordham's new location was actually a straight shot down Southern Boulevard from where John Josef and Katie were living in 1913, in a new five-story building that had running water and may even have had steam heat. They were living right by several busy intersections and, for five cents, could hop on the subway at the nearby Interborough Rapid Transit station. The Collinses would have been surrounded by first- and second-generation immigrants—fellow Irish but also Germans, Italians, and people referred to in the census as "Russian Yiddish." They had married only a few years earlier, in 1909, when they'd both been in the country less than a decade, and quickly started a family. By the time Katie went into labor with her third child, she might have walked to Fordham, as some laboring women did in this era, or perhaps she took the hospital's newfangled "automobile ambulance."

The year little Jim was born, the population of the Bronx was six hundred thousand, and to keep pace with growing demand for the borough's only municipal hospital, Fordham had just built another ward to separate medical from maternity cases, rented a nearby house to accommodate its growing staff, and would clearly need another expansion very soon, its trustees declared. Its nurses would likely have been specially trained in how to care for the inhabitants of its maternity ward, which was small but growing by the year. According to pamphlets from the Bellevue Training School for Nurses, which appears to have been serving Fordham and several other hospitals in this era, nurses learned about the stages of labor, Cesarean sections, and post-labor complications.

They learned about newborns—caring for their eyes and for the dressings on their umbilical cords, the ins and outs of breast-feeding and infant formula. They seemed to offer poor women state-of-the-art care.

And yet, when I spoke with historian Sandra Opdycke, who wrote a book on the city's public hospitals called *No One Was Turned Away*, she told me Fordham was very likely being largely staffed by students still learning their craft. "Pupil nurses" provided the bulk of the nursing force and many of the interactions with patients. Old pamphlets from the Bellevue Training School indicate these pupils worked twelve-hour shifts every single day except for "an afternoon during the week, and half of Sunday." They were paid in food, lodging, and education, with the tiniest of stipends to cover uniforms and textbooks. In between all this, they had lectures and exams—though Opdycke told me she never could figure out how these students fit their studies around their work shifts. Under the nursing staff, which was typically spread thin, there would have been "hospital helpers," poorly paid, poorly trained, and poorly regarded staff with fantastically high turnover, who would also have helped with the newborn babies, Opdycke says. In other words, the people in charge of the hospital experience of someone like Katie Collins and her baby were likely tired, overworked, inexperienced, and underpaid.

Nearby Fordham University was not formally connected to the hospital, but the two institutions had a relationship, and university medical students trained there. In 1911, a promising young man received his medical degree from Fordham University. His name was Jerome McSweeney, and he graduated with the Dean's Prize, for the highest average throughout the entire course. Later, as an officer of the medical reserve corps during World War I, he'd be sent to Army Medical School in Washington, DC, return and buy a house on Bathgate Avenue in the Bronx, eventually open an office on Park Avenue, and die at the age of fifty-eight, still living in the borough. But in September 1913, McSweeney was a young doctor working in the Department of Gynecology at Fordham Hospital, which was where his path, along with those of a number of overworked nurse-trainees in peaked white caps, would intersect with that of an Irish immigrant giving birth to her third child.

* * *

By 2013, Alice and Gerry were buried in the genetic segments they hoped to follow like breadcrumbs on the trail of history. The sheer volume of their data was astonishing.

Over the years Alice and Gerry were working their case, the databases yielded a total of 6,912 DNA cousins for the seven Collins siblings, who altogether had 311,467 "segment matches"—that is, segments along the chromosomes that overlapped with those of the Collins siblings. Each one of these slivers of genetic material was a potential clue, as were the surnames associated with the matches, all 14,182 of them. The sisters imported all this data into vast Excel spreadsheets that showed an array of information along rows and columns of cells, including the number and length of the overlapping chromosome segments, and the locations of those segments along the chromosomes.

For instance, for a particular genetic segment match along chromosome 16, one cell might indicate the starting point was "11376029" and the end point was "17065473"—a detail of profound granularity that would mean nothing to you or me, but that Alice understood could hold the key to understanding who her father was, if she could just deduce what these numbers were trying to tell her. Sorting the data various ways, Alice and Gerry went down the list, Gerry following Alice's methodical approach and sometimes suggesting refinements to make their search more efficient. They examined match after match, chromosome after chromosome. Separated by 1,700 miles, the two sisters sent each other updates by email when they found promising relatives to pursue, and every day, Alice searched the spit databases to see if new cousins had turned up, and when they had, she downloaded their data and manually copied-and-pasted it to her spreadsheets. She tracked big trends and granularities, noticing what countries and surnames were showing up over and over, trying to make sense of it all. "If you're look-ing for relatives, you're looking for patterns," she told me.

Keeping all this data on spreadsheets became so unwieldy that Alice and Gerry's brother Jim the younger, the retired software engi-neer, created an iPad app to manage it all. Within the app, Alice could manipulate the data in different ways, searching and sorting by length and by chromosome, and in a notes field beside each DNA cousin, she kept track of pertinent information, like the names of the villages a cousin's grandparents hailed from or the fact that one cousin matched another along the eighth chromosome. She asked her brother to add cer-tain functions to the app, and in those instances when automation failed her, she used brute force—spending weeks, for example, just moving data from one field to another so that she could search it.

Meanwhile, to gather all this data, the sisters were reaching out to DNA cousins to ask them to share their genomes and other informa-tion. Over time, they sent at least a thousand such requests through 23andMe. While some people ignored these messages, plenty were intrigued by the mystery the sisters told them about, and eager to offer Alice and Gerry genetic genealogy techniques, or to scour the Internet for clues into the family trees of other relatives. Some became friends. The long history of intermarriage among Ashkenazi Jews meant that these helpful cousins were almost always not as closely related as the databases suggested they were, which made Alice and Gerry's job much more difficult. "There are only a few people I know of who really know how to do a Jewish unknown parent search—it's just incredibly mud-died by endogamy," says Jennifer Mendelsohn, who runs a Facebook group for Jewish genetic genealogy. "If you pull any two Jews off the street, the chances are pretty high that they're going to share some seg-ments." For example, Alice shows up as related to Bennett Greenspan in FamilyTreeDNA's database; I show up as related to his business partner, Max Blankfeld. This meant that not only were many of the sisters' sup-posed relatives essentially false ones, reflecting distant ancestry many times over, but they also had many more people to comb through. And that's not all of the challenges Alice and Gerry faced. The search for ancestral records is difficult in places where many Jews historically lived. "There's this incredible complication of provenance because borders changed so much," Mendelsohn says. "The place where my grandfather was born was part of the Austro-Hungarian Empire, and then it became Poland, and then Ukraine."

In the beginning, when she didn't know any better, "I'd look at people who were probably related to us through Moses," Alice says. And yet, once upon a time, far enough back, the Collinses and these distant cousins *had* descended from the same people, had shared soup and heartache, had slept in the same bed. A 2014 study in the journal *Nature Communications* found that all modern-day Ashkenazi Jews are likely descended from what researchers describe as a population bottleneck in the late medieval times. The population was very small, perhaps as few as 350 people, but more likely somewhere in the thousands. Something like ten million Ashkenazi Jews descend from this tiny group, which lived, according to several studies, around twenty to thirty generations ago. After centuries, Alice and Gerry were connecting again with their people through the technology of DNA testing, and although the genetic material of these distant cousins might not help the Collins family solve their case, perhaps their research efforts might.

* * *

Alice's problem was complicated enough that she knew using autosomal DNA alone might not be enough. So, even as she and Gerry worked through those thousands of database cousins, she convinced her brother Bill to take a Y-DNA test through FamilyTreeDNA, which offers a database of matching customers. When the results came back, she began something CeCe Moore calls "fishing in the Y chromosome."

The idea was to find other men closely related to Bill Collins along the male sex chromosome and look at their surnames. This works because a father typically gives two things to his son that are of interest

to genealogists: a last name and a Y chromosome, which stays virtu-
ally unchanged as it gets passed down. Within many populations, the
Y chromosome and the surname can be associated going back many
generations, with occasional interruptions for things like adoptions
and those "non-paternity events." So, a man fishing for the identity of
his father can look at his matches in a Y-DNA database, hook a likely
surname—let's say fourteen instances of the last name "Ogle" show
up among men with similar Y chromosomes—and then go back and
search among close autosomal DNA matches for folks with "Ogle" in
their family tree. If the method works, he can trace his way into the right
branch of the Ogle family.

Alice figured this technique was worth a shot, even though she
knew that once again, the unique history of the Ashkenazi Jewish people
might pose a challenge. Most Jews in eastern Europe only acquired fixed
surnames starting at the end of the eighteenth century, in response to
laws forcing them to adopt them. As a result, says Moore, Ashkenazi
Jews "with the same direct paternal line two hundred to three hundred
years ago are unlikely to carry the same surname now." Yet, incredibly,
when Alice examined Bill's results, she found a man with a very close
match to his Y chromosome. The man was unreachable, but Moore
somehow figured out his full name, and Alice's brother Jim managed to
find the man's aunt listed online at a synagogue. Alice got in touch with
the synagogue's rabbi, who connected her to a relative, who connected
her to the aunt, who agreed to take an autosomal DNA test to see if she
was related to the Collins siblings. Alice bought a kit and had it sent,
and the woman spit in the tube, and they waited and—

There was no match. Well, they were related, of course, in the
way that Ashkenazi Jews are often related—"pull any two Jews off the
street," as Jennifer Mendelsohn would say—but very distantly, many
generations back.

It was disappointing, but Alice plodded on. One helpful DNA
cousin suggested she try another helpful DNA cousin, who passed her
on to a third helpful DNA cousin, who happened to be a genealogy

scholar, who roped in another scholar, and the two men began brain-storming ideas for how to help Alice and Gerry. Finally, one of them suggested to Alice that the answer to her mystery might lie in the last name "Cohen." There were a few reasons why this idea made sense, including the notion that if the nurses at Fordham Hospital had relied on an alphabetical system to organize their babies, they might have confused a little Collins with a little Cohen. Alice reached out to genetic genealogist Gaye Sherman Tannenbaum, who was able to look at a copy of the New York City Birth Index of 1913 and find one Cohen who fit: a Seymour Cohen, also born in the Bronx, and on the same day as Jim Collins, September 23, 1913.

Alice began to do research on the family of that little boy. She found a birth announcement and figured out where Seymour and his parents had lived, not too far in the Bronx from John Josef and Katie and their children. She traced where they'd come from in Europe. Was this Cohen family her family, too, the answer to so much hard work and search-ing? She reached out to Moore, who looked at Bill's Y-DNA results and noticed a fairly good Cohen match, which added extra credence to the theory. "It just felt right," Alice said.

It felt so right that Alice and Gerry decided to announce their news to the family. Alice prepared an email. It was May 2013, just under a year since she'd sent away for that first AncestryDNA test, and while that seemed like a long period to be immersed in this consuming and painstaking work, it was, she felt, actually pretty short, given the degree of difficulty of what she and Gerry were trying to do.

"Our search for Dad's parents is most likely over," she wrote, and revealed to her siblings their Cohen grandparents' names. It was an astonishingly gratifying moment.

To confirm what they already knew was true, Alice and Gerry needed to compare their DNA against someone from that little boy's family, so helpful DNA cousins fanned off across the Internet and helped track down a professor in North Carolina who was a descendant of Seymour Cohen's sister. Alice wrote to her and asked if the woman would be willing to help them solve an incredible genetic detective case, and the woman said she'd be delighted. Alice paid for the kit and had it sent, and the woman contributed something completely free and absolutely priceless. She sent it in, they waited, and—

Well, they were related a little bit, if you went back far enough. "Were we ever wrong," Alice says.

13

THE AMERICAN FAMILY

Stephanie Talley learned she was donor-conceived when she was twelve, but it didn't mean too much to her then. It wasn't till later, when she got to the stage when "your parents are suddenly embarrassing and not cool—that's when I started to wonder about that person who contributed 50 percent of my body," she says. "I used to look in the mirror and try to subtract my mom." Her parents divorced when she was a teenager, and after that she wasn't close with her dad, who mostly disappeared from her life. Over time, Stephanie romanticized her anonymous donor, imagining him as an "artistic beatnik guy" responsible for her love of the arts and photography, and wanting to find him. But she couldn't get very far until consumer genomics came along.

Stephanie started testing in 2012, and got her first big break two years later when a half-brother showed up in the relative database. Then she found another half-brother and a half-sister, all of them donor children, and they began to get to know each other. She was using genetic genealogy methods to piece together her donor's identity, and was about to solve her case in 2016 when her biological father showed up in the AncestryDNA database. By then, there were five donor-conceived siblings, and together they crafted a letter to send him over Facebook Messenger. They went back and forth, fine-tuning the tone. Talley told me she understands why so many seekers fear rejection when they reach out: "I was so worried about being good enough to be wanted."

But the donor did not shut them down. He had raised two daughters, and he felt the decision about whether to know these donor children—his daughters' half-siblings—was more theirs than his. He talked to them, and they were open to it, so he and they and some other members of his family agreed to meet up in New York with his donor-conceived children. Stephanie came up from the Tampa Bay area and was so anxious before the meeting that she threw up. What struck her during this visit were the commonalities that genetics seemed to have bestowed on her and her half-siblings, beyond even their physical likenesses: shared sarcasm, outgoing temperaments, artistic inclinations. It turned out her teenage imaginings had been right: Her donor father has a degree in fine art and, after retiring from his career as a civil servant, now works in ceramics. "I feel whole," she confessed over dinner with her biological family, tears springing to her eyes as she looked at one of her donor father's daughters. Her half-sister teared up, too, and they hugged.

But the story didn't end there, because the siblings kept showing up in the DNA databases. When I first interviewed Stephanie during the summer of 2018, she told me her donor father had a total of fifteen biological children "so far," including the two he'd raised. When we spoke again less than a year later, there were nineteen, the last having shown up within the previous forty-eight hours. I talked to their donor father, Robert Jenkinson, and he told me he figured he'd about neared the full tally of his donor-conceived children, although "I wouldn't be surprised if there's another one out there." (Months later, the count had gone up to twenty-three.) Most, though not all, of the donor-conceived siblings were forging relationships within this new family. A few had spent Thanksgivings with Jenkinson and his daughters, and gone camping with them at Virginia's Shenandoah National Park. One of Stephanie's half-sisters had come down to visit her in Florida. For an only child, Stephanie told me, having this abundance of family in her thirties was wonderful, and seeing some of her own qualities reflected back in her siblings had given her "permission to be myself more." And yet, aware of how many other donor-conceived people have had bad experiences in their search for genetic family, Stephanie says, "it's hard for me to completely bathe in how good it is." She told me she feels as if she's won some sort of seeker's lottery and worries about trumpeting her good fortune, not wanting to belittle other people's experiences or gloss over the complications of being donor-conceived.

Some donor-conceived people are deeply troubled by aspects of sperm donation, including instances in which people are not told the truth of their origins until later in life—or until DNA testing uncovers it. As with parents of adoptees, parents of donor-conceived children are generally more open with their kids about how they came into the world than they were decades ago, when, as a piece in the journal *The New Atlantis* explains, the shame of male infertility and the "stigma of illegitimacy" encouraged parents to keep the practice quiet, including from their own children. (The tradition of secrecy goes back to the 1880s, when, in one of the first documented uses of artificial insemination by donor, a doctor sedated and impregnated his patient with the sperm of a good-looking medical student, without telling her or her husband.) Once upon a time it was not uncommon for doctors to report mixing donor sperm with the sperm of the intended father in order to "circumvent legal or psychological difficulties" by keeping the child's paternity ambiguous. This was sometimes known as "confused artificial insemination," and I found a 1990 paper stating, less than reassuringly, that it had by then "almost died out." Plenty of people in their forties, fifties, and sixties are just discovering now how they came into being, and often their parents are no longer alive to have conversations about what they knew and how they felt about it.

But even when donor-conceived people are told the truth of their paternity in childhood, some of them describe the way it is practiced in the United States as "unethical" and have found the truth of their conception destabilizing. "Conversations about this topic happen about us—without us," Joanna Rose, an activist for the rights of the donor-conceived, has said. Stephanie Talley told me she's conflicted by the technology that brought her into being; on the one hand, practices like sperm and egg donation enable gay couples, single women, and those with fertility problems to have children, and she supports that. On the other hand, these practices seem to place a parent's need for a biological connection to their child by means of assisted reproduction above the child's need for that same connection to the person who contributed half of their DNA.

Donor-conceived people have been seeking their genetic families for a long time. In 2000, Wendy Kramer, urged on by her donor-conceived son, started a Yahoo group that became the Donor Sibling Registry, a nonprofit established to help people conceived by sperm, egg, and embryo donation make contact with their genetic families. Kramer's organization also conducts research and advocates for the rights of the donor-conceived, as well as for better regulation of the industry. The registry has more than sixty-five thousand people in paying family memberships in more than one hundred countries, attesting to the deep desire on the part of the donor-conceived—as well, as in some cases, the donors themselves—to seek out their biological kin. But this precise moment, in which so many people are discovering who their genetic fathers are through DNA testing, is at times exacerbating tensions between people who are strangers in all but genes. Most donors gave sperm with the understanding that they'd remain anonymous, and Stephanie has seen seekers receive legal documents, warning them that additional attempts to contact donors will be considered harassment. (In one case, reported in the *New York Times*, an Oregon mother who contacted her daughter's biological grandmother after finding her on 23andMe was threatened with $20,000 in penalties by the sperm bank she'd used to conceive her child.)

Genetic testing is also showcasing what many see as inadequate regulation of the industry that brought them into being. Some want limits on how many children a donor can produce. In contrast with a number of countries, there are no legal limits to this in the United States, though the American Society for Reproductive Medicine, a professional society, recommends no more than "25 pregnancies per sperm donor in a population of 800,000" to prevent what it calls "inadvertent consanguinity"—as in half-siblings having a child together. Within the Donor Sibling Registry there are a number of groups with at least 100 genetic half-siblings, with one group hovering around 200, Kramer told me. "Inadvertent consanguinity" is a real concern for the donor-conceived, especially in cases where many live in the same city; Kramer has said she advises people "to memorize their sperm or egg bank name and donor number, and to share that information with potential dates."

The DNA age is also bringing to light more cases in which fertility doctors appear to have secretly substituted their own sperm. There are cases like this from all over the United States and the world, but one of the more well-known involves Indiana doctor Donald Cline, who, according to the *The Atlantic*'s Sarah Zhang, told his patients in the 1980s that he was using medical residents as donors and limiting their contributions to a small number. Instead, *he* was the donor. At least sixty-one people say they've discovered Cline is their biological father from genetic tests, the *New York Times* has reported. He pled guilty to obstruction of justice charges for initially denying the allegations, but received only a suspended sentence. Cases like this have led to lawsuits and calls for new laws, and in 2019, Indiana passed a fertility-fraud law initiated by several of Cline's donor children.

If assisted reproductive technologies have changed how we can make nuclear families, it has taken DNA testing to demonstrate how far beyond "nuclear" the definition of "family" can extend. How do you adapt language to account for the kind of family that someone like Stephanie Talley now finds herself in? People have invented new nomenclature for these relationships, with some in these expanded networks referring to genetic half-siblings as "halfies" or "sperm sib-lings" or "diblings" (for donor siblings). Stephanie, who dislikes the "arm's-length quality" that a term like "dibling" implies, and simply uses "brother" and "sister," struggled with what to call Robert Jenkin-son, feeling "donor" was sterile and "dad" too loaded. They talked it over and she eventually arrived at "Pops."

As for Pops himself, he told me he never expected any of this; he couldn't foresee the future any more than the people who ran the fertil-ity clinics could. He says he donated for about two years in his twen-ties, when he was newly married with a new daughter, looking for a job, with money tight. It was the 1970s. His wife at the time had given him the idea for how to make a little extra cash, at twenty-five dollars a contribution, if memory served. He never thought he'd meet a child he'd helped conceive; instead, he thought of it as helping someone else expand their family.

And then, several decades later, his daughters gave him an AncestryDNA kit as a Christmas gift, as happens in so many families, and he never realized it would give him anything beyond that little pie chart. ("Surprise!" his current wife said when they recounted this story.) Jenkinson says the biggest gift from this revelation has been the relationships forged between all his biological children. "I hope that continues after I'm gone," he told me.

When we spoke, Stephanie and two of her donor siblings had just returned from the camping trip with Pops and his wife and the two daughters Pops had raised. "It was cool, it was a family thing," she said; even so, she'd been so anxious during the trip that she'd thrown up again. Sometimes, she felt like she didn't deserve this newfound, uniquely modern family. She knew some donor-conceived people who could only dream of a camping trip like that. And on the other hand, she knew many people who were not donor-conceived and who took for granted the warm, loving relationships they enjoyed with their genetic families.

"I feel so privileged, but should I?" Stephanie asked. "Why should I feel like I don't deserve it?"

* * *

Even though it was clear that a long-ago baby named Seymour Cohen was not the solution to Alice and Gerry's puzzle, the sisters believed they'd been right in one crucial sense. Some baby, born so long ago that he almost certainly was no longer alive, was the missing piece, if they could only figure out who that baby was. At Fordham Hospital on or about September 23, 1913, their infant father had been switched with another, and the two had gone home with the wrong families. It had to be. The idea that their father had been switched at birth was the only theory they had left.

Alice being Alice, she started hammering away at proving this theory. She figured that the name of this baby was surely somewhere in old genealogical records. So, she began working with the New York City Birth Index of 1913. The Bronx in 1913 saw 14,679 births, and for Alice this meant 159 pages of baby names. Unfortunately, the names of these babies were not organized by what date in 1913 they'd been born; nor did the records specify which births took place in the hospital—that would have helped narrow it down. Instead, the names were alphabetized, which was thoroughly unhelpful, because Alice didn't know the name of the child she was looking for. Undaunted, she put all those baby names into a white binder, and then she set about combing through every single page, squinting at the old type, which was sometimes blurred into illegibility.

What Alice was looking for were a few demographic factors: a child, likely male, with a last name that sounded either Jewish or ethnically neutral, born within a day on either side of her father's birth date, September 23, 1913. Although Katie Collins had spent thirteen days in the hospital, not unusual for the time, Alice figured only those three days mattered because only just-born children were relevant. No mother who'd spent days holding her infant could mistake some other child for hers.

She searched methodically as best she could, marking all the names that interested her, which were a lot. She tracked the birth certificate numbers associated with these baby names, looking for ones close to her father's birth certificate number, figuring a baby born in geographic and temporal proximity to her dad would likely have a number close to his. Her white binder gradually became littered with turquoise stickies with names and numbers written on them.

Then, she began to play with her data. Over the previous year, Alice had collected an awful lot of family surnames from DNA cousins: so many Levines and Cohens and Friedmans, not to mention Slutzkys and Franks and more. Now, she attempted to see if the last names of any of the babies-of-interest from the index showed up within the pool of surnames she'd collected. If a close DNA cousin had one of these surnames in his or her lineage, they might be related to the woman who'd given birth to her father—back when he was intended for a name other than Jim Collins.

About thirty-five names seemed especially promising, and Alice made herself a spreadsheet of them, then searched systematically in alphabetical order. She started with Appel, then Bain, then Bamason (or was it Bamson? Or Ban_son, with an illegible letter?), Bohrm, Bresky, Brody, Cherkunes, Dazey, Edelman, Efferenn . . . and all the way up the alphabet to Paltzold. She was getting hits, but not on any close cousins. She'd spent enough time in old censuses and the like to know that officials sometimes had atrocious handwriting and a flagrant disregard for original spellings, so she did wildcard searches, in case names were misrecorded. Through the summer of 2013, Alice searched this way, scrawling handwritten notes to herself on a copy of the spreadsheet she'd printed out, writing down when a baby surname she was interested in showed up associated with people predicted to be DNA cousins.

And she got nowhere. It appeared to be another dead end.

* * *

Back in the early 1990s there was a pretty crummy movie starring Brendan Fraser (remember him?) as a frozen caveman who gets thawed by some high school kids and gets to live among us moderns, and that's what I think of sometimes with respect to consumer genomics—all those secrets frozen and brought back to life in a vastly different era. These days, we place a high value on transparency and truth-telling, but the truth as revealed by our saliva is at odds with certain historical high stakes. Infertility, donor conception, adoption, extramarital affairs, and interracial relationships once could have threatened a family's integrity or its standing within a community. Some of those stigmas seem so dated that you'd think these secrets couldn't hurt anymore, but they do. Because, in fact, out of those secrets came people—people who can sometimes still sense the shame that surrounded their hidden identities.

Genetic testing makes clear that the past was a foreign country. Once upon a time, as Rosario Castronovo discovered when he was researching his grandparents, Vermont convicted and imprisoned people for the crime of adultery. Once upon a time, young women who became pregnant when they were unmarried were sent to places like the Veil Maternity Hospitals in Missouri and Pennsylvania. Promised protection of their "reputation and social standing," the women gave birth, sometimes under assumed names, and typically never saw their babies again. I heard about one of those babies, now in her eighties, who spent years searching for the identity of her mother before DNA cracked the case.

Such late-in-life revelations have an extra poignancy. A Florida TV station reported on how DNA testing had for the first time brought together four elderly siblings, the product of a man who'd divorced his wife, changed his name, and started a new family. "To learn at this time in my life that I have a sister—and my sister's 102?!" said one of the men, grinning and clutching his cane, while the sister beside him smiled behind her walker. I talked to two sisters, both in their eighties, for whom DNA testing had become the means to reconnect after forty-five years of estrangement. They never addressed the reasons for their split—"We purposefully don't talk about any bad things, only good things," one of them told me—but this late-in-life reconciliation was deeply meaningful.

And in another case, the *New York Times* reported on a woman who was unmarried when she gave birth in a hospital in 1949. She was told her baby had died. In fact, that baby was alive and was adopted out; somebody, as CeCe Moore would say, was playing God. The mother was eighty-eight years old when she met her sixty-nine-year-old daughter. What do you make of such a story? That the loss is heartbreaking? Yes. That the revelation is a lucky one? Yes, that, too. The gratitude and the sorrow are woven together. DNA testing is a race against time.

Part of what interested me in talking to seekers was the question of how they processed the pain of the past: the unfinished business, the unanswered questions, the unresolved anger and shame and resentment and grief. And how did they make sense of lives with huge people-shaped holes in them? So often, the secrets themselves loom as a kind of negative space in those timelines: the teenage girl who went away for months, or the father who disappeared, or the relatives whose names are listed together on a ship's manifest, yet for some reason scattered in the new country. With the truth written into visibly overlapping DNA segments in commercial databases, we are all now left to make something out of those negative spaces.

"There are secrets people don't want to talk about and then they pass, and there's no way they can talk about it," Krista Driver says. Driver, a therapist now in her late forties, was adopted. When she met her biological mother at twenty, her mother told Driver who her father was, and Driver went on to meet this man and to believe he was her genetic father for twenty-six years. And then, in 2018, she took a DNA test, and that blew everything apart. When she figured out who her real biological father was, she texted her mother his name and asked, "Who is this?" Her mother never replied. "She hasn't stopped crying" since you texted, Krista's half-sister wrote her. When I spoke with Driver in the late winter of 2019, she told me her biological mother had passed away just three weeks earlier. She had taken her tears and explanations to the grave.

Driver told me she worries about the way the press focuses on happy reunions when they cover genetic testing surprises, failing to account for the emotional complexity of such situations. Driver is CEO of the nonprofit Mariposa Women & Family Center in Orange County, California, and since her own DNA revelation, she has started counseling groups for those who've experienced NPEs (which she defines as "not parent expected"). "There's nothing really in the mental health community addressing this," says Driver. She's seen others express frustration: "I went to a grief specialist, but they didn't really understand." Driver doesn't directly treat people who've experienced NPEs because her personal experience makes her too close to the topic, but she trains other counselors who do, and she says there's often a panoply of emotions accompanying the discovery that one or both of your parents are not who you thought. One is a particular grief at "losing a life that you didn't even *know* you didn't have," she says; she was herself flattened by this emotion after a reunion organized by her genetic father's family at which she saw four cousins who grew up less than a mile from her, and with whom she could have had lifelong relationships if only she'd known they existed. There's also sometimes anger at the parents who didn't tell the truth, and the ensuing guilt over being angry at one's parents.

And there's anxiety at the loss of identity. "Human beings have an instinctive need to connect with each other; we're pack animals," Driver says. "For somebody to suddenly be disconnected from their pack, from what they were connected to, it's like being unmoored, it's being untethered." Her words made me think of Jacqui Ochoa's description of the alienation she felt when she discovered her newfound siblings were not her genetic siblings, and the language Alice used when she learned her beloved cousin Pete Nolan was not her genetic cousin— she'd felt "adrift."

At its core, Driver says, a DNA revelation like this can create profound trauma, made all the more difficult to process at first because the cycle—from surprise to meeting new genetic family—is often so short. She told me she didn't spend a lot of time on a Facebook group she'd joined for those experiencing misattributed parentage because "a lot of people are so broken and angry." She's seen cases in which people in sobriety have relapsed after discovering genetic surprises.

A few other specialists have emerged to help people through the particular landscape of DNA revelations. Farther north of Driver in California, therapist Jodi Klugman-Rabb has been offering a treatment for what she terms Parental Identity Discovery. And across the country outside Charlottesville, Virginia, Brianne Kirkpatrick, founder of Watershed DNA, believes she is the only genetic counselor who helps people not only with the implications of medical genetic testing but with family surprises as well. She runs several secret and intentionally small Facebook groups for people who very often "just want to know that what's happened to them is normal."

Kirkpatrick preaches proactivity for the keepers of genetic secrets, both because people on the outside of a secret have a right to know who they are and because, in the age of DNA testing, they will eventually find out anyway. It's not a matter of *if*, but *when*. For parents who haven't yet told their children that they are, for instance, donor-conceived, she offers resources and helps them plan for these conversations. And for those who've discovered a surprise by way of a DNA database and ask, "Can I keep this a secret?" Kirkpatrick usually explains to them why this is impossible, even if they were to delete their account at MyHeritage and say nothing to anyone. Do they have children who might someday test? Do they have siblings who might? Time is the enemy of secret-keepers, because with time comes the inevitable revelation, one way or the other.

Kirkpatrick also works with the children of incest, a revelation sometimes made through DNA testing by adoptees, and one that can be incredibly painful. "Once you know it, you can't un-know it," she says. She prefers to use less stigmatizing language, describing these people as having "high ROH," which means "high runs of homozygosity," or stretches of identical DNA inherited from both parents. There's a tool on GEDmatch called "Are Your Parents Related?" that shows what this looks like, with long segments of identical DNA on both of a pair of chromosomes indicated by blue bars. In an information sheet Kirkpatrick produced for those with high ROH, she offers resources and explains the health implications—in some cases, being the child of incest can pose medical issues—and emphasizes that these results are not the fault of the person making the discovery, a point I saw made elsewhere. "You cannot control your own birth!" blogged one adoptee who wrote about confirming that his biological grandfather was also his biological father. "The important thing is to not let it control you or your life."

Among the groups she runs, Kirkpatrick started a group for women married to "birth fathers," men who discovered by testing that they had children they didn't know about. These women call themselves "birth wives." According to the accounts of these birth wives, there is "a very unusual dynamic that happens when it's an adult daughter coming into the life of her birth father," Kirkpatrick says. "The way the wives have described it, in some of the situations it turns into almost a romance," with the husbands secretly texting their daughters, buying them flowers, and primping before seeing them. "In some situations this is how GSA can happen," Kirkpatrick says, referring to something called "genetic sexual attraction," which has not been the subject of much research. GSA is not necessarily always acted upon, Kirkpatrick says, so it can't be equated with incest, but it can be troubling and painful for the people involved. Kirkpatrick says it has cropped up as an issue within the adoptee world in the past and worries it is now a looming matter of concern for the growing universe of people affected by DNA surprises.

In coming years, as history collides with the present, we may grapple with other issues stemming from such surprises, ones we can't even imagine now. Krista Driver argues that America needs to have a broader cultural conversation about DNA revelations: about how common they are, about the wide range of emotions that accompany them—not only on behalf of all the people who've already experienced their genetic surprises, but on behalf of those whose surprises are to come.

"If we start having a conversation now, and start taking the shame out of it, that day of reckoning for people may have less of a sting," Driver says. "We just need the past to catch up to where we are now."

* * *

Is it always better to know the truth? For seekers, I've found, the answer is almost always yes. CeCe Moore told me she's found the same thing, and it surprised her: "I really do expect more people to say I wish I never tested." That's despite the fact that, as Krista Driver and many others have pointed out, reunions aren't always "happy" ones. Sometimes, seekers are disappointed once they get to know the people they were looking for. Sometimes tensions arise over whose narrative of the past will prevail. It may take time for this to play out; I've checked back in with seekers after many months or a year and found their perspectives have shifted. For instance, the seeker I call Nancy, who got to know her father after taking a DNA test, began over time to doubt his account that he'd tried desperately to have a relationship with her when she was a child. Nancy said her father repeatedly avoiding discussing what she called "the pink elephant in the room, which was why the hell were you not in my life for the last forty-five years?"

And yet, Krista Driver says, "I wouldn't want to hold people back from finding their truth because they heard someone say how horrible their experience was." Because that truth is worth it. "I don't want to live with secrets. Secrets make you sick."

Brianne Kirkpatrick says she watched this sentiment echoed in one of her Facebook groups, when a man posted that he'd figured out his wife's biological father was someone other than she thought. "Should I tell her?" he asked. The people in the group who'd experienced similar surprises said yes. "There's a really big push from the NPE side of things that people are owed the truth about their genetic origins—even if it means a painful process," Kirkpatrick says.

To be clear, it's hard to extrapolate across the human spectrum from a pool of people who are invested in uncovering the truth. How much of this is the result of self-selection? "Maybe those are the ones I don't see because they don't pursue it any further," Moore says. And indeed, there are surely many people who receive surprising results after testing on a whim and don't immediately decide to devote hours upon hours to solving that mystery. In some cases, they might decline to look further because the news is destabilizing, whereas in others, they might not have the time or interest. Many testers are like Jessica Benson, a stay-at-home mom in her thirties living in Raleigh, North Carolina, who tested in 2014, hoping to find out more about her late father's side of the family. When her ethnicity results didn't line up with what she expected to find along her paternal side, she spoke with her dad's sister, who was equally mystified, and then threw up her hands.

"Somebody did something with someone they shouldn't have," Jessica concluded, and that's where she left it, until the truth came looking for her later. There are surely many more people with discrepancies like this lurking in their results, discrepancies that did not turn them into seekers.

Other things matter, too, in a person's experience of a revelation. Like the way the news unfolds for you. A philosophy student, writing in *The Guardian* about how genetic testing had uncovered that his father had lied about his ethnicity, rupturing his lifelong identity as an American Indian, resented "the insistence with which my aunt had pushed my sisters to take these tests," and the fact "that I couldn't discover who my father was on my own time." As psychologist David Brodzinsky puts it, "those who feel in control generally feel better about what's going on, and if you're doing the searching, you're the one in control." Some of the way we process these disruptions comes down to a sense of agency. It is one thing to be invested in a search, another to have the truth thrust upon you.

And of course, the stories of Laurie Pratt's father, who deleted his AncestryDNA account after she found him, and Jacqui Ochoa's half-siblings, who question the science behind DNA testing because her existence implies their mother abandoned a child, have demonstrated that it is a vastly different thing to be the *seekee*, especially when the news suggests something shameful or troubling about you or the parent who raised you. Genetic genealogist Debbie Kennett says there's a great need for research into the cultural and personal impacts of DNA testing and the secrets it uncovers. "We've got all these stories coming out, but certainly in our community, we're seeing a very one-sided view. We're always seeing the perspective of the people who are searching, and we never seem to hear about the effects on the people who are contacted," she says.

But for seekers, revelations about the past and how they came to be can offer context for otherwise inexplicable childhood memories. In some cases, they can answer lingering lifelong questions. "My entire existence was a cognitive dissonance," one seeker told me. The truth colors countless interactions, tiny and weighty, stretching back through a life and before it. Ricki Lewis, who learned through genetic testing that she'd been conceived by a sperm donor during an era when some doctors mixed sperm, has been waking up with old memories she hasn't thought of in years, as if her subconscious is rewriting her childhood in light of new facts. She looks back and wonders whether her father knew he wasn't her genetic father, and if so, might this help explain the depression he struggled with? She wonders if the fact that she was not related to her father explains the almost instinctive affinity she felt as a child for her mother's family, but lacked for her father's.

The need for certainty and clarity helps explain why people can become obsessed with their genetic mysteries, their investment building during the painstaking process of discovery. Adoptee Linda Minten told me what drove her search was "the truth, just the truth, just the truth," beating the phrase like a drum. "I need the truth because I can deal with that head on. It's all this underlying gray—secrets, hiding—that stuff messes you up." The secrets were no foundation, she said; they were like quicksand. Cultural psychologist Steven J. Heine observes that uncertainty can be more difficult to handle than even sure knowledge of bad news.

Knowing the truth is about claiming a complex, clear-eyed self. Seekers told me they needed to know their own beginnings in order to write the rest of their stories. A woman I'll call Sara told me about her journey: about discovering through a spit test that her father was not genetically her father, about unraveling that her genetic father was her mother's first cousin, about calling her mother, who cried and said that her encounter with this cousin had only just started coming back to her and that what she remembered was that she had been raped.

Sara told me her upbringing had been deeply troubled and unstable, marked by her mother's emotional unavailability and abuse in the family home. In that single phone call, things shifted. "It was the most honest and real that I felt she had ever been with me in my whole life, and I felt more connected to her after that whole conversation than I ever felt," Sara said. "It was like all of a sudden, all of these pieces made complete sense." The news changed how Sara related to her mother, for whom she had more empathy and understanding. She encouraged her mother to see a therapist so that she could process what had happened.

Sara told me she was under no illusions that her family was somehow fixed. The truth was no magic balm. Her revelation had upended her mother's life, and she did not know how that would play out for her. But for Sara, at least, this tumultuous journey had been a profoundly good thing. Knowing the truth had freed her somehow, even though the truth itself was ugly.

* * *

In the summer of 2013, the Collins family held a reunion in San Francisco and celebrated the upcoming hundred-year anniversary of Jim Collins's birth. The siblings spent days together, talking about their dad and the mystery still at the center of his life. Alice and Gerry updated their brothers on all the work they'd done and all the techniques they'd tried, fielding questions and suggestions. Gerry had made beautiful stained glass Christmas ornaments for everyone, each a Jewish star of David with an Irish shamrock in the middle. Alice's son, Gidian, hung his from a button on his shirt, like a necklace.

The culmination of the reunion was a picnic in Marin County, at Mount Tamalpais. Alice remembers two things in particular from that reunion. She remembers the siblings discussing a theme that occupied many of their conversations during this time: that they would not and could not be bitter about what had happened to their dad. Without his case of mistaken identity, however precisely it had come about, Jim the elder would not have met Alice the elder, and the seven Collins siblings would not be. The vagaries of history had led them here, so they had to make peace with it.

The other thing Alice remembers is lying. She told her siblings she was certain that if she just kept plugging away, she could eventually find their biological grandparents. In truth, she felt no such certitude.

After the reunion, Alice went north to Washington State and kept working her case. She hoped that by Jim's birthday on September 23, she'd be able to solve it, but that was not to be. Instead, on the day after his birthday, she and her best friend, Kathy Long, set off for Europe. The birth index may have been blurry, and surnames and segments might mislead, but Alice knew the geographic origins of her DNA. Her closest genetic cousins had told her where their grandparents and great-grandparents had lived before they came to America, and she'd written down the names of all those towns. They went to Poland, Latvia, Lithuania, Estonia, and Russia.

It was an emotional trip for Alice, who little more than a year before had not known she had any connection to this part of the world. The friends visited a Jewish ghetto in Krakow and a memorial to Jews who'd lost their lives in Warsaw. When Alice and her siblings had lived in Germany in the mid-1950s, the reconstruction of the country that had murdered so many Jewish people had been the backdrop to their lives. Now, at the Warsaw Jewish memorial, Alice thought, *These are my people*. She began to cry.

In Lithuania, she and Kathy visited the capital city, Vilnius, and toured a museum that had been made out of a former Soviet Union KGB headquarters. Inside were photographs of Lithuanian people; terrible things had happened to some of them. And among the faces, Alice glimpsed a familiar one. It was a little boy who looked so much like her brother Brian had looked when he was young that her heart stopped. Alice stood there, riveted, unable to get over the idea that she'd come so far to see a face from home.

Alice flew back to Washington and kept plugging away at her detective work. She continued with her segment triangulation, day after day. She put the data from each new promising DNA cousin into the app Jim had made for her, and sorted and studied the results. She chased down leads, and was disappointed, and then did it again. In time, she and Gerry would buy at least twenty-one DNA test kits for themselves, relatives, and strangers suspected of being relations. And she tried to get close to her grandparents by homing in on where they'd once been. She watched a seminar intended for people researching ancestors affected by the Holocaust, and came away with a new technique, the fruits of which she showed me when I visited.

On her Mac, Alice navigated to a spot within Google Maps she called "ancestor pins." We watched as an image of the earth spun around to show the landscape of eastern Europe come alive with yellow dots. They were all over Belarus, Latvia, and Lithuania, but they clustered, like raindrops, over the Lithuanian city Vilnius, where she'd seen her brother's face.

She could see where her DNA was coming from, right there on the screen, but she was still wandering without a map. She was Alice in the wilderness.

14

YOUR GENES ARE NOT YOURS ALONE

In 2012, right around the time that Alice Collins Plebuch was trying out Ancestry's beta version of an autosomal DNA test, Yaniv Erlich was a fellow at the Whitehead Institute for Biomedical Research in Cambridge, Massachusetts. He was studying genetic privacy, a concept that fascinated him. But not everyone in his academic community was thrilled with his choice.

"Several people came to me and told me, 'Why even study this type of thing?'" Erlich told me over the telephone from his home in Ra'anana, Israel, in early 2019. In many ways, 2012 was a long time ago. "This is pre-Snowden, pre–Cambridge Analytica," says Erlich. It was before Russian bots on Facebook, before the election of Donald Trump. "2012 was just the dawn of big data," Erlich recalls. "People thought, *That's going to be great for everything.*"

Erlich was not sure big data was going to be great for everything. Inspired by the story of Donor Sibling Registry founders Wendy Kramer and her son Ryan, who'd managed to track down Ryan's anonymous donor father using Y-DNA testing and genetic genealogy all the way back in 2005, when Ryan was just fifteen years old, Erlich and colleagues had been working on a paper showing that they could fully identify men who'd anonymously donated DNA samples for research. They did this by taking advantage of the same correlation between surnames and Y-DNA that Alice had been hoping to exploit when she tested her brother Bill, and that genealogy researchers had been tracking in surname projects within public databases. Combining the power of those databases with the Internet, genealogical websites, and the limited "nonidentifying" information available about five research genomes, Erlich, his student Melissa Gymrek and their colleagues were ultimately able to identify not only those five men but many of their family members as well, altogether close to fifty people.

Erlich says there was some perception among his colleagues "that this is just fear-mongering," that he was pointing out a privacy loophole that only "boring academics" would bother to exploit. And at the same time, there was concern that his work would discourage the sharing of genetic data and make scientific research harder. "Some senior people told me that this might destroy my career," Erlich says. He was looking to move to an academic appointment once his fellowship was over, and this made him nervous, but he believed that what he was doing was important—that by alerting everyone to vulnerabilities, he might prompt solutions that would make things safer. At the very least, people participating in studies should know what they were getting into.

He submitted his paper, and it was accepted by the journal *Science*. On the day it came out, he told me, he went out to a celebration dinner with his wife and had to force himself to eat.

* * *

In April 2018, a major news event demonstrated the power of genetic genealogy. Major outlets reported that investigators in California, matching crime scene DNA against the genetic profiles of ordinary people in the free, publicly-accessible genealogy database GEDmatch, uncovered the suspected identity of the notorious Golden State Killer and arrested him. Thought to be responsible for at least thirteen murders and fifty rapes in California in the 1970s and '80s, the Golden State Killer became known for his brutality and sadism. "As his victims lay terrified," the *New York Times* reported, "he would pause for a snack of crackers after raping them. He placed a teacup and saucer on the bodies of some of his victims and threatened them with murder if he heard the ceramic rattle." Victims' descriptions of him—a white man with blond hair—generated multiple sketches, while his crimes inspired terrifying nicknames: the East Area Rapist, the Original

Night Stalker, the Visalia Ransacker, the Diamond Knot Killer. Yet for decades, the culprit could not be found.

Police announced that the accused was Joseph James DeAngelo, a seventy-two-year-old former police officer living in a suburb of Sacramento. The notion that the Golden State Killer might be a cop had come up over the years, but according to the Sacramento district attorney, DeAngelo's name never had. "We weren't even close," Paul Holes, previously a cold case investigator with the Contra Costa County District Attorney's Office, told an interviewer.

Some months later, after hesitating over concerns for her safety, the genetic genealogist who'd assisted police stepped forward. She was Barbara Rae-Venter, a retired patent attorney with a PhD in biology who was trained by and then volunteered with DNAAdoption.org, the same organization Alice Collins Plebuch had turned to in order to hack her genes. Rae-Venter had not meant to be solving crimes—"I'm retired; I'm supposed to be working on my own family history," she explained during a lecture in Dublin, Ireland—and had essentially fallen into this work. Some years earlier, at the request of a detective, Rae-Venter had used segment triangulation (filtered through Alice's brother's iPad app, as it happens) in an enormously complex and time-consuming case to uncover the true identity of a woman who had been kidnapped, abused, and abandoned as a small child. Rae-Venter then helped determine the identity of the woman's deceased abductor, who police believe was a serial killer responsible for a number of deaths. Holes, who was nearing retirement, heard about Rae-Venter's work on those cases and contacted her to see if she might be able to help with the Golden State Killer, a case he'd been working on since the 1990s. Rae-Venter got on board and began to learn the details of her newest case. She thought Holes's criminal, whoever he was, made her previous serial killer "look like a choirboy."

Here's where the story gets thorny. The account that was given by authorities and reported by the media when the Golden State Killer case broke was that Rae-Venter, Holes, and other law enforcement investigators solved it by uploading a profile made from crime scene DNA to GEDmatch. From genetic cousins found in that database, the account went, they used segment triangulation to build family trees and identify a most recent common ancestor, and then traced forward in time to candidates descended from that ancestor, ultimately homing in on a man named Joseph James DeAngelo.

Then, as in the many cases like this since that time, police followed the suspect until he discarded his DNA in a public place—in DeAngelo's case, in a tissue in a trash can—and matched it to crime scene DNA. DeAngelo was arrested three weeks after Holes retired. Rae-Venter told me many of the serial killer's victims had continued to live in fear of his return; he's believed to have terrorized some by phone years after he brutalized them. When she watched the news of DeAngelo's arrest, Rae-Venter told me, she was struck by the thought that genetic genealogy meant not only would the culprit see justice, but his victims might finally have some peace of mind. "They don't have to wonder that he's going to come through the door," she said. In 2020, DeAngelo pled guilty to thirteen murders, and, after a hearing filled with days of anguished testimony from survivors and their families, was sentenced to multiple life sentences without the possibility of parole. It was only after DeAngelo's sentencing that a more complex—and for privacy scholars, more troubling—account of the Golden State Killer case was uncovered, involving even more databases. But we'll come back to that.

If the Golden State Killer case was a wake-up call for many Americans about the unforeseen applications of genetic genealogy, there were a number of people who were not surprised: members of law enforcement who'd already dabbled in trying to solve case through public databases, and certain people in the field of genetic genealogy, including CeCe Moore, who'd known this kind of approach was possible for years and had been thinking through the ethics of it. "The law has been lagging behind technology for a very long time," says Moore, who remembers speaking at the International Symposium on Human Identification several years before the Golden State Killer case was solved, and walking her audience through what she could do as a search angel. "They were dumbstruck. They couldn't believe what we could do."

And then there was Yaniv Erlich, who had published that paper in *Science* just five years earlier, when the experiment of using public genetic genealogy databases to identify anonymous research subjects struck some of his colleagues as unwise. In the end, that paper didn't damage Erlich's career. The *New York Times* ran a story on his "surprising" discovery, and he was invited to give a lecture and to give testimony. He became a professor of computer science and computational biology at Columbia University, and later took a leave of absence to become the chief science officer at MyHeritage. In addition to his work on other things—including the construction of an astonishing thirteen-million-person family tree offering insights into migration patterns and genetic longevity—Erlich continued to work on genetic privacy and, in 2014, published a paper in which he surveyed all the ways that privacy could be breached, warning about a then-obscure hobbyist website called GEDmatch. At the time, he allowed, the threat was mostly theoretical, but the world of genetic genealogy moves so fast that it took only four short years to get from theoretical to arresting a man in one of the most notorious cases in California history.

To understand why police were so excited to have access to a site like GEDmatch, it helps to understand how they usually use DNA to solve crimes. For many years, forensic DNA testing by law enforcement has relied on a different kind of technology than the one companies like AncestryDNA use. Instead of looking at SNPs, police rely on STR testing—"the number of times certain short, known sequences repeat at various locations on the genome," explains NYU law professor Erin Murphy, author of *Inside the Cell: The Dark Side of Forensic DNA*. The genetic markers law enforcement uses usually "reveal little to nothing about a person's medical or clinical history"—in contrast to the many SNPs employed by genetic genealogy companies, which are typically not used in a criminal justice context "in part due to these privacy concerns."

Police search for DNA matches by uploading the DNA profiles from crime scenes to local, state, and national databases that are known colloquially under the name CODIS. Looking at thirteen to twenty of these genetic markers known as STRs, the system compares crime scene DNA profiles to the millions of profiles of offenders, arrestees, and others already in the system. If the suspect is not already in the system—if he or she is not someone already known to law enforcement, in other words—police may be out of luck as far as this technology can help them. That's because "long-range" familial searching—trying to piece together the identity of someone who left their DNA at a crime scene by using relatives—is very difficult to do using the DNA profiles in CODIS, which are helpful only for finding the closest of relatives, like siblings, a parent, or a child. And besides which, not all states permit or practice familial searching, and those that do may impose restrictions.

To be sure, police are able to use CODIS databases to assist in solving crimes—hundreds of thousands of them, according to FBI statistics. But there has long been interest in taking advantage of the public databases filled with the research of genealogical hobbyists. Genetic genealogist Colleen Fitzpatrick consulted on the Golden State Killer case for California's Irvine Police Department as far back as 2011, using Y-DNA techniques to try to suss out the identity of the perpetrator. Reporting by several outlets shows that in 2017, investigators attempted to find the identity of the Golden State Killer by tracing DNA through a small, now-defunct public genealogy database for Y-DNA called Ysearch.org, run by FamilyTreeDNA. Law enforcement was able to get a subpoena to force FamilyTreeDNA to turn over some information about the account—"I explained to them why it wasn't going to help them, because the data they had was too puny," company president Bennett Greenspan told me—and that led to the swabbing of an elderly man in a nursing home. (The man turned out to be an extremely distant relation to the Golden State Killer, Paul Holes told me.) But, as Rae-Venter pointed out to me, relying on Y-DNA alone often leads to mistakes and misidentifications. Neither that approach nor searches in CODIS had anywhere near the power and reliability of harnessing autosomal DNA gathered for genealogical research.

"We've had the Golden State Killer's DNA up in CODIS since 2001: no hits," Holes told a news station afterward. "We've done familial searches here in California every single year in which it's been legal: no hit. We've done international searches. We did everything without success. Law enforcement has been chasing this guy for forty-four years. . . . Once we started down this genetic genealogy road, it took us four months."

Using consumer DNA profiles meant law enforcement didn't have to worry about whether or not a suspect or his brother happened to be in CODIS. So long as there were cousins, that could be enough. What's striking, population geneticists Graham Coop and Michael "Doc" Edge write, is that because of the richness of the DNA genotyping approach used by ancestry companies, GEDmatch's roughly one million people was bound to be vastly more helpful to investigators than the thirteen million DNA profiles then in CODIS.

* * *

At the time Rae-Venter and investigators unraveled the identity of the Golden State Killer, few of the hobbyists who'd placed their genetic information into the GEDmatch database could have anticipated that it might be involved in solving a crime. For years, genealogists like Alice Collins Plebuch who'd had their saliva or cheek cells processed by consumer DNA testing companies had been downloading their raw data files and uploading them to this third-party site.

Founded in 2010 by a genealogist and retired businessman named Curtis Rogers and an electrical engineer named John Olson, GEDmatch was free and run by volunteers, though it charged a small amount for some advanced tools. A small private company developed in the spirit of open sharing and research, the site's policy was loose in 2018, warning users of the possibility that it could theoretically be used for purposes other than genealogy. And while it was much smaller than the databases of major DNA testing companies, the appeal of GEDmatch to investigators was that it was accessible to them: 23andMe and AncestryDNA, for example, don't permit uploads of DNA processed elsewhere, and say they don't permit law enforcement to access consumers' genetic information unless they're required by law to comply with things like subpoenas and warrants.

Uploading crime scene data to GEDmatch required investigators to create a fake profile on the site. "An undercover profile," corrected Holes when a journalist asked about the ethics of this. "There's plenty of case law that allows law enforcement to pursue undercover activities to effect an investigation. So this is something that is lawful." Indeed, Erin Murphy told me, the only oversight of police using a genealogy database this way would come from the database itself. The weakness of that oversight would become clear only years later, when investigative reporting revealed that GEDmatch was just one of several databases used to solve the Golden State Killer case. But in the weeks and months following Joseph James DeAngelo's arrest, it was only tiny GEDmatch that had to answer the thorny questions about whether to continue cooperating with police, bearing both the backlash from those horrified by this crime-solving approach and the burden of expectations from those who cheered it.

GEDmatch co-founder Curtis Rogers had not wanted his site used for criminal investigations. He later said that when he first learned it had been accessed by investigators, this felt at first like a violation. But he was surprised by the reaction GEDmatch got: In emails, people were largely supportive and grateful. He came to feel the use of the platform in this way was a good thing. He and Olson alerted users and changed the site's policies to say that law enforcement could upload DNA for the sake of solving murders and sexual assaults and for identifying remains. Law enforcement officers across the country became aware that they had a powerful new tool that could radically improve their ability to solve some of the worst crimes.

Within weeks, a DNA technology company outside Washington, DC, called Parabon NanoLabs made an announcement. Parabon, which had previously offered law enforcement agencies a service that creates a composite sketch of a suspect or victim based on DNA, now had an investigative genetic genealogy unit helmed by CeCe Moore. Less than a month after DeAngelo's arrest, Parabon had already uploaded genetic information from about a hundred crime scenes into GEDmatch, and Moore had begun to unravel them, at a typical cost to police departments of about five thousand dollars per case. I spoke with CeCe Moore that summer, and she told me she'd long wanted to use DNA in this way, but for years had turned down requests from detectives, coroners, and district attorneys. At first, she declined in part because GEDmatch's database was too small to be useful. But more broadly, she had ethical and legal concerns. She couldn't do this kind of work behind the backs of the genealogy community, she felt, especially since she was responsible for encouraging many people to test. So, for a long time, she sat on her hands, and *not* doing this work was its own kind of burden. If a perpetrator was still out there actively committing crimes, was her inaction resulting in more victims?

The Golden State Killer case, and the publicity around it, had blown Moore's concerns away. "I couldn't have asked for a better advertising plan to get the word out," she said. Anyone who didn't want their data to be used that way could take it down. Besides which, the case seemed to make clear that she was legally protected if she did this kind of work, and, as she gauged the reactions of bioethicists who weighed in, she came to believe her gut instinct was vindicated.

"I just think it's the right thing to do," Moore told me. She frames investigative genetic genealogy as akin to the work she's done with adoptees and the donor-conceived. "My work is about family and trying to bring closure and answers and peace, and it's exactly the same thing—it's just a different avenue to do that." She told me she was still consulting to *Finding Your Roots*, but had scaled back on her search angel work to make room for a second full-time job at Parabon. Sixteen-hour shifts had turned into twenty-four-hour shifts. "I've had cases where I just could not walk away," she said.

By May 2019, a year after Parabon had launched this service, it announced it had helped solve fifty-five cases. Many of these were cold cases involving crimes that took place well before commercial DNA databases and genetic genealogy existed, not to mention the Internet; the oldest had been committed more than fifty years earlier. Working from her couch, Moore was able to tease answers from the old crime scene DNA and give the names of possible suspects to police as tips for them to confirm. Many of the cases were jaw-dropping in their horror. They included a man in his fifties arrested for raping and strangling an eight-year-old Indiana girl in 1988 (he pled guilty and was sentenced to eighty years), and a man in his thirties arrested for the rape just a few months earlier of a seventy-nine-year-old woman.

Meanwhile, Barbara Rae-Venter has also been working cases, offering a sliding scale and even working pro bono when local agencies' budgets are tight. The FBI flew her to Texas to teach federal agents and police her methods. And Colleen Fitzpatrick has been helping with murders and unidentified remains, adding autosomal DNA techniques to her old methods involving Y-DNA. In the months and years after DeAngelo's arrest, the faces of long-ago victims were revived again in news stories: mugging for the camera or smiling from school photos. Most were women and girls, and the crimes perpetrated against them often involved sexual assault. Now, these gendered crimes were being solved by female citizen-scientists, who were pointing police toward men who, if they were guilty, had lived for decades without any repercussions for the most atrocious crimes. It prompted a kind of reckoning, a sense for some that this is the way the world is supposed to work and that genetic genealogy was making it work that way.

Sometimes, Moore says, she sweats while looking for a culprit, cries upon finding him. These discoveries can be heavy things because once she pieces her way to a likely suspect, there is a period of time when she is perhaps the only person in the world who knows his identity, other than the criminal himself. In June 2019, one of Moore's solves became the first case involving genetic genealogy to go to trial and lead to a conviction after defense attorneys didn't challenge Moore's techniques, agreeing to treat her work as "just a scientific tip," as she puts it. "That's really good news for us." After a jury pronounced William Earl Talbott II guilty of murdering a young Canadian couple in Washington State in 1987, a local investigator said genetic genealogy was "an amazing tool, and we'll be using it again."

But a number of privacy experts are uneasy about the way that Moore, Rae-Venter, and others have been helping police make arrests. In a paper published in *Science*, legal scholar Natalie Ram and others argue that while "there are few legal roadblocks to police use of genetic databases," this may nonetheless violate the spirit of American law. "The Fourth Amendment's protection against warrantless searches and seizures generally does not apply to material or data voluntarily shared with a third party"—in this case, a platform like GEDmatch. Yet part of the Fourth Amendment's civil rights protections involves "a commitment to protecting privacy or freedom from government surveillance," Ram and her colleagues wrote. "Police cannot search a house without suspecting a specific individual of particular acts—even if doing so would enable the police to solve many more crimes." In other words, solving crimes is not the only thing American law values; our right to privacy isn't—or shouldn't be—subservient to this interest. A Supreme Court decision handed down a few months after DeAngelo's arrest has further unsettled this area of the law, potentially reinvigorating privacy protections for those identified though investigative genetic genealogy.

"The fundamental order that's set by our Constitution and by the protections in the Fourth Amendment and elsewhere . . . is one that says people should be left alone by the government in its policing capacity unless the police have a suspicion, a reason to think you've committed a crime," says NYU's Erin Murphy. "Something as simple as your name can't be forced from you by police without suspicion." DNA databases, Murphy argues, are part of a larger trend in which new technologies involving big data and surveillance upend that presumption. "DNA, when used to find suspects or to put people under suspicion who otherwise would not have reason to be under suspicion, is an inversion of how we actually do things." And the shared nature of genetic data means that "this idea of consent or this idea of third-party disclosure doesn't really work," she continues. After all, when a man uploads his genetic data to GEDmatch, can his sister be considered to consent to this? Or, as search angel Laurie Pratt put it after pulling her genetic data from GEDmatch: "Do I have the right to be your genetic informant?"

But Paul Holes argues that these new techniques can be less invasive than a traditional investigative strategy relying on "the thousands of
tips that have come in because somebody looked like a composite drawn
back in the 1970s." With previous methods, he told the *New York Times*
podcast *The Daily*, "we gathered DNA from hundreds and hundreds of
people because of these tips and because of circumstantial evidence
within the case file." Besides, he pointed out, "we aren't accessing anybody's actual genetic information. We are just seeing how closely or
how distantly related some people are to who our offender might be."
Rae-Venter told me she thought it illogical for people to object to her
using the same techniques for crime-solving that she'd long employed
in her search angel work. "Nobody's jumping up and down about anybody using the database for identifying people who were sperm donors,"
she said, and plenty of those donors were actually promised anonymity,
unlike criminals leaving their semen in the act of committing rape.
Most Americans appear to agree with Holes and Rae-Venter; one study
of about 1,500 people surveyed online in the United States found that a
sizable majority supported police searching of genetic genealogical databases to identify perpetrators of violent crimes, perpetrators of crimes
against children, and missing persons.

What's needed, observers on both sides of the debate told me, is
oversight. Thomas Murray, the Hastings Center president emeritus,
told me he supported the use of genetic genealogy "to prevent the most
extreme sorts of harms," so long as there are "good guardrails" in place
to make sure police don't use these tactics for lesser crimes. A number
of scholars believe there need to be legislative fixes that at the very least
delineate when and why police are allowed to utilize genetic genealogy
databases, whereas a state legislator in Maryland proposed a law ban-
ning the use of genealogical databases for crime-solving altogether.
And yet, not long after I spoke to Murray, news outlets reported on
the use of genetic genealogy leading to the arrest of a South Dakota
woman "charged with murder for allegedly leaving her newborn in a
ditch thirty-eight years ago." Several genetic genealogists expressed
concern about a slippery slope, about the kinds of crime this powerful

new technology was being used for, and about the implications of such cases for women's reproductive and bodily autonomy.

Another concern among genetic genealogists, who know the perils of making mistakes in their work, is that no one is making sure these new crime-solving techniques are executed properly. "I trust somebody like CeCe—she's really good—but we don't have any protocols for genealogists doing this type of work and what the best practice is, and the police themselves don't know who to use," Debbie Kennett told me. Kennett has heard of cases in which adoptees were reunited with the wrong parents because of mistakes in interpreting data, and she worries about questionable genetic genealogy techniques leading investigators astray.

In fact, this has happened, most famously in 2014, when, according to a story in the *New Orleans Advocate,* Idaho investigators trying to solve the brutal 1996 murder of a woman named Angie Dodge became interested in a New Orleans filmmaker in his thirties. His name was Michael Usry, and he became a suspect because Y-DNA from the crime scene was partially matched to the DNA of his father, who years before had submitted a sample to the Sorenson Molecular Genealogy Foundation. Ancestry, which had acquired Sorenson's assets, received a court order to reveal the father's name, and investigators traced their way to the son. They got a search warrant ordering Usry to provide his DNA, and after about a month it turned out that he was not a match. Shortly afterward, Ancestry made the Sorenson database private to protect customers' privacy, a loss Moore mourned as akin to burning a library. "Civil rights are being taken from us at an alarming rate," Michael Usry told the *Los Angeles Times* a few years later.

And yet, autosomal DNA tracing techniques in the hands of a skilled and experienced genetic genealogist are much better than the techniques employed in the Usry case, which Rae-Venter told me shouldn't even be considered investigative genetic genealogy. And besides, advocates say, this is why police confirm the tips that Rae-Venter and Moore give them. (Other apparent problems in the Angie Dodge investigation underscore the fact that DNA is not alone in its potential for police misuse. In 2017, another man who'd been implicated by the investigation was freed and would eventually be exonerated following two decades in prison after giving what he now says was a coerced confession to aiding and abetting in Dodge's rape and murder.)

In 2019, as if in vindication of investigative genetic genealogy's promise, police in Idaho announced the arrest of another man for Dodge's murder. This time, they'd homed in on their eventual suspect using GEDmatch, through the work of CeCe Moore.

* * *

In late 2018, Yaniv Erlich published another paper. This one contained the revelation that, from the genomic data of 1.28 million people, it was possible to find a third cousin or closer for 60 percent of Americans of European descent, the primary population engaging in genetic testing. This meant that all those Americans, the vast majority of whom had never spit into a tube or swabbed their cheeks, could potentially be identified with additional research. Consent truly didn't matter anymore. Most Americans would soon be consented for, whether or not they knew it.

This time, Erlich told me, he didn't get pushback on his findings. So much had changed in five years that his colleagues didn't question his premise that genetic privacy was profoundly compromised. They didn't suggest that by showing what was possible, he was acting unethically. "I think this is part of the maturation of our society that we now are well aware that big data can be abused," he says.

The statistical implications of Erlich's paper were that once a public database contained about three million people of European descent, 99 percent of the greater population of these Americans would be potentially identifiable to law enforcement, at which time the United States would essentially have what legal scholar Natalie Ram describes as a de-facto national DNA database. At the time Erlich pointed this out, it seemed the prediction wouldn't be a reality for some time, till the database of a third-party website like GEDmatch got big enough. But a few months after his paper was published, news broke that almost overnight, the population that law enforcement had access to had doubled, to about two million.

BuzzFeed News reported that FamilyTreeDNA was allowing the FBI to search its database for relative matches of crime scene DNA to identify perpetrators of certain violent crimes, as well as the remains of John and Jane Does. It was early 2019, and the company told *BuzzFeed* that it had been processing crime scene samples for the agency in its lab and uploading them "on a case-by-case basis" for several months by that point. A number of genetic genealogists, who had regarded Bennett Greenspan's company as one created by a genealogist for genealogists, said they felt betrayed. Judy Russell, a former lawyer and newspaper reporter known for her blog *The Legal Genealogist*, pointed out that FamilyTreeDNA was giving law enforcement access, in many cases, to genetic relatives' full names, email addresses, and family tree information.

In the days and weeks that followed, FamilyTreeDNA unleashed a blizzard of statements and policies, eventually announcing that consumers in the United States could opt out of having their DNA used for law enforcement purposes if they wanted to. And Greenspan offered a confusing explanation as to what in his company's terms of service had permitted him to give law enforcement access to his database—even as Russell told me the contractual language appeared to say the opposite. I'd spent months researching privacy policies and clickwrap contracts by then, and this seemed a striking example of the problems around informed consent in the context of genetic data. If even Russell, with a law degree, couldn't make heads or tails of a company's policies, what hope was there for the average consumer?

It turned out the process of becoming involved in genetic crime-fighting had been rocky not just for Greenspan's customers but also for Greenspan himself. At least, that's what he told me. When I spoke with him a few months after the *BuzzFeed* story broke, he told me he'd had to grow a new skin over the past few months and that he'd given his customer service representatives "combat pay" for the days following the article. And he claimed that his hand had been forced by the fact that, unlike 23andMe and Ancestry, FamilyTreeDNA permits people who've had their DNA processed by other companies to upload their data to its site, which meant law enforcement could theoretically upload crime scene data so long as it was compatible. He said that his company discovered that someone was attempting to upload files in a strange format. Greenspan told me he believed this someone was the FBI, and that this, combined with several subpoenas he'd received for information in his databases led him to conclude that law enforcement's presence there was inevitable. The best approach, he decided, was to work with law enforcement and regulate the situation by subjecting cases to approval. But he was cagey about certain details and the account he gave me differed from one he'd given a publication called *Forensic Magazine*. Months later, in an interview with the *Wall Street Journal*, Greenspan offered a third version of events. He now said his company's relationship with the FBI began when he willingly cooperated on two criminal cases—including the Golden State Killer case—because it was the right thing to do.

And then, two and a half years after Joseph James DeAngelo's arrest, and months after DeAngelo's sentencing, the public's understanding of the Golden State Killer case changed, underscoring again how much power over consumer genetic data rests in the hands of individual companies and the protections they do or don't extend. Paige St. John of the *Los Angeles Times* revealed that the investigation was actually more extensive that previously understood, and that it was not tiny GEDmatch's publicly-accessible database that solved the case after all, but a company with an ostensibly private database that had never before been implicated in the official version of events. The *Los Angeles Times* found that after searches at GEDmatch and FamilyTreeDNA failed to yield anything closer than third cousins, genetic genealogist Barbara Rae-Venter uploaded the killer's crime scene DNA profile to MyHeritage without the company's knowledge and in apparent violation of its privacy policies. That search, which Rae-Venter says was sanctioned by an FBI attorney, yielded a closer relative, which helped break the case. Afterward, the FBI tried to keep the role played by FamilyTreeDNA and

MyHeritage secret, invoking "a legal privilege that protects the names of confidential sources." MyHeritage says it only learned of the breach to its own database years after it happened, shortly before the Times article came out.

Natalie Ram, law professor at the University of Maryland, described herself as outraged but not surprised by the revelations in the *Los Angeles Times'* reporting. "We've been told from the beginning to trust law enforcement, that law enforcement is going to use this responsibly, for the most serious cases—that this will be limited. But if they've already told an untrue story at the outset, why should we believe anything they've said about the limits they're putting in place now?" she asked. "How can any of us hold law enforcement accountable within reasonable bounds when we don't even know what they're doing in our name?"

The use of genetic information to solve crimes, however worthy, has to be balanced with legitimate concerns on the side of protecting privacy, MyHeritage's Yaniv Erlich told me. We spoke before the news of his company's unwitting involvement in the Golden State Killer case came to light, though his words would take on new weight after that reporting. The problem is one of trust, Erlich argued. "Genetic information is essential if you want to advance precision medicine," he said. "We have families, some of them I know personally, that genetics was able to help, to identify the cause of serious conditions." But what if people were to stop trusting researchers to keep their information safe? "If we cannot recruit tens of thousands, maybe hundreds of thousands of people for these studies, we cannot help these people," Erlich said. "We cannot use the power of the genetic revolution to empower our understanding. And that's a huge missed opportunity."

This erosion of trust may already be happening. In early 2019, 23andMe co-founder and CEO Anne Wojcicki said the market for her product had slowed, speculating that this might have to do with broader privacy concerns stemming from things like the Golden State Killer case and Facebook's dicey data-collection practices. Some observers have suggested that the lack of privacy protections around genetic data may disproportionately affect minorities, and I heard a number of accounts of African Americans quitting GEDmatch after it became clear that police were using it.

And, of course, concerns about how we protect our genetic information extend beyond the context of recreational testing. Some experts worry about the increasing use by police of Rapid DNA machines, which can process samples in ninety minutes, and which they fear poses the possibility of errors and privacy violations. There are serious concerns about DNA testing on migrants. And all of this is taking place within the broader context of what legal scholar Andelka Phillips calls "ever-increasing monitoring, where we are all subject to complex data mining and profiling." The problem, Phillips says, is that it's difficult to imagine how data from direct-to-consumer testing may in the future be linked with other data and used in ways we can't anticipate.

It might, in theory, be used by some future totalitarian government to discriminate against its citizens. This may seem implausible, except that the day I spoke with Erlich—and asked him if he could imagine a future world in which our ancestral backgrounds could be used to hurt us, and he said he didn't need to project into the future because "I can go to the history"—on that same day, the New York Times ran a story about how Chinese authorities were using DNA as part of a "campaign of surveillance and oppression" against the country's ethnic minority Uighurs.

Journalist Kristen V. Brown has thought a lot about big data and unintended consequences. In 2015, she covered the hack of the website Ashley Madison, an online dating service geared at people who wanted to cheat on their partners. "Life is short. Have an affair," the site's slogan went. Brown wanted to know how the hack and subsequent exposure of the site's users' personal information had changed their lives, so she interviewed more than one hundred people impacted by it and learned about divorces, blackmail, and suicides. It's easy to assume that if you don't cheat on your spouse, you don't need to worry about this sort of thing, and that if you do cheat, well, you deserve what you get. But Brown thought the incident had broader implications. She was fascinated by a concept from Georgetown Law professor Paul Ohm that we are all at the mercy of the massive troves of data that businesses collect and keep on us, and that somewhere amid all that information, every one of us has a devastating secret. Ohm called this eventual, interconnected treasure trove of information the "database of ruin," and he urged in a 2012 *Harvard Business Review* article, *Please don't build this.*

This idea stuck with Brown, and when she started covering consumer DNA testing, she saw how this new technology fit the paradigm. "I was like, 'Oh my god, our genomes have now become databases of ruin.' It's just another piece of data that can be incriminating," she says. "I think we're at the beginning of living in a time where you can't really keep secrets anymore. And how does that change how we go about the world and our daily lives?"

How does it? We're such bad prognosticators. Historian Melvin Kranzberg once wrote that technology is neither good nor bad, nor is it neutral, by which he meant technologies play out in vastly different ways depending on the context, and that we often lack the ability to anticipate how they will change our lives. "He realized that the impact of a technology depends on its geographic and cultural context, which means it is often good and bad—at the same time," columnist Christopher Mims writes in the *Wall Street Journal.* The pesticide DDT, heralded for knocking out malaria and then reviled as an environmental threat, was the example Kranzberg used; Mims invoked Facebook. Consumer DNA testing seems a pretty good example, too.

Just over a year after Joseph James DeAngelo was arrested, the rush to that de facto national database abruptly slowed. After a seventy-one-year-old woman in Utah was attacked and choked into unconsciousness as she was practicing the organ in her church, GEDmatch co-founder Curtis Rogers made an exception to the site's policy that law enforcement access the database only for murders and sexual assaults, and allowed police in to help solve a case he described as "as close to a homicide as you can get." In the ensuing controversy over the site unilaterally making an exception to its terms of service, GEDmatch decided to expand the definition of violent crimes that law enforcement matching could be used for—and to automatically opt all its users' information *out* of being available to law enforcement. If people wanted their genetic information used in this way, Rogers decided, they needed to proactively choose it. Overnight, the database of people that CeCe Moore and Barbara Rae-Venter could access to help police narrow in on suspects went from about a million to zero. Moore called it "a setback for justice."

When I spoke with Rae-Venter a few weeks after GEDmatch's decision, she told me she was hopeful that a growing opt-in movement would eventually make the database a destination for law enforcement matching again. By then tens of thousands of people had logged into GEDmatch to request that their genetic information be used to solve crimes. In the meantime, she said, she was still able to use FamilyTreeDNA for investigations. Some months later, FamilyTreeDNA announced a new investigative genetic genealogy unit to rival Parabon's, headed by Rae-Venter. And then the *New York Times* reported that a Florida detective had obtained a warrant giving him permission to search *all* of GEDmatch's database, including data from the majority of users who hadn't consented to being involved in criminal investigations. The judge's decision to grant this order was a game-changer, experts told the paper; it would likely embolden other agencies to seek similar warrants for huge companies like Ancestry and 23andMe, potentially "turning all genetic databases into law enforcement databases."

DNA revelations mean that a sadistic serial murderer named Joseph James DeAngelo is behind bars, but they also mean that you could be inadvertently involved in the arrest of a relative. It means that an adoptee can find her parents, and it also means that a family can find out that decades before, a man cheated on his wife and produced a child—which could be good for that child and bad for that wife and a mixed blessing for that man. Perhaps consumer genomics means we are all on guard waiting for the other shoe to drop, or perhaps it means that we are all forced to be more honest with each other. One of the central conundrums of spitting into a tube is the way one person's rights so often collide with another's after the tube is sealed and sent in.

15

LATE NIGHT

By early 2015, two and a half years had passed since Alice had first spit into a tube. She and Gerry had folded a platoon of people into their effort—from a rabbi in New York to a professor in North Carolina, from genetic genealogists in California and Uruguay to distant cousins in Massachusetts and Minnesota. But in the end, it was not any one of these people who provided the crucial bit of evidence the Collins sisters needed. The missing link came in the form of a genetic stranger who was not looking to solve a mystery. This stranger did not know, when she got her own results, that they were much of anything to ponder or that they would profoundly alter her own family as well as another.

Alice and Gerry were monitoring quite a number of DNA kits by then. Mainly, they checked on their own results and those of their brothers, looking to see if any new DNA cousins had shown up who might help them unravel family relationships and solve their case. Alice was also managing the account for her first cousin Pete Nolan, the son of her father's beloved sister, Kitty, but she would check on it only occasionally, as it rarely turned up any new close cousins. The significance of Pete was that he *wasn't* genetically related to the Collinses—as Alice and Gerry had learned after testing him at 23andMe. Pete's mom, Kitty, had been Jim's sister, but they were not genetic siblings, and it was this crucial bit of evidence that allowed them to understand that Pete was the biological grandchild of John Josef and Katie Collins, whereas they and their siblings were not. This, in turn, had allowed them to discard several of the theories they'd come up with to explain their unexpected ethnicity, and to conclude that their father must have been confused with another child, most likely when he was born at Fordham Hospital. But they had no conclusive proof of this theory and no inkling who that other child might be. If their father had been taken from another

family, what family was it? So far, none of their efforts to unravel the
truth had worked.

One day in January, Alice decided to send Pete an email updating
him on what was happening or, more precisely, not happening, with
her search. She logged onto his 23andMe profile, not expecting anything interesting.

But something interesting had appeared. A new close cousin had
shown up at the top of Pete's list of relatives. The significance of this
anonymous woman and her relationship to Pete were not clear, but
Alice was intrigued. Since Jim and Kitty were not biological siblings,
Kitty must have had a genetic brother she never met—the baby boy
born to her parents, Katie and John Josef Collins, and switched with
Jim. And that baby boy was presumably raised by the *other* family, the
one Jim should have gone home to, and might have grown up and had
children of his own—and one of his descendants might be interested in
genealogy and might have sent her saliva into 23andMe. And this person might be showing up as Pete's closest cousin at this very moment.
Alice, staring at her iPad in the family room, began to shake. She had
a prescient feeling.

Alice wrote the woman through 23andMe's internal messaging
system and asked if she would be willing to compare genomes with
Pete Nolan. She explained that she was Pete's cousin and that she was
trying to solve "a hundred-year-old mystery concerning my father."
Then she waited at her iPad, worried that the stranger wouldn't reply,
as so often happened with these requests. But a short time later, the
stranger did reply, mutely signing into her account and allowing for
genome sharing. Alice studied all the places where the stranger's DNA
overlapped with Pete's and concluded that the stranger was probably a
first cousin once removed based on the centimorgans—an extremely
good match. But was the stranger related to Pete in a way that would
be helpful to Alice? She wrote again and asked the woman what she
knew about her lineage.

The woman on the other side of the exchange studied the screen.
Her name was Jessica Benson, and she was that Raleigh, North Carolina,

stay-at-home mother whose ethnicity results through her paternal side hadn't been what she'd expected, but who had not been transformed into a seeker by the discrepancy. Jessica's dad and his parents were no longer alive, so she'd gone to her dad's sister with her unexpected ethnicity estimate, but her aunt couldn't shed any light on it, and Jessica had thrown up her hands. Yet one lesson of the consumer genomics age is that even when you don't go looking, the truth can come looking for you. In this case, the truth took the shape of Alice.

"I was actually expecting to be much more Ashkenazi than I am," Jessica Benson wrote back. The weird thing was, 23andMe was telling her "that I am actually Irish, which I had not expected at all."

Alice got chills. She wrote again. Her father had been born at Fordham Hospital on September 23, 1913. Had anyone in Jessica's direct ancestral line been born on that date in that place? Jessica didn't know. She didn't even know where Fordham Hospital was. She googled it and learned it had been located in the Bronx, where her dad's family had lived. She pulled up the Social Security Death Index and wrote back that it appeared her grandfather, Phillip Benson, had been born right around that date.

Alice began to cry. She went downstairs and told her husband, Bruce, she thought she'd found something big. She started looking through the pages of the New York City Birth Index of 1913, trying to figure out how she'd missed the name Phillip Benson, if indeed such a baby had been born in the Bronx around September 23. If this child were real, he had to be in there.

It was looking increasingly likely that the birth certificate Alice had for "James Collins" was, in fact, not properly her father's birth certificate. It had become his when he'd been handed to the wrong mother and gone home with the wrong family. But when he was born, before capricious fate confused him, Alice's father had been "Phillip Benson," and Phillip Benson had been James Collins, and everything that followed had flowed from this single mistake. It was a wild thing to contemplate, and yet it was the simplest explanation given everything Alice knew.

It was nearly midnight when she finally found the right page among her 1913 baby names. In her thick white binder, there was one line that she had circled and put an asterisk next to. The name was blurry. The birth certificate number was completely illegible. The first name was Phillip, albeit misspelled, and the last name was something—Bamson or Bamason or Ban_son. During those months combing the index years earlier, Alice had struggled with this name and placed a turquoise sticker with a question mark next to it.

But now she knew: "Philip Bamson" was really Phillip Benson. She sent Jessica an email with a screenshot of the page: "I FOUND HIM!" Alice barely slept that night.

* * *

The next day, Alice emerged from sleeplessness and began to tell people about what had happened: her siblings, CeCe Moore, and her DNA cousins, who immediately started scouring the Internet looking for archival information on the family of Phillip Benson.

Jessica Benson began sending photos of her father's side of the family. Alice was in a sewing class when she got one of her emails and saw, for the first time, an image of a man named Sam Benson, who had been Phillip Benson's father—and who (it was now becoming clear) was Alice's genetic grandfather. She was stunned by how much Sam looked like one of her brothers. Meanwhile, Jessica put Alice in touch with her aunt Pam. Pam Benson is the daughter of the late Phillip Benson, which is to say she was like Alice on the other side of the looking-glass—a woman in the process of learning that her father was not who she thought, and neither was she.

In retrospect, the conversation between Alice Plebuch and Pam Benson has to have been one of the stranger iterations of a modern-day game of telephone. After Alice's long search for something else entirely—for her genetic cousins—she had found the daughter of the man her father had been switched with. Only, she didn't realize it; she assumed that Pam had married into the family. She had never intended to contact a direct descendant for fear of disrupting someone else's life as hers had been disrupted.

Meanwhile, Pam, who lives in Lawndale, California, was not really prepared for this conversation, if it's possible to prepare for such a thing. The news that her dad had been switched at birth with another child unfolded in a rapid and confusing fashion. Her first inkling that something was strange was that after her niece Jessica had taken a DNA test, Jessica had contacted her saying, "Auntie Pam, do you know we're Irish?" To which Pam had replied: "You're nuts. The test is wrong." Pam told me she went through "this whole spiel" explaining why Jessica's results were impossible—her father had always said his family were Russian Jews, and the genealogical paperwork confirmed it. Then, some time later, Jessica's mom contacted Pam again with a crazy story involving a woman named Alice, and could they give Pam's phone number to this woman?

And so Pam and Alice spoke, and Alice laid out her theory, and Pam's reaction was "You've got to be kidding me," and for some time Alice didn't realize she was talking, in essence, to herself. She was telling another woman that her father had been swapped in the hospital and originally destined for a different family, a different culture, a different religion, and a different life. When Alice put everything together, she felt terribly guilty about the way that first conversation had gone with Pam.

Pam ordered her own DNA test, but well before the results came back, she sensed in her bones that Alice was right. She saw a photograph of John Josef Collins, the man Alice had always known as her grandfather, and it was clear that he was, in fact, Pam's grandfather. He looked just like Pam's brothers. And it was becoming clear why Pam's own grandparents, who'd died before she was born, looked nothing like her father. Pam's grandfather Sam had been five foot two, and her grandmother Ida four foot nine. Their son Phillip had been six foot four. In a photograph, Sam barely cleared his son Phillip's shoulder, and they looked absolutely nothing alike. "I always wondered, 'How could Grandpa be so short and Dad be so tall?'" Pam told me. She'd asked her dad once for an explanation, and he'd said, "recessive genes."

Phillip Benson, center, with his father, Sam, at right, and a relative

Now, Pam was where Alice had been, only Alice had had years to digest the news of a disrupted identity and make sense of it. She had been an agent in her own discovery. Pam had had the news thrust upon her, and it was jarring. She was in her fifties by then, nearing retirement. How could she suddenly "go from this to this"? Pam asked me. "It's, like, nuts," she said, using her favorite word. There were times when Pam's daughter would find Pam crying. "What's wrong?" she'd ask her mom. What was wrong was—well, everything. How was Pam to reconcile her new identity when she'd lived so long with an old one? She was *of* her father, and now her father was someone else. Pam had not been raised Jewish—her mother, Phillip Benson's

247

second wife, was a Lebanese American Catholic—but "that was what I knew my dad was," she says. Growing up, she celebrated Hanukkah and Christmas; her parents sometimes went to friends' houses for Passover Seders. To learn now that "my dad was really Catholic," and Irish—this jarred Pam. It was as if she had to let go of something on her father's behalf.

And she felt sad for him. She imagined her tall, blue-eyed father sensing all his life that he was out of place in his family, towering over his parents, looking not a whit like his sister. It was too crazy-making to have this knowledge and imagine Phillip in ignorance of it. Somehow, the revelations had "altered the course of time," Pam said, and made the past present. More than three decades after her dad's death, she spoke of calling questions up to him in heaven. She pondered how he'd made sense of his difference—if he'd questioned whether his parents were truly his parents or wondered whether his mother had cheated on his father.

Phillip Benson in his later years

And so, when Pam's daughter asked what was wrong, Pam would say, "I'm just wondering what my dad would have thought. Would it have answered questions for him?" Within that question was embedded another: Who was Phillip Benson, really? Who was Jim Collins? What makes us who we are? Blood? Family? Culture? At one point, Pam thought back on her father's character, the generosity he showed throughout his life, and concluded that this was innate. "The family that you're given can alter your environment and the way you're raised but it's not going to alter your soul," she said. "Your soul is given to you by a higher power."

Things might have unfolded differently if she'd made the discovery herself, Pam told me. It might have eased her initial sense of disbelief and befuddlement. "You have more control when it's you," she said. Indeed, "there was a split second when I wished that the DNA wouldn't have been done, and I would have gone on happily ever after." But then, as Alice sent old photos and began to fill Pam in on her father's family of origin, and told her she had a first cousin she'd never heard of—a man named Pete Nolan—the truth became fascinating to Pam.

"I believe that it is better to know," Pam told me. The truth answered so many questions. "I mean," she added, "I still say it's nuts."

* * *

The incontrovertible proof of her theory, Alice learned following her 23andMe breakthrough, lay in the DNA of a woman named Phylis Pullman. Everything pointed to Phylis being the biological cousin she'd never known.

When Alice's father was born—intended for the life of Phillip Benson, but transformed by an accident of fate into Jim Collins—he already had a sister, a toddler he never met. That girl, Miriam, grew up—well, sort of; she never quite reached five feet—and had a child named Phylis, who is also small in stature. Alice got in touch with Phylis's daughter and laid out what she allowed was an "absolutely bizarre" tale, and

Phylis's daughter reached out to Phylis and said, "Mom, I got the strangest email."

The first time Phylis and Alice spoke by phone, Alice was struck by the way Phylis used a phrase that had been one of Jim Collins's favorites. "Such is life," she said, and in those words was so much: a shrug and a laugh and a shaking of the head in wonder at the strange diversions of fate. Phylis sent away for a DNA test and began to share family stories. As Alice would learn, Phylis, now in her eighties, did not know her uncle Phillip well, because he'd moved to California when she was young. But as a little girl she perceived him to be "a giant" in contrast with her own petite family, even if no one around her appeared to make much of this. Yet others may have been struck by Phillip's differences. Phylis told Alice how, when Phillip was courting his first wife, her observant Jewish parents didn't believe he could possibly be a member of the tribe. "He had to bring his birth certificate," Phylis said. "Little did we know it wasn't *his* birth certificate."

Through stories and genealogical research, Alice began to fill in the story of her grandparents' lives. Like the Collinses, the Bensons were immigrants during the first decade of the twentieth century. They came over as young adults from the same part of the world as each other, part of a large wave of Eastern European Jews escaping poverty and discrimination who settled in large cities like New York. Sam's and Ida's documents generally listed "Russia" as their birthplaces, but Alice is a digger, and eventually she was able to figure out that they came from two towns in Lithuania with sizable Jewish populations. About thirty-five years after Sam and Ida left, the Jews in these towns were murdered during World War II German occupation. The places Sam and Ida emigrated from are within about an hour of Vilnius, the city that had been blanketed in yellow "ancestor pins" when Alice had mapped out her DNA cousins' ancestral villages, and now that pattern made sense. She'd known her people came from there; it just took finding Jessica Benson to figure it out, one accident of fate correcting another.

According to accounts, Ida was sweet and nurturing, whereas Sam could be taciturn and cold. When Phillip was born, the Bensons appear to have been living in a five-story apartment building that was just a ten-minute walk from the Collins family; Miriam Benson and Kitty Collins, both toddlers, could have played with each other at the big park just up the road. The Bensons then moved around before eventually settling in Brooklyn. Sam started in the position of "button hole maker," then became a machinist in a factory. Like his biological son Jim, Sam loved to play cards; unlike his son, he lost the family paycheck more than once, forcing his wife Ida to borrow money from family. Eventually, Sam ran a shop where he repaired, of all things, Singer sewing machines—another tiny accident of fate that parallels his granddaughter's love of high-tech sewing machinery. What was his shop like? Was it like Alice's office, crowded with machines and other things, always barely on the safe side of chaos? And what was his mind like? Could he, like his son, follow all the moves of an imaginary game of pinochle in his head? Did he have an eye for detail so that he could—as Alice did with computer programs—look at a sewing machine and rapidly pick out the "nit" that was making it break down?

Or is it facile to assume such a parallel? After all, Phillip Benson also took after Sam, the man he knew as his father. Like Sam, Phillip had a talent for taking things apart and putting them back together. When he was sixteen years old, he was listed on a census as a mechanic working in an automobile garage. Later, he helped run a company that built custom airplane parts.

When Phylis's DNA test results came back, it confirmed what she and Alice already knew: They were first cousins. When Pam's DNA test came back, it confirmed that she was first cousins with Pete Nolan, and she began to email with him. Alice and Pam, whose fathers had been switched in the hospital, began calling each other "swapcuz." And Alice and Pam and Phylis, making up for whole lifetimes, embarked on what I call "The Great Exchange," swapping visits, photographs, family stories, and memorabilia. Alice flew to visit Phylis in Florida, and the first time

Phillip Benson and his first wife, Esther Abolafia Benson,
sitting with their son Kenny. Behind them are the parents
Phillip grew up with, Ida and Sam Benson.

the first cousins sat across from each other on a couch, inadvertently holding the same pose, Phylis's daughter laughed and told them they looked like mirror images. They have the same lips, the same nose, the same eyes, and the same build. They were born with identical gaps in their front teeth.

Phylis told me she was thrilled to discover she had seven first cousins and all their spouses and children—she'd long considered herself rich in family, but now, her wealth had doubled. She told me that getting to know "the Collins-Bensons," which is what she calls Alice and her family, "has been the most exciting experience of my life." But the discovery was tinged with sadness that she'd gotten this news so late in life. When Phylis was sick, Gerry went to Florida and cooked her meal after meal, the way you do for family (including family you didn't know existed till you took a DNA test). "I've only got so much time with her," Gerry told me.

And, in early 2016, Alice flew south to California to visit her swap-cuz Pam, who, at five foot ten, takes after her father, Phillip, in height.

Alice's head barely clears her shoulder. The swapcousins were not related by blood or marriage but they were related in a way, by a mistake a century earlier, one that only DNA testing could have uncovered. "It was wonderful," Alice told me. "We essentially had experienced a lot of the same things, just from the other side." Pam felt moved to give Alice things that had belonged to her grandparents: the Bensons' naturalization papers and Sam Benson's Social Security card. Alice sent her swapcuz the 1918 war zone pass that had belonged to John Josef, the man she thought was her genetic grandfather but who was, in fact, Pam's.

It took Pam a while to get her father's birth certificate from New York State, in part because Phillip's last name had been misrecorded as "Bamson." But when it arrived, she and Alice could see for themselves exactly how one moment long ago in a hospital that doesn't exist anymore had set two men off on mistaken courses they'd live and die without ever knowing about, and shaped generations to come.

There was the birthdate, the same as Jim's: September 23, 1913. There was the same stamp, skewed at a diagonal: Fordham Hospital. There was the signature of the promising young doctor, Jerome McSweeney, who delivered both babies, assisted by nurses, nurse trainees, and hospital helpers whose identities are now rendered invisible by time. And there was the birth certificate number: 10942. It was one digit off from the number on Jim Collins's birth certificate, 10941.

The little Irish boy, the son of Katie and John Josef Collins, had presumably been born first, and at some point later in the day, the little Lithuanian Jewish boy, the son of Sam and Ida Benson, came along. At a hospital with fewer than a child a day born on average, they were quite possibly the only two babies to arrive at Fordham Hospital that day. And yet somehow, the two newborns had gotten confused.

* * *

In the months after finding Pam and Phylis, Alice searched online and found a book with the ponderous title *Brought to Bed: Childbearing in*

REPORTED DECEASED: DOD 12-11-97

STATE Ca D.C. # 5005

THE CITY OF NEW YORK.
DEPARTMENT OF HEALTH.

STATE OF NEW YORK.

No. of Certificate

CERTIFICATE AND RECORD OF BIRTH 10942

OF

Name of Child *Philip Benson*

Sex	*Male*	Father's Occupation	*Benson Cole mach.*
Color	*white*	Mother's Name	*Sen Benson*
Date of Birth	*Sept 28 1913*	Mother's Name before Marriage	*Sen Bott*
Place of Birth, Street and No.	*Fordham Hospital*	Mother's Residence	*1185 Union Av*
Father's Name	*Ben Benson*	Mother's Birthplace	*Russia*
Father's Residence	*1185 Union Av*	Mother's Age	*23*
Father's Birthplace	*Russia*	Number of previous Children	*1*
Father's Age	*24*	How many now living (in all)	*2*

I, the undersigned, hereby certify that I attended professionally at the above birth and I am personally cognizant thereof; and that all the facts stated in said certificate and report of birth are true to the best of my knowledge, information and belief.

Signature, *J M Sweeney*

Residence, FORDHAM HOSPITAL

DATE OF REPORT, 19

07/19/2016

R 0 0 3 6 3 0 5 7 3

THE CITY OF NEW YORK.
DEPARTMENT OF HEALTH.

STATE OF NEW YORK.

No. of Certificate,

CERTIFICATE AND RECORD OF BIRTH 10941

OF

Name of Child *James Collins*

Sex	*Male*	Father's Occupation	*Driver*
Color	*White*	Mother's Name	*Katie Collins*
Date of Birth	*Sept 28 1913*	Mother's Name before Marriage	*Katie Kennedy*
Place of Birth, Street and No.	*Fordham Hospital*	Mother's Residence	*1285 So. Blvd*
Father's Name	*John Collins*	Mother's Birthplace	*Ireland*
Father's Residence	*1285 So. Blvd*	Mother's Age	*32*
Father's Birthplace	*Ireland*	Number of previous Children	*2*
Father's Age	*37*	How many now living (in all)	*3*

I, the undersigned, hereby certify that I attended professionally at the above birth and I am personally cognizant thereof; and that all the facts stated in said certificate and report of birth are true to the best of my knowledge, information and belief.

Signature, *J M Sweeney*

Residence, FORDHAM HOSPITAL

DATE OF REPORT, 19

DATE ISSUED AUG 21 1997

DOCUMENT NO. D 231156

The birth certificates for Phillip Benson (misspelled)
and James Collins

America, 1750–1950, by the childbirth historian Judith Walzer Leavitt. She bought it and was struck by a picture inside.

The caption read, "Babies on the cart at the Manhattan Maternity and Dispensary, 1912." The black-and-white image showed about fifteen babies piled together, some leaning on each other, like so many cabbages on display at a grocery store.

"Every time I show it, when I give lectures, the whole audience gasps," says Leavitt, a retired professor at the University of Wisconsin–Madison. "You can understand how possible it was to switch babies inadvertently."

Because hospital births were still unusual in 1913, procedures to identify babies were changing and inconsistent. It's hard to know what practices were in place at now-defunct Fordham Hospital, but in 1913, it's entirely possible that Fordham didn't use any kind of identification system at all, that "they just depended on mothers' recognition or nurses' remembrance," Leavitt says. If they did have a system, it may have been crude. I found a number of scholarly papers and patents from the 1920s attempting to address the problem of keeping babies straight in maternity wards: aluminum tags tied around babies' necks, bracelets with lettered beads, slips of paper tied to umbilical cord stumps, and this memorable one: "The child's name written on its chest with indelible pencil." Alas, none of these approaches was foolproof, wrote one prominent obstetrician in a 1928 issue of the *Journal of the American Medical Association*. A "69 could be read 96, or the adhesive plaster labeled J. Jones get on J. James' back, or a necklace intended for baby Flanagan get round the neck of baby van Dorn, if the two babies should be placed together on a table while their beads were being strung," he wrote.

Leavitt told me that it wasn't till the 1930s or '40s that it became standard for hospitals to give both babies and their mothers identifying wristlets or anklets so that they could be matched to each other. Even then, as a few baby-switch cases from the 1940s attest, the methods were far from perfect. Complicating the situation, Leavitt added, was the fact that while some hospitals kept babies sleeping in cots by their mothers' beds, other hospitals kept them in nurseries, increasing the chances of a mix-up.

I couldn't find any pictures of the babies at Fordham from this era, but a 1907 report on obstetrics practice from Bellevue Hospital, with which Fordham was allied and appears to have shared pupil nurses, gives a sense of what giving birth in a hospital could have been like back then. New mothers were wheeled out of the delivery room and into a "lying-in ward," while their babies were weighed and measured and dressed and put into cribs. Newborns were not brought to their mothers to nurse for the first time till they were about six hours old. Could a mother, woozy and exhausted after childbirth—one that may have taken place under the influence of chloroform or ether—reasonably be expected to know that a baby she barely glimpsed six hours earlier was hers?

And let's say she received the right baby. This first period of time, right after childbirth, would not be the last time her baby's identity might be confused. Bellevue's nurses appear to have been handling the babies a lot, given that the hospital enforced a strict feeding schedule of nursing every two hours during the day and twice during the night at precisely 1 A.M. and 5 A.M. And throughout a mother's stay, nurses daily took her baby's rectal temperature, counted his pulse, gave him a bath with white castile soap, dusted him with talcum powder, and washed out his little eyes to prevent infections. Assuming the baths weren't given at the mother's bedside, each one of these daily journeys with the nurse increased the chances of a mix-up.

In their efforts to keep babies healthy, clean, and well-fed, it appears that early maternity wards were unaware of the very real danger they posed. At some point—perhaps while putting the babies into cribs or while parceling them out from a cart—an exhausted hospital employee could not tell a James from a Phillip. Two mothers were handed the wrong babies, perhaps even as they lay near each other in the same large, light-filled ward. Two mothers were handed the wrong babies, and they nursed those babies, and admired their tiny faces, and then took those babies home.

The repercussions of that single event, the act of setting one baby down in one place and another baby down in another place, meant that two men wound up living lives intended for each other. A century later, Jessica Benson would write that this knowledge was humbling. She and her family, and Alice's family, were children of a whim. We are all, of course, children of whims, and we have always known this to be true, long before the DNA age brought to light tales of one-night stands and donor sperm swaps and the mix-up of two babies in a hospital. We are whims of our mother deciding to attend a certain dinner party and meeting our father. We are whims of the sperm cell that beat out millions of other sperm cells to reach our mother's egg. And as we come into being, we are still whims: of childhood accidents, and the precise location of the house we grew up in; of a college admissions officer; of an email never opened; of a drunk driver we never saw coming. As destabilizing as it is to admit, life is just one grand unfolding of accidents—accidents or God's will, depending on how you see it. Yet there is something poignant and deeply disturbing about the idea that an act of momentary carelessness by the professionals charged with caring for the most vulnerable of human beings could have had such profound consequences.

Alice emailed Leavitt and told the historian her tale. Alice doesn't often get emotional, but she does when she asks herself one of the questions she asked of Leavitt then. She wanted to know whether or not Jim ever enjoyed the comfort of his biological mother Ida's arms. "I wanted—at least one time—to have him held by his mother," she told me.

The question has special resonance for her. Alice has three sons, and one day over lunch she told me about the birth of her eldest, Gidian. He's a radiation technologist, and he and his wife and Alice's three grandchildren, whom she adores, live only two houses down. But when Gidian was born back in 1973, Alice told me, she did not know if he would make it. He had to be delivered prematurely by an emergency C-section because Alice was having complications, and afterward, the newborn would sometimes stop breathing. Alice was weak, and the hospital would not tell her what was going on with her son. She remembers finally gaining the strength to walk down the hall to the nursery only to watch in horror as staff crowded around her son to perform some sort of procedure that no one would explain. Bruce was teaching and she could not get ahold of him. The doctor, when he came, could not tell her if her son was going to live or die.

Gidian was three or four days old by the time Alice was finally able to hold him, and this memory of that early separation helps explain why the not knowing and the sadness of that loss—of Ida's loss, of Jim's loss—haunt her.

It is a question without an answer, a puzzle not even DNA can solve.

16

ALICE REDUX

What role did nature play in the lives of Jim and Phillip, and what role nurture? Who were Jim and Phillip really, in their souls, as Pam would put it? And who would they have been if not for that mistake?

"His story is one of a lot of tragedy," Jim's son Jim the younger says of his father. Jim the younger ticks off the losses: his dad's mother dying when he was a baby, his dad going into an orphanage, his dad leaving the orphanage and being on his own before the age of fifteen. What would have been if Jim Collins had lived the alternate life, the one intended for him? The Collins kids can't regret what happened—"If he wasn't switched, we wouldn't be here," Gerry points out—but that doesn't mean that Gerry can't feel sorrow for her father, for what he had at the very beginning of his life and all he was forced to give up.

It's hard to explain this mix of gratitude and regret, Jim the younger says. "Because you loved this person, you wish things would have been a little bit better for them, or a little bit easier," he says. "I'm happy that we know where his family came from, and we met them, and they are good people. It's a little sad, too, that *he* never had that opportunity, and I don't know how he would have felt—years later, after being raised one way, you find out you're somebody different. It's sort of a perplexing situation no matter how you look at it."

Among the photographs Pam showed Alice during The Great Exchange was one of her father, Phillip Benson, as a child. Phillip is about three. His shiny hair is trimmed into a neat bowl haircut, and he is sitting on the back of a pony, lightly gripping the saddle, with a faint smile playing on his lips. Alice scanned and restored the photo and sent it back to Pam, and meanwhile she thought about the image, which provoked in her a pang of jealousy. She did not begrudge Phillip the pleasure of that pony ride, but she thought of her father, given an orange for Christmas, and she wished for him the pleasures of the life he was meant for.

Phillip Benson as a child

It's impossible not to wonder what would have been. Bill Collins wonders if his dad could have had a better education if he'd stayed with his biological parents. "He was a very smart man and he would have been in a different situation," he says. Would Jim have chosen something other than work in regimented institutions that so mirrored life in the orphanage—the army, then correctional work? Might he have done something else entirely, Alice asks, "something where he used his mind?"

Alice and Gerry also wonder what affect the baby swap had on the family that took little Jim into their home. It's difficult to know exactly what was going on during those nine months of turmoil, the nine months between when John Josef and Katie brought their youngest child home from the hospital and when Katie died. Was the turmoil because Katie was sick, or because she and John Josef had separated, or because John Josef was "intemperate"? Were these things related? Most crucially, why did Katie and John Josef separate? For Irish Catholic immigrants to break up in 1913, just after having a child, seemed surprising. Alice and Gerry and Pam all speculate that the separation may have resulted from Katie bringing home a child who did not look like the rest of their family, and John Josef becoming suspicious.

"He had two children already and this was not his child, and he knew it—like a mother seal knows her baby pup," Gerry told me. If Jim had not been swapped, Gerry went on, she believed the Collins family would have remained intact. The theory is impossible to prove, of course, but if it were true, it would mean that the baby swap was not just the engine by which Jim was given to another family, but the means by which that family was dismantled.

It is tempting to wonder what might have been if Jim had known about the switch when he was alive. Would he have been better off? Would it have helped him reconcile with the past, or would it have been a burden for him to contemplate what he'd lost, and threaten the loss of more that was important to him: his memories of his siblings, his sense of himself as an Irish Catholic man with an Irish Catholic family? These *what-ifs* are the existential questions of the DNA age, the ones that loom over revelations of misattributed parentage and donor conception and secret adoptions, over the stories of people like Linda and Laurie and Jason and Rosario and so many others. Who are we, really? Who would we have been if we'd known the truth earlier?

The Collins kids have wrestled with who Jim Collins "really" was and who they, in turn, really are. They have wrestled with whether he would have wanted to know. "He was raised Irish but he was Jewish—that's what I say," Alice says. But her brother, Jim the younger, says, "He was Irish in every sense of the word other than having been born to an Irish couple." And they have wrestled with whether or not they would have told him the truth if they could have. *Would there have been a purpose in telling him?* Alice wondered early on. But eventually, she came around to Gerry's thinking, which is that it would have been the right thing to tell him. "I feel like it would have answered questions for him," Gerry says. "I feel like he would have finally understood." Alice imagines the Collins kids telling their father and Jim saying, as he sometimes did, "Such is life." Still, perhaps it was a mercy that the Collins siblings never had to make that call. "My dad would have lost his identity," Alice says. "He's been kind of spared that."

Are you what you live? Are you what you were born into? As Rosario Castronovo wondered, to what extent can you claim a heritage you didn't grow up with? And how do you decide what to claim? Experience gets a lot of weight, of course, but so do history and intention. Rosario's black grandfather lived and loved as a black man, and paid the price of a black man. The erasure of black identity in his descendants is arguably yet one more loss. Similarly, Jim Collins's biological mother never intended for her son to be raised in another family, in another culture and religion. But if the philosopher Françoise Baylis is right—if identity is something we work out in constant conversation with the rest of the world, a kind of story we tell, shaped by our beliefs and desires and fact-checked and validated by others—then the only person who could have said who Jim Collins truly was was Jim Collins himself.

And likewise, what the Collins kids feel themselves to be varies; it is a question with a seven-sided answer. One brother told me it was a little strange at first to look at himself in the mirror and realize that his own history, and the generations of people leading up to his existence, weren't what he'd believed all his life. John Collins, a practicing Catholic, told me the news had not changed his religious practices, but for Alice and Gerry, I sensed that the revelation felt more personal. Gerry told me a relative had objected when she described herself as half-Jewish, saying "No, you're not, you're Irish," and reminding Gerry that she'd sung an Irish blessing at a wedding. "This doesn't have anything to do with you," the relative told her. But it did, Gerry felt. It had everything to do with her. Gerry told me the news that her father was Jewish made sense, confirming an orientation to the world she had long felt.

What can we make of the fact that some people say this is true for them after DNA revelations, and others don't? Are some people more sensitive to the tiny clues that an out-of-order identity can drop? Or is this a matter of us storytelling human beings looking for patterns to explain the world, the way that Jason saw his own temperament reflected in his father—only to find out later, when DNA set him straight, that this father wasn't his?

I don't know. I do know that Alice says the same thing as her sister.

"One almost wonders whether you have genetic memory for your people," Alice says.

* * *

In January 2017, an unusual group took a cruise to the Caribbean. It was a family reunion, though it was more than that; this assembly of more than thirty vacationers amounted to an illustration of the wonders of consumer genomics, as well as a lesson in how the past lives on in us. It was Alice and all her siblings, plus assorted Collins family members, along with their newfound cousin, Phylis Pullman, and members of her family, and the Collins clan's new swapcuz, Pam. Two years before, many of these people had been strangers to one another; now, they were eating at the buffet, shopping at the ports together, and dressing up in suits and gowns for a huge family photo.

It was oddly comfortable, said Phylis, who didn't want the week to end: "None of them felt like strangers to me because they all looked like they belonged to my family." Someone made T-shirts that said "A Family United: Benson-Collins," and on the back, two trees twisted around each other with a heart in the middle, containing the names "Phillip" and "James" and their shared birthday. One night, Phylis told me, she and some of the family climbed up a whole lot of steps, and the cruise ship turned off its lights, and they stargazed right there on deck, in the middle of the water. "All these stars—you felt you could touch them," Phylis said. The points of light that come to us from years gone by seemed so present at that moment, as if the past had finally arrived.

"It was wonderful," said Pam Benson. "It was like we were all one big swap family." At breakfast one morning, Pam watched Alice's brother John walk into the room and was stunned. There was her grandfather Sam Benson, a man she'd met only in photographs, come to life. "Tears welled up because he looks so much like my grandfather," Pam said.

Even so, Pam told me, when we first spoke about the cruise in 2017, she wished her genetic first cousin Pete Nolan, the man whose DNA had been so crucial to Alice's case, were there. Pam had not met Pete yet at that time, though they'd been emailing. It was hard for her watching Alice find Phylis and those two families come together. She longed for that same experience, what she called the "family bond." And in the years that followed, she told me she also struggled with how to reconcile with a sense that in the swap "my dad got the better end of the deal." That version of events began to bother her. Her father had not asked to be placed in the wrong crib, after all, or on that pony in the park. And his life had not always been easy. He, too, had had difficulties, as well as triumphs.

Phillip Benson had married a woman who was one of four sisters, and the four husbands of the four sisters all went into business together, moving to California to build parts for airplanes. Then Phillip's wife died of cancer, leaving behind two young boys. Phillip remarried and was nearly fifty by the time Pam was born. Pam told me her dad had loved tinkering with cars; he'd buy old Ramblers just so he could fix them up. They lived near Los Angeles International Airport, and Phillip knew the flight schedules in his head, organized by the type of plane. And he kept homing pigeons, a hobby he'd loved since he'd been a child hanging out on a New York City rooftop. Pam remembered driving with her dad to LAX to pick up pigeons he'd bought from all over the country, and she remembered the pretty tumblers, who seemed to fall out of the sky. Her father would put bands on their ankles and drive them miles and miles to free them, and take bets on which birds would come home first.

"He loved them," Pam told me. No matter how far he drove them, the birds always returned. "It was amazing. How did they know to come home?"

Pam told me her father was generous. He had done well in business. A relative moved out to California and Phillip rented the guy an apartment and gave him a job. Every two years Phillip bought a new car, and the old cars went to people who needed them. "Don't worry about it; buy me a beer," he'd say. And then his fortunes turned. He had a stroke when Pam was eleven and stopped working, and the restaurant where Pam's mother had waitressed closed down. Phillip never regained the use of his right hand, and he had to walk with a cane. Phillip lost his oldest son in 1985, and Pam said it seemed that her father never recovered from having to bury his own child. He lived only two years after that, and he was a changed man—quiet, heartbroken.

It had been a few years since the DNA discovery when Pam finally flew out to North Carolina to meet her first cousin Pete Nolan in person. She stepped out of her rental, and Pete was outside taking a stroll, and time stopped while she looked at him. There was so much to see. Her father's nephew had Phillip's height and coloring. And he had those blue eyes she hadn't seen since 1987, when her father had died. Before anyone said "hi," before Pam told Pete to come give her a hug, they were both silent, looking at each other. He was family. She told me she could feel it with her whole being.

They sat at Pete's table and shared stories. Pam pulled out a photograph of her father and told him, "Pete, look at this, you have your uncle's ears." And afterward, Pam kept thinking about what her father's life would have been like if not for the switch. What would his career have been, his loves, his children? "I mean, your mind can just travel and travel," she told me. But there had to be an explanation for it all, right? Pam was raised Catholic, and now she said she could not help but think there was a guiding hand in all of this—in the switch and in its discovery—because, well, because. Maybe because the alternative seemed unthinkable. "This was supposed to happen, because that's what did happen," she said.

Pam told me she'd enjoyed the shock of seeing how very alike Alice and Phylis looked, and she was glad, at last, to have her own version of that story with Pete. She was struck by the fact that two families had been united by a single moment a century before, had taken separate paths, and had wound up back on the same road all these years later.

She told me seeing Pete had given her a sense of closure. She already wanted to see him again. She had trouble putting into words how good it felt to know him after having traveled so far. There was a kind of inevitability to this reunion—as if, like her father's birds, she'd always known where she was going. From the great distance of time, she was finding her way home.

* * *

The trouble with Alice's primary method, using genetic segments to reverse-engineer her family's DNA into the right tree, was not that it wasn't sound. It was that in the years she was using the technique, the genetic testing revolution hadn't yet reached the right people. The databases were too small. In all that time, combing through cousins who appeared to be more related to her than they actually were because of the influence of genetic interrelatedness, and stymied by a preponderance of last names that often didn't follow family relationships in an expected way, Alice never found a cousin close enough to allow her to unravel a relationship.

But in early 2017—about two years after she'd found Jessica Benson and unraveled the mystery of her father—Alice's long years of work with genetic segments were vindicated. A man showed up as a relative at 23andMe. She looked at his chromosomes, and they did not look like those strands of endogamous spaghetti CeCe Moore had talked about. The man shared a whopping 217 centimorgans of genetic material, suggesting a closer relationship than any other Alice had found, and, instead of being arranged into many short, chopped-up segments reflecting relationships far back in time, most of them were large. A big chunk of the shared genetic material was along the X chromosome, suggesting this man was related to Alice through his mother. And that's how Alice found her way to a woman named Ruth Klein—her father's last living first cousin. She flew to New York to meet her.

Looking back on how she solved her case before Ruth came along, Alice is always struck by how it happened. There were so many barriers along the way, and so many accidental connections that almost didn't happen. What if Pete had never decided to go looking for his uncle's family? What if Pete had not written that letter to the orphanage at the precise moment that he did, so that his letter and Alice's letter could cross the desk of someone who read both of them and realized that these two letters were requesting information on the same family? Without that, Pete never would have found Alice, and Alice never would have forged a relationship with the man she still considers her cousin in everything but blood. And without that, she never would have been able to find her biological cousins, because she would not have found Jessica Benson and learned of the existence of Jessica's grandfather, the other baby at Fordham Hospital.

"I keep on thinking, what are the odds?" Alice says. Alice is not really religious, but she can't help but think "there's some overriding force—God? Who knows what?—trying to set this world straight."

I'll add to this pile of improbabilities, although Alice never mentioned these: The whole notion that a person could, by spitting into a vial, uncover the truth of something that happened over a hundred years earlier to two people who are no longer alive—this also seems utterly preposterous, or would have, just ten years ago. Consumer genomics moves so fast that we take for granted now that such a thing *is* possible.

This speed underscores another point, which is that the herculean efforts that Alice had to put forth to find Pam Benson and understand what had happened to both of their fathers would not be necessary now. Because she first tested in 2012, the hunt for understanding took two and a half years. But were Alice testing for the first time just a few years later, she would have been able to journey from the shock and confusion of that unexpected ethnicity estimate to unraveling that her dad had been switched in the hospital and figuring out his identity in just a few days. She would not have needed Pete and Jessica; she would not have needed to test her whole family or send kits out to strangers or have her brother create an app to track 311,467 genetic segments. Finding Ruth Klein's son in the database would have been enough to set the world straight.

17

WHERE WE'RE GOING

In the years between when Alice first spit into a tube and now, the landscape of consumer genomics, and genomics research more broadly, has changed radically. On Twitter, I followed academics, lawyers, and scientists debating many issues touching on genetics research and use. I read books and scholarly articles, knowing that the landscape was changing drastically even in the time it took for those works to be published.

A few months after I flew out to spend time with Alice in Washington State, a huge controversy unfolded over a Chinese scientist who announced he'd created the world's first genetically edited babies, a move that scientists worldwide denounced as unethical, unnecessary, possibly dangerous to the babies, and potentially impacting on future generations. "Ever since scientists created the powerful gene editing technique CRISPR, they have braced apprehensively for the day when it would be used to create a genetically altered human being," the *New York Times* reported. "Many nations banned such work, fearing it could be misused to alter everything from eye color to IQ." One of the inventors of the CRISPR technology, who years before had told a magazine she'd had a nightmare in which Adolf Hitler himself asked her for a tutorial on "this amazing technology," pronounced herself "physically sick" when she learned the news about what had happened in China.

As I was reporting and writing, I followed many growing debates: about so-called three-parent babies, created by a technique called mitochondrial transfer to avoid the passing down of certain genetic disorders; about the potential and ethics of growing efforts to conduct genetic sequencing of newborns in hopes of spotting otherwise undetectable health issues; about research attempting to link genetics to IQ, to educational attainment, and to income; about work being done within the field of epigenetics. On that last topic, there are even debates over how the term "epigenetics" itself is used; the epigenome, according to the National Human Genome Research Institute, "is a multitude of chemical compounds that tell the genome what to do"; it can be affected by the environment, and in some cases, epigenetic changes can be inherited. But how much can the epigenetic influence of trauma, for instance—experiences like war and poverty—be passed down? Some studies have suggested "that we inherit some trace of our parents' and even grandparents' experience, particularly their suffering," the *New York Times* reported in late 2018, but critics "contend that the biology implied by such studies simply is not plausible."

And meanwhile, as I was talking to seekers, the prospect of whole genome sequencing was becoming ever more possible for everyday Americans. The Human Genome Project took thirteen years and cost $2.7 billion, but in recent years the cost of sequencing has dropped to about one thousand dollars or less, and by the time I was writing this book, Illumina was talking about a not-too far-off future in which it will cost one hundred dollars, about the same as the current cost of consumer tests like AncestryDNA. In *She Has Her Mother's Laugh*, science journalist Carl Zimmer writes about having his own genome sequenced a few years ago. "When I started writing about science, if someone had said, 'Do you want to get your genome sequenced?' it would have been like saying, 'Do you want to go to another galaxy?'" Zimmer said in a talk. He took his raw data around to different scientists for analysis, and was struck both by what this data could tell him and by what it could not. Studying the actual Neanderthal genes he'd inherited was amazing, Zimmer said: "I can say that I got that gene from somebody sixty thousand years ago, maybe, and that does give you this profound connection to the past." But at the same time, for so many genes he was looking at, "we don't really know what most of them actually do."

"What if: Everyone Had Their Genome Sequenced at Birth?" was the title of a panel discussion at the World Economic Forum Annual Meeting at Davos in early 2019, conjuring up images of a future that several geneticists and marketplace experts had been describing to me in which genomic data is as integral and accessible to doctors as a person's health history and BMI. The audience was split on whether such a thing should happen now, but nearly unanimous that it was inevitable that most infants would ultimately be sequenced. The panelists debated the benefits and drawbacks. In cases of children known to be ill, a strong case could be made for sequencing, but were we ready to start uniformly sequencing healthy babies, and what to do when genetic issues cropped up for which nothing could be done? What were the implications for consent, privacy, economics, and health inequities?

I watched as the use of genetic genealogy to narrow in on criminal suspects went from a controversy to a novelty to a fairly normalized aspect of crime-fighting, even as bioethicists, lawyers, and legislators were still debating it. A woman named Brandy Jennings, who lives in Alice's town of Vancouver, Washington, found out the genetic file she'd uploaded to GEDmatch and forgotten about had been used to locate a murder suspect, a second cousin once removed whom she'd never met. She said she was glad her DNA had been able to help, even if she never could have anticipated it.

And I followed genetic genealogy blogs explaining the new techniques unveiled for "smashing through brick walls" by companies like Ancestry and MyHeritage at RootsTech 2019, the massive Salt Lake City conference put on annually by FamilySearch, which boasted twenty thousand attendees that year. The features do things like show which common ancestor likely connects two DNA cousins, and sort DNA matches into clusters. In short, they make easier the kinds of breakthroughs Alice could only dream of when she first started trying to unravel the mysteries of her DNA.

The pace of change was so rapid that I became something of a Twitter junkie, nervously checking many times a day to make sure I wasn't missing anything. I worried sometimes that writing this book was a fool's errand—that whatever happened in the realm of genetic privacy, forensics, consumer protections, and family relationships that seemed fresh and revelatory as late as 2019 would seem familiar and even obvious to readers in 2020.

But then I would remember that I was in a bubble. The people who were getting DNA kits for Christmas had no idea what was coming for them. And the ramifications of what they might find would not be short-lived; rather, they amounted to a fundamental reshaping of the American family. It was something they would deal with for the rest of their lives and pass on to the generations that followed.

* * *

Every morning, Alice has her routines. She wakes early and puts on MSNBC, plays solitaire, does several crosswords on her iPad, and checks the DNA databases, to see if any new DNA cousins have shown up. She makes notes of how much genetic material she shares with each one, and how they might be related, and reaches out to the closest new cousins. Her curiosity and her desire for connections means she's constantly emailing and calling and driving and flying to visit with an ever-expanding set of relatives. Because of her love of genealogy, Alice's family is bigger, and her world smaller, than many people's. Among the things her father passed on to her is a sense that family is everything.

Alice being Alice, she is still searching and has built a huge family tree she calls Twisted Branches, which incorporates both the Benson and Collins lines, both the family her father knew as his, and the genetic family he never knew. The tree has 1,331 people in it, with the earliest ancestor born in 1536. Alice has made contact with a relative in Israel, and found cousins in South Africa, and discovered links to a family of Estonian rabbis. She used her genealogical knowledge and her trick of thinking sideways to figure out her likely connection to a DNA cousin through a tiny village in Belarus. When George H. W. Bush died, Alice sent me a photo taken of his casket—the forty-first president was a seventh or eighth cousin of her mother's, something she uncovered by tracing back to the Plymouth Colony—flanked by a marine who happened to be her dad's great-grandnephew. "We never thought we'd see a state funeral where both sides of our family were represented!" she wrote. The genealogy she was doing, evoking long-gone towns in eastern Europe, and long-dead family members with names like Freyda and Yetta and Mendel, meant that the past was not a static thing, any more than Jim's identity was. It was a puzzle forever being solved. In asking all those questions, in combing through old archives, Alice was continually putting her father's story together, keeping him with her.

Alice fields questions from people trapped in genetic mysteries, and skipped a class in the finer points of serger sewing to tell her story at a nearby genealogical society. She has helped Pam unravel her genealogy, and together the swapcousins have been trying to get New York State to annotate their fathers' birth certificates, to reflect their true parentage. She continues with her sewing projects and made a holiday runner for Hanukkah, which she has added to the long list of holidays her family celebrates. She told me that sometimes she goes to a reform synagogue for Friday-night services. She never had been a "good Catholic," by her own reckoning, and finding out her father was Jewish by birth has given her another means to feed her spirituality.

One afternoon, Alice showed me something amazing. We were sitting in her office eating peanuts and M&M's. She went onto GEDmatch and into a tool called Lazarus. She showed me how, using the DNA tests she, her siblings, and father's cousins had taken, she was able to get a sense of what her father's genome would have looked like. She didn't need his saliva; Jim had seven living children, and each had inherited genetic material from him, some of it the same and some different. Alice verified which of the Collins siblings' genetic segments came from their father by matching them against known paternal cousins, and, by putting it all together, she could approximate a good portion of what Jim's chromosomes looked like, effectively raising him from the dead.

"God*damn*," I said, wondering for the umpteenth time if I'd ever met anyone so nerdy and so cool.

* * *

The fundamental lesson of the DNA age is that the past is not over. We may feel we are leading modern lives, having left behind certain tragedies and injustices, certain mistakes and anachronisms, but it is all still there. It is etched into us, and it only requires technology to be revealed. It is revealed in the segments of DNA that demonstrate how contemporary African Americans carry the legacy of white men who forced themselves on the women they enslaved. It is revealed in the discoveries of those who were donor-conceived at a time when such reproductive methods were kept secret, even within families. It is revealed in cases of adoptions and NPEs, some covered up when doing so was a matter of survival. DNA reveals decisions made in the aftermath of events of all kinds: acts of passion, acts of violence, acts of love. We get to decide who we are, but DNA allows us to inform that decision by understanding the truth of how we got here, of how we came to be. It forces us to look back and examine the circumstances under which difficult and life-altering decisions were made. It gives us context.

When I interviewed Bennett Greenspan of FamilyTreeDNA in Texas, he told me he could not have envisioned what his business would become. During that weekend in August 1999 when he had the idea of using DNA to get around a brick wall in his own research, he was not imagining how DNA could be used to explain how we all fit into the massive family tree of our species, or anticipating the FBI uncovering the identity of suspects based on the genetic information of fellow hobbyists in his database. He was not thinking that one day genetic testing could be used to show that two babies had been switched in the hospital back in 1913. He was just a seeker. He was "looking for my *emes*," he said, using a Hebrew word. "My real truth."

But of course, your tree is my tree is his tree is her tree. Your emes is not my emes, and sometimes, in a genetic database, they collide.

Things have changed so much since the advent of consumer genomics. Alice's search encapsulated a brief moment in time when the technology existed, but the databases were relatively small, meaning that the techniques of genetic genealogy made unraveling connections just barely possible, if combined with enough work and luck. But in the future, we may not even remember this moment in time, even as it is truly the moment when everything changed, and when a woman like Alice could make the choice whether or not to search. Because in the future, and I'd guess it's a near one, enough people will have undergone consumer testing that solving a genetic mystery will be in many cases almost instantaneous. And nearly everyone related to the seeker—which will be not a special class of people willing to learn complex genetic genealogy techniques, but, effectively, anyone who sends in their saliva—will be rapidly implicated in any discoveries that arise, whether or not they choose to test. When the future catches up to the past, and there are no genetic secrets, we will need to rethink many things—starting with how we talk to our children.

In the summer of 2018, CeCe Moore told me that since Alice's discovery, she'd assisted in solving five more hospital baby-switch cases. Alice's father's switch had occurred the furthest back in time, and the most recent had taken place in the 1950s. News stories have recounted two instances of baby swaps uncovered through recreational testing, both involving girls born in the United States in the 1940s. Both were solved quite rapidly because the databases pulled up a biological sister in one case, a mother in the other. There was no quest, as there had been for Alice—just a sudden, searing revelation.

"Sometimes I had that sense that I didn't quite fit in with them," said one of the women, who was switched at a Minneapolis hospital in 1945, reflecting back on the family she was raised in. She was echoing themes I'd heard some seekers give voice to, a lifelong sense that not all the puzzle pieces of their nuclear families fit together, even when they didn't know why. But women in both stories also talked of an expanding of their sense of family once the revelation was made, rather than a replacement of it—not nature over nurture, not a zero-sum game. It was language I'd heard from other seekers, like Laurie Pratt, who never told her father that she didn't share DNA with him because her dad was her dad, no matter what a test said.

Throughout the years I was researching and writing this book, my interest in DNA was not just professional. My parents both began researching their roots in earnest, helped along by growing DNA databases and genealogical archives. My mom was able to confirm the identity of her great-grandfather. My dad met the son of a first cousin he never knew along one branch of his tree, and along another, he found a genealogist relative in Sweden, and we decided to embark on our own DNA heritage trip. In southern Sweden, we visited the farm where my dad's grandmother Sigrid grew up. We walked the lanes Sigrid walked and saw the church where she worshipped and the school she attended, learning about the entrenched poverty and family dynamics that must have contributed to her decision to emigrate at sixteen. We shared meals with my dad's second cousin and her daughter, people we never would have known but for ancestry testing. We became, as a family, seekers, the way so many Americans are becoming. I fielded messages from a woman trying to help an adoptee who wondered if my mother's DNA might hold the key to the adoptee's identity, and I scrutinized the overlapping segments with this distant cousin, and thought endogamy might be confusing the issue. I scrolled through my match lists, pausing at the close cousins with unfamiliar names, wondering when the roulette wheel of some unexpected revelation *would* stop at our family, because by now it seems inevitable that such a thing will happen to someone in our familial orbit, eventually.

I spoke again with Rosario Castronovo, who described his search for the truth as an effort to reconcile with decades of his life during which the most important things were left unsaid. I spoke again with Linda Minten, who told me she's been helping other seekers find their genetic families, and she warns them to hope for the best but be prepared for the worst. "Some people bite and some are open-armed." I emailed with Ricki Lewis, the geneticist and science writer; less than two months after our first conversation, she'd discovered she had not just one half-sibling through the sperm donor she'd never known was her genetic father, but six. AncestryDNA kept spitting out new ones. She was a "DC-NPE" in the lingo of this world—the product of both donor conception (DC) and a "not parent expected" event. Lewis is the author of a college textbook on human genetics, and she recently revised the thirteenth edition to include a chapter on NPEs, informed by her own experience. "Bioethics seems to be lagging behind the science, dumping a mess onto the mental health community," she wrote in a recent column on ancestry testing.

And meanwhile, the headlines kept coming—"New Siblings, Old Secrets," and "Your Father's Not Your Father," and "My DNA Test Revealed a Genetic Bombshell," and "Woman's DNA Test Revealed a Shocking Family Secret"—and countless tales about surprises for "adopted Dover woman" and "Tuscaloosa dentist" and "Houston-area mother" and "Minnesota man." And the seekers kept flooding the Facebook help sites, and the children of NPEs kept searching social media for clues about their biological fathers, and the adoptees kept crafting careful letters of introduction, and the genealogy addicts kept building out their trees, and a whole swarming mass of Americans found themselves converts to a search many never planned to embark on, a question they'd never known they were asking:

The question of themselves.

SELECTED BIBLIOGRAPHY

I relied on many sources to understand the broader scientific and historical themes I was interested in exploring in *The Lost Family*, most notably five excellent books: Carl Zimmer's *She Has Her Mother's Laugh*, David Reich's *Who We Are and How We Got Here*, Christine Kenneally's *The Invisible History of the Human Race*, Siddhartha Mukherjee's *The Gene*, and Adam Rutherford's *A Brief History of Everyone Who Ever Lived*.

François Weil's *Family Trees* offered the history of American genealogy I relied on for the book, and *Hearts Turned to the Fathers* recounted the great influence of the Church of Jesus Christ of Latter-Day Saints on the field.

For the early history of genetics and the rise of eugenics recounted in chapter two, the following books by historians of science and other scholars were invaluable: Peter J. Bowler's *The Mendelian Revolution*, Nathaniel Comfort's *The Science of Human Perfection*, Nicholas Wright Gillham's *A Life of Sir Francis Galton*, Daniel J. Kevles's *In the Name of Eugenics*, Mark A. Largent's *Breeding Contempt*, and two books by Paul A. Lombardo: *Three Generations, No Imbeciles* and *A Century of Eugenics in America*. I also relied on research and work by Gregory Radick, Susan Rensing, and Alexandra Minna Stern.

To understand the cultural meaning of DNA, as well as our cognitive biases about it, and how popular portrayals of it intersect with our thinking about race, I drew on several books, including *The DNA*

Mystique, by Dorothy Nelkin and M. Susan Lindee; Steven J. Heine's *DNA Is Not Destiny*; and Alondra Nelson's *The Social Life of DNA*.

Thomas H. Murray's *The Worth of a Child* offered a fascinating exploration of timeless moral and philosophical questions about the relationships between parents and children.

Genetic Secrets, edited by Mark A. Rothstein, explores themes around genetic privacy we've been grappling with for decades.

Erin E. Murphy's *Inside the Cell* delves into privacy, accuracy, and justice in the use of forensic DNA.

For a history of early maternity wards, I relied on Judith Walzer Leavitt's history of childbearing from 1750 to 1950, *Brought to Bed*. Sandra Opdycke's book *No One Was Turned Away* helped me understand New York's public hospitals, and Evelyn Gonzalez's *The Bronx* fleshed out the social and historical context for immigrants living near Fordham Hospital after the turn of the twentieth century.

Dani Shapiro's memoir *Inheritance* movingly recounts the experience of a genetic surprise from the inside out. Other books that shaped my understanding of the power of DNA testing and genetic genealogy both to disrupt and to reunite are *Finding Family*, by Richard Hill; *The Stranger in My Genes*, by Bill Griffith; and *The Foundling*, by Paul Joseph Fronczak and Alex Tresniowski.

In addition to other research and reporting, I consulted Ricki Lewis's comprehensive and comprehensible college textbook *Human Genetics*, and for pure pleasure reading on the topic, I recommend *The Cartoon Guide to Genetics* by Larry Golnick and Mark Wheelis.

Those hoping to understand how to use genetic genealogy to find their biological families—as well as how to reach out to those families, and what to expect—should seek out the very fine guide *The DNA Guide for Adoptees*, by Brianne Kirkpatrick and Shannon Combs-Bennett. Blaine T. Bettinger's *The Family Tree Guide to DNA Testing and Genetic Genealogy* offers a broad and exhaustive overview of the field.

ACKNOWLEDGMENTS

I am deeply grateful to Alice Collins Plebuch, her sister, Gerardine Collins Wiggins, and their brothers Jim, John, Bill, Brian, and Ed for allowing me into their lives and helping me understand their emotional trajectories, as well as the experiences of their father, Jim Collins. Alice's incredible mind and perseverance made it possible for her to solve the mystery of her father; her record-keeping and meticulous memory—not to mention her time and patience—made it possible for me to recount her efforts. Gerry offered important context and emotional perspective, and had records and recollections that deepened my understanding of the family and its search. I am also deeply grateful to Pam Benson for her time and thoughtful insights into her own journey and the life of her father, Phillip Benson.

I want to thank the "seekers," many of whom spent hours telling me their stories of heartbreak, wonder, and love. It is a leap of faith to tell a stranger your shock, your shame, your anger, and your fears, a leap of faith to let a stranger into your most intimate musings. I found myself grateful for their trust and determined to tell these stories with as much sensitivity and emotional nuance as possible, in hopes of doing justice to a phenomenon that is increasingly a part of our American story. One of the seekers I interviewed said she hoped that accounts like hers will make genetic surprises, and the secrets they reveal, "not so taboo." That's my hope, too.

This book began when my editor at the *Washington Post*, the wonderful Amy Argetsinger, came to me with the idea of writing a feature story about DNA revelations. She and others at the *Post* gave the story the space, design, and platform to catch the attention of readers—a lot of readers. As hundreds of emails poured into my inbox with captivating stories that made me gasp, smile, and cry, I knew I wanted to write a book that would attempt to capture the scope of how this technology is changing how we relate to one another. *The Lost Family*'s other fairy godmother is the wise and thoughtful genealogist Jennifer Mendelsohn, who first suggested to Amy that she commission a story on this important topic. Jennifer's guidance about and enthusiasm for the world of genetic genealogy have been invaluable to me throughout the process of writing this book, as was her reading of my manuscript early on. I am grateful for her foresight, patience, and feedback.

A number of experts have generously given their time to help make this book as accurate and compassionate as possible. Genetic genealogist CeCe Moore first told me Alice's remarkable story in early 2017. Over many subsequent conversations, CeCe shared countless insights, much technical knowledge about her field and her work as a search angel, and a big-picture, sociologist's-eye-view of the changing state of the American family. I am exceedingly grateful to geneticist Ricki Lewis for reading through the manuscript to make sure the science was correct, comprehensive, and comprehensible. Her book *The Forever Fix: Gene Therapy and the Boy Who Saved It* is a model of compelling science journalism. Genetic counselor Brianne Kirkpatrick counseled me on the science and ethics of her field, fielded countless questions from me about everything from *BRCA1* mutations to high runs of homozygosity, and spot-read certain science-heavy paragraphs. Genetic genealogist Laurie Pratt McBriarty-Sisk acted as a sounding board and sensitivity counselor, helping me understand experiences like hers from the inside out, and spent an inordinate amount of time making sure I understood the techniques involved in an unknown parent search. Genetic genealogist Debbie Kennett has always found time to answer my questions about her field and the home DNA testing industry, and she's a font of knowledge about ethics, privacy, and international issues.

I also want to thank Kathryn Garber, Emory University professor of human genetics, who read and critiqued several DNA explainer paragraphs that appear early in the book, as well as the American Society of Human Genetics for facilitating this assistance. I am indebted to Johns Hopkins science historian Nathaniel Comfort, whose guidance about the early history of genetics (recounted in chapter two) was invaluable, as were his recommendations about the best historians and sources to consult. I'm grateful to University of Leeds historian of science Gregory Radick, who coached me through the nuances of technical language and aided in my understanding as I worked on this same section. Many more people contributed to *The Lost Family*; in cases in which I've quoted someone at length, you can usually assume that meant one or more long initial interviews, numerous follow-up emails, and another epic fact-checking phone call at the tail end. I also want to mention those who don't appear in the book much, if at all, yet lent a special expertise or were especially supportive of the book and its mission. These include Megan Allyse, Alexander Beider, Benjamin Berkman, Shai Carmi, Ellen Wright Clayton, Ken Cobb, Michael Coble, Robert Cook-Deegan, Graham Coop, Michael "Doc" Edge, Colleen Fitzpatrick, Dov Fox, Tana French, Ariella Gladstein, Evelyn Gonzalez, Jason Harrison, James Hazel, Erika Hutchcraft, Razib Khan, Maarten Larmuseau, Thomas Murray, Sandra Opdycke, Itsik Pe'er, Andelka Phillips, Julia Robbins, Judy Russell, Alicia Semaka, Dani Shapiro, Alexandra Minna Stern, and Carl Zimmer.

My agent, Jess Regel of Foundry Literary + Media, believed in this book when it was just an idea and has offered me savvy counsel and support at every step. I have been thrilled and honored to have as my editor Jamison Stoltz, whose unerring sense for narrative and pacing helped me see how I needed to shape the manuscript. Like *The Lost Family*'s protagonist, Alice, Jamison can see the big picture at the same time as he can tell you where the "nits" are. I am grateful to the whole team at Abrams Press for their support and guidance, including Lisa Silverman, Gabby Fisher, Kimberly Lew, and Sarah Robbins.

Many friends have been with me through this marathon, cheering me toward the finish. In particular, I want to thank Ilene Wong, also known as the young adult author I. W. Gregorio, who since becoming my friend in college has encouraged my ambitions and pushed me to take up more space. Ilene's *why-not?* attitude is infectious. Karen Pasternack knows me so well that at times I'm certain we share a brain. She has comforted and counseled me countless times during the course of this book, making me feel braver, calmer, and wiser. And Nina Morgenlander's faith, love, and optimism have always made me feel like I can do anything. Many other friends, family members, colleagues and wise people offered encouragement and advice, including Angela Giuffra, Monica Hesse, Mark Kuniholm, Katherine Reynolds Lewis, Henry Miller, Suz Redfearn, Gina Scharoun, Lyn Traverse, Neely Tucker, and Gustavo Stolovitzky.

Writing this book has caused me to think a great deal about inheritance, genetic and otherwise. I am deeply grateful to my mother, Sarita Eisenstark, whose curiosity, encyclopedic knowledge and intellectual rigor made me a journalist before I was one. And I want to thank my father, Charles Copeland, whose emotional honesty, generosity and optimism have helped shape my compassion and moral compass, invaluable skills for a writer.

To my children, Olive and Lev, thank you both for your wisdom, sensitivity, and love. As a family we worked together to make this book real. And to my husband, Dan, thank you for believing in me before I did; thank you for your calm and certitude and soulfulness; thank you for always making time for this project, even as your own load grew heavier. I wouldn't want to attempt to write a book (or do anything else, for that matter) without you.

INDEX

THE

GODSTONE

A NOVEL

VIOLETTE MALAN

DAW BOOKS
New York

For Paul

Acknowledgments

I started this book while we were still living in Canada, so my first thanks have to go to our friend Patti Groome, who housed us for three months until we left the country, and still gets bombarded by ridiculous amounts of our mail. You are the easiest person in the world to live with. Thanks also to my friend Stuart who always helped me with my computer needs. You're going to be hard to replace. To the girls in Cahucholas Servicios Gráficos, who did all my printing and copying for me until my printer arrived in Granada.

Final thanks to my wonderful agent, Joshua Bilmes, and my Hugo-award-winning editor, Sheila Gilbert, without whom none of this would have been possible.

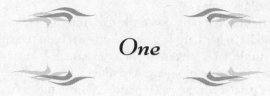

One

ARLYN

THE LETTER CAME on a low day. Bad timing. I set it aside on the table, next to the dirty breakfast dish. I wandered into my workshop, stared at the tools. All clean, neatly put away. No project on hand. Could have sharpened chisels. Chisels always need sharpening. Seemed like a lot of trouble. Could go down to the baths. Maybe later.

Usually I'd go to Fenra when I got this bad, but she'd been busy.

Might as well open the letter. It could be a commission, or maybe a payment. Impossible as it might seem at the moment, I could one day be interested in building something—or making money. A single sheet of parchment was folded into the sealed cover.

I had to read the letter over twice; I was too numb at first to take it in.

"Oh no. No, no, no." My stomach sank away, cold as the abyss, but my head was on fire. I spread out the cover and grabbed a piece of the dark chalk I use for marking pale wood. "On my way," I wrote. Refolded it, resealing it by warming and reapplying the original seals, one white, one red.

Before I could talk myself out of it I trotted down to the mill. The courier was long gone, but Ione Miller promised to hold my letter for him to pick up on his way back to the City in the evening. The Road doesn't go much farther from here. Used to, but not anymore. My answer would get to the City much faster than I could.

My burst of energy deserted me on the way back to my workshop. I stopped, rubbed my face with my hands. The thought of all I'd need to do smothered me, the planning, the details, the organizing, the execution, the decisions needing to be made, contingencies to consider—where would I even start? What if I *didn't* go? What was the worst that could happen?

A little girl came hopping down the path toward me, jumping from puddle to puddle in her bare feet, splashing dirt onto the embroidered hem of her tunic. She looked at me, concerned. I knew her name, couldn't think of it that moment. My smile must have been crooked, the smile she gave me back was tentative. Forced myself to smile better. Went down on one knee in the wet and clapped my hands, holding them up palms toward her, began to whistle.

Giggling now, she ran up to me and smacked her palms to mine. The clapping pattern was a little complicated for her, but we finally made our way to the end. She hugged my neck, kissed my cheek, skipped away, splashing into every single puddle as she went.

Felt a little better. Knew that later, once I'd been leveled, I'd be pleased I'd done this, that I hadn't hurt little Garta with my lowness.

Watched her until she turned off down the alley to her home. What was the worst that could happen, I'd asked myself. It had already happened to me. But Garta, and others like her—everywhere—they enjoyed their lives. If they knew what I knew, they'd want me to think of them. They'd want me to stop what was coming. Or at least try.

Ducked into my place, picked up the letter, left the door wedged open for the cat, set off back the way I came. If Fenra wasn't home, I had a pretty good idea of where I should look.

But she was home. As usual, except in the worst weather, her door stood propped open with an old chunk of marble. I saw light, movement within. Fenra glanced at me, half smiled, not really paying attention, turned back to what she was doing. A sudden flare of anger burnt through me. What? Too much trouble to look at me? I wanted to grab her arm, but Fenra isn't someone you can grab safely, despite her lameness. She spoke without looking up.

"Can I leave Terith with you?"

This was so completely unexpected that my surge of rage vanished. "What? Why?"

"I have to be away for a while."

That was when I noticed she'd been sorting her things into packs and bags and boxes.

"But why now? It's too late to do your fall rounds." Twice a year

Fenra visited other towns and villages in the area. Now she stopped what she was doing, looked at me. I saw fatigue in her face, shadows, the remains of tears in her eyes.

"The Ullios' little girl died this afternoon."

I'd been with Fenra the day they'd brought her the girl. Sitting outside this very door on the wooden bench I'd made, enjoying the last warmth of the sun. The child—Jera? Jena?—had obviously been ill for days, vomiting, from the smell of her, incontinent as well. The parents had probably tried every other remedy before bringing her to a practitioner, someone they'd have to pay. By then it was too late. "Surely they don't blame you?"

"Not today, no. Today they cry in my arms and thank me, knowing I did everything I could. But in a few days, or perhaps a few weeks, their own guilt will have them looking for someone to blame. They will forget I promised nothing and asked no fee." She rubbed at her forehead, leaving a streak of dirt. "The easiest one to blame is the one whose practice failed them, so best I am not here for a while."

No argument there. In this Mode it's a short step from wise one to witch. The farther out the Mode, the shorter the step.

"You'll need the horse," is what I said to her.

"Not going that far." She shook her head, lips pressed together. She loved the horse.

"I can't keep him." I felt guilty but relieved to have an excuse. "I have to go to the City."

All this while she'd continued sorting through her spare clothing, her scrolls and tablets, her medicines and potions. Though she'd looked at me, she hadn't seen me. Now she stilled, looked me straight in the face. "You are low. Why did you wait so long?"

I shrugged. "Well, first you were busy with your friend from the City —"

"Medlyn Tierell was my mentor at the White Court, not just my 'friend.'"

"Whatever." If the man was just a mentor, why was he still coming to visit her so often? "Then the Ullio girl got sick . . ."

She made one of those noises that say you're exasperated, looked around the room with more purpose, finally waved me over to a four-legged stool I'd made her, moved the pile of parchment sheets that sat on it, set them on the floor. She knelt in front of me, took my hands. I made a half-hearted try at pulling them away; Fenra's strong when she wants to be.

"You're too tired." But I stayed where she put me.

"Never mind that now," she said. "Half the practice is learning what can and can't wait."

Her power flushed over me and through me, cold and bubbly and clean. My skin twitched and my ears buzzed. Plants must feel this way after a rain. The deep breath I took didn't release itself in a sigh, rather it filled me completely. I could feel it in my toes and my fingers, I swear even in my hair. Muscles I didn't know were tense relaxed, my mouth quirked into a smile.

I saw for the first time the toll of this year's fever season. Fenra's face was thin and drawn and she had a little tremble in her left eyelid. Her hair hadn't been brushed out for days, her braid hung half untwisted, and she'd kept the tangles off her face with combs shoved in any which way through the blue-black curls. Her normally dark skin was sallow, almost ashy, only her gray eyes the same color as usual—too light for the darkness of her skin and hair. I was ashamed that I'd taken her strength—but not so much that I wouldn't do it again.

Fenra sat cross-legged on the floor in front of me, agile despite being tired. "Best I can do for now," she said. "So tell me, why would you need to go to the City? In all the time I have known you, you have never gone anywhere. Your customers all come here to you, to the village."

"Here, read this." She took the letter from me, stood up in one move. As I said, agile. She crossed over to the lamp on her table, held it at an angle. I watched her eyes flick back and forth.

"I am sorry about your cousin. Were you close?" There must have been some change in my expression because she added, "Are you telling me you have no cousin?" Smiling, tilting her head, she held up the letter. "Is this some scheme to get you to pay for an inheritance that doesn't exist?"

"Oh no, Xandra is—was—definitely real. But there is no chance, none whatsoever, that he would have named me executor of his testament. In fact, I happen to know he left no testament at all."

She raised brows so perfect they could have been drawn on her face. "And so?" she said when I didn't continue.

"And so I'd better go and see what's really going on."

She nodded slowly, refolded the letter, tapping it on the palm of her hand. "Hence you cannot look after Terith."

My turn to nod. "Hence I can't."

Her hands stilled. "So what is it about your cousin's death that has you in such a swivet?"

I sat up straight, tilted my head back to look her in the eye. "I'm not so sure he's dead."

"Where is he then?"

"I think he just went walk-about. It's not so common now I hear, but practitioners used to—"

"Wait." She held up the letter like a baton. "He was a practitioner?"

"Didn't I say?"

"No, you did not. But that certainly explains why someone is so anxious to have the man declared dead. He left artifacts?"

I shrugged. "I imagine." A practitioner's properties and tools could carry quite a lot of power in them, to say nothing of their usefulness as already practiced objects.

"So they need you, someone of the same blood, to release any seals they find." She made a face. "I have never seen it done." She tilted her head to one side, narrowing her eyes. "What are you not saying?"

I rubbed at the back of my neck. "There's something in Xandra's vault, besides the ordinary artifacts. Something dangerous. That's why I know this testament is fake. He would never want the seal released."

She tilted her head, studied me with narrowed eyes. "Stay here then. Without blood kin they cannot break the seal."

I scratched at my elbow. "There are . . . others, cousins, somewhere, who don't know the full story." Fenra was the type to think of others. It'd seem a good reason to her. Still, she looked at me sideways. "And if I don't go, they'll just send for me again until I do. You know what the Red Court is like."

Tapping her lower lip with the folded letter, she looked at me through narrowed eyes. "Are you going to tell me what is this thing?"

I tasted the words on my tongue. If I told her she wouldn't believe me. No one would. Except perhaps whoever had sent the summons. If they knew, that is. If they weren't just curious and greedy.

While I hesitated Fenra had shut her eyes. She swayed a bit and I got to my feet, prepared to catch her if she fell.

"I have to go," I said. "Officially deny my permission to break the seal. In person." A thought struck me that wouldn't have when I was low. "You should come with me." I gently took the letter out of her fingers. She still looked at me sideways.

"Look, you think the Ullios are going to blame you, and if you rush off—out of season and within hours of the little one's death—Jera, wasn't it?—they'll think they have reason to. But if you leave because you need to accompany me to the City, no one will think twice. Everyone knows how much better I've been since you came." She couldn't argue with that.

"You were bad at first, that's true."

"If you leave with me, come back with me, everyone will accept it." Her eyebrows rose. "You think you will come back?"

"I certainly intend to." And that was true.

She looked around her room as though seeing its contents with different eyes. She turned toward me, but she wasn't looking at me. "I have been happy here. I have been useful."

I made her sit while I gathered together the things needed to make tea. The heavy iron pot with the wooden handle I'd carved her, two clay cups glazed only on the inside, the larger clay jar with its wooden stopper where she kept tea, the smaller one where she kept honey. "And if you come with me, we can take your horse."

"Now there's a reason I cannot disagree with."

I was relieved. Fenra was the only person I'd ever met who could level me. I hadn't known it could be done. A remarkable gift considering her leg. Though come to think of it, healing others is always easier than healing yourself. Many practitioners have found that out.

⟜

FENRA

There are days when what I would most like is to shut my door in Arlyn Albainil's face. Knowing that I am not about to do it only adds to my irritation. I am not likely to forget what he was like when I first came to the village. Old Boose Miller, the headman before his daughter Ione took over the business, brought me to Arlyn almost as soon as I arrived and they knew what I was. Before I came and leveled him, they had been taking turns feeding and bathing him, though his money had run out long before. He lived in the workshop even then, but once, when old Boose had had more to drink than he should have, he told me Arlyn had not made anything since anyone could remember. I did not believe him, of course. Not then.

I had neither seen nor heard of an example of lowness before Arlyn. Even my mentor Medlyn Tierell, who is the Lorist for the White Court, has never found mention of such an illness in any book or scroll. From what I have seen, it consists of an increasing detachment, both from the world and from one's self, a sinking into apathy so profound as to be possibly fatal. When I first saw him, Arlyn was completely unresponsive, though when I touched him I could tell there was still a person inside. Old Boose pointed out the scarring on Arlyn's wrists, old enough to be very faint indeed. Unmistakable, however, as the marks of suicide attempts. In fact, from the extent of the scarring, it was incredible that he had not died. Perhaps he *had* known another practitioner before me. I have given up trying to find out.

At that moment, however, I wondered what could be in the sealed vault, just how dangerous it was, and to whom. Still, I knew that if I pressed him I would only get a plausible lie.

Early the next morning Ione Miller helped me lift my packs on to Terith's rump. The saddlebags carrying our money I had managed myself.

"It's good you go with him."

I looked at her with eyebrow lifted, but I let my smile die away when I saw she was serious.

"Maker-touched people like Arlyn are lucky for villages—look at the business he brings," Ione said in response. Ione, like her father before her, was the unofficial mayor, so she would know. "People come to him for his furniture. They have to stay somewhere, eat somewhere, and buy supplies from us. Besides, the children like him."

I patted the woman on the shoulder. "He wants to come back," I reminded her. "And generally speaking, what Arlyn Albainil wants, he gets."

"Well, remember, we need you just as much as we need him—and the children like you, too."

Where the Road passes by the village it is little more than a track. In a few places two ruts show where wheels regularly pass. In this, as in any Mode, soft spots on the Road are reported by couriers, and strengthened with rocks and stones by what the White Court calls Wayfarers, apprentices doing solo tours of all accessible Modes before they take their class exams. The Road is their special business. Couriers are apprentices also, but their special business is information. A few practitioners, like me, do not return to the City from their tours.

I love to ride, but I always tell people, if I do not walk, I could lose the use of my leg entirely. It's my favorite bit of misdirection. Terith can be as happy on the move as in a stable with his nose in a bag of feed—such bags as he was himself carrying, as it happened.

"Why carry so much food anyway?"

Arlyn's question surprised me. "I know you have not strayed from the town since I have known you, but you must have arrived there somehow, and if not by horse, how?"

He shrugged. I sighed and rubbed my eyebrows with the thumb and middle fingers of my right hand. "The grazing that can be found by the roadside is minimal." I indicated the trees growing close to the Road. "Every animal that has passed this way has been eating it."

"But the fields—"

"Are owned by someone likely in a position to make their complaints heard."

He nodded then, brows drawn together as if he had never considered this before.

A few hours later we came to one of the Wayfarers' rest stops, in this Mode no more than a small lean-to with a fire ring made of carefully placed rocks. These shelters are also for mundanes to use, not just practitioners, so learning the differences and matching the Mode is part of an apprentice's course of study. I unloaded Terith and left Arlyn to make a fire while I tied the horse around back of the structure, making sure he could reach what little grazing there was. I told him I would break out a feedbag for him later. He wrinkled his nose in what horses use for a smile.

I found Arlyn standing with his thumbs hooked in his belt, staring at the pile of twigs and dried leaves as though it would light itself.

"Can you light the fire?" he asked. "Without a sparker, I mean."

"I know what you mean." I lifted my chin and examined him. His eyes looked into the present and his facial muscles were relaxed. It was too soon for him to be low again.

"So *can* you?"

I squatted on my heels, most of my weight on my left foot. Though I had not bothered for years, I remembered how it was done. I reached my practitioner's hand through the nest of twigs Arlyn had created and touched one particular dry leaf with the tip of my finger. Everything works better if you know where to start.

But apparently not today.

"Isn't calling fire the first *forran* practitioners are taught?"

"Calling light is, actually. But even the easy *forrans* need practice," I said, accepting the sparker he handed me.

Supper consisted of two strips of dried venison, a large potato, and three carrots, all simmering in a small travel pot on the fire as we shared a cup of mint tea. Between us Arlyn and I had food for about three days, what we had been able to assemble out of our own stores. We would use the money in my saddlebags to buy more supplies when we reached a town that had them to spare.

As we were cooking Arlyn finally asked the question I'd been waiting for.

"Why did you help Jera, if you knew she would die?"

I poked at the stew with my ashwood spurtle, grimacing at the heat of the fire. "I might have been wrong. I did my best to prepare the parents, though I knew in the end it might bring me grief."

"How so?" Arlyn took a sip from the cup of tea.

"They will remember it later, and they will believe I did not try hard enough, having already decided there was no point." I sat back on my heels.

"You sound as though this has happened before."

I turned my head to better look him in the face and lifted my left eyebrow as high as it could go. He shrugged and nodded at the same time. "Sorry. Stupid thing to say."

As we ate the stew a young badger rustled through the underbrush, curious, but beyond the light of the fire.

"I'm no hunter, but if you call the animal to you, I think I could manage to stab it. For tomorrow," he added when I froze, my spoon halfway to my mouth.

"It doesn't work that way."

"What doesn't?"

"The animals come to me because they trust me. If I abuse that trust, they will never come again."

"That can't be true." He waved his spoon at me. "I've seen practitioners do it repeatedly at court."

And when were you at court, that you saw all this? "It may work that way for them, but not for me."

After staring at me for a moment longer, he lowered his eyes to his bowl and started eating again.

ARLYN

"You might as well tell me what it is." Fenra was riding, so her voice came from just over my head.

I opened my mouth, shut it again. How much to tell her? How much would the cabinet maker she thought I was reasonably know? At least she'd waited until daylight to ask. "It's a dangerous artifact," I said finally. Absolutely true. "I don't know where he found it, or how he took charge of it." Not true at all.

"How dangerous?" From her tone, she'd been doing a bit of thinking since we'd set out.

I was ready for that. "Xandra said it could bring irreversible damage to the world, perhaps even un-Make it." Now I looked up at her face. "What?"

She shrugged. "Just a little surprised to hear the village superstitions coming from you. You seem a little sophisticated to be referring to the Maker of the World."

"Yes, well, maybe I was once." I could hear a tautness in my voice, so I looked up at her and smiled.

Her eyes clouded. "Can the artifact be destroyed?"

My mouth opened before I meant it to. It's the schoolteacher tone, we all respond to it. "If it could, wouldn't Xandra have destroyed it himself?"

"Something so powerful is a great temptation for a gifted practitioner."

That worried me a little. I didn't think she'd recognized the name. "How do you know he was gifted?"

"Because I have never heard of him, and it's only the really gifted ones who choose to keep themselves out of the records. Only they have reason—such as something to hide."

That part was truer than she knew. "You think he was tempted, then?" I kept my eyes on the track ahead. Apparently Fenra didn't notice that I moved us along at the horse's best walk—respectable on the smoother surface at this part of the Road. Trips to the City have been known to take months, but we didn't have that kind of time.

"Of course, anyone would have been. He would have found a dozen excuses for doing as he wished. That's what they are like. The powerfully gifted ones." She must have seen something in my face, because she changed the subject. "How can you be certain your cousin did not make a testament since the last time you saw him?" She shifted in the saddle, made a small disgusted sound, stopped the horse and slid off, landing on her good leg. She looked over at me, lifting her left eyebrow in that irritating way she has.

"Even if he did, he would never have named me executor. Never. No one in my family would have, I can be certain of that." I looked away; let her think I was studying the fields to the left of the Road. Farmers would be cutting the hay soon, if the weather held.

"And why would that be?"

When I didn't answer right away she just waited. People use that technique because it works. "I wasn't always a furniture maker."

"I know you have scars that woodworking won't account for." She pointed toward my right wrist with her chin.

"I was . . ." I smiled. I was about to tell her a true story. "I was a highwayman."

"What?"

The look on her face was priceless. "I robbed people on the Road—"

"I know what a highwayman is." Her voice snapped. "When did you learn to make furniture?"

"Oh, the furniture came first. I ran away from my apprenticeship with my uncle—Xandra's father—and I swept the roads for . . . actually, I'm not sure how long exactly."

"And then?"

"Then I was caught. I was to be hanged. My cousin came and rescued me." I said this as matter-of-factly as I could. This was the tale-as-I-wished-it, not the tale-as-it-was.

"How?"

"There's money in the family." I looked sideways at her, shrugged. From the look on her face she knew firsthand that money provided a lot of solutions. "He was the only one who hadn't already disowned me. Xandra would never have made me executor if there was any chance I'd have to deal with the family again. Never."

"I have a great many questions," she said.

"I imagine so." She wouldn't get any more answers just now. There's a limit.

"Unless we are to go our separate ways, I will need to know what to expect. You will have to tell me everything, soon or late."

"Yeah, well, let's make it late."

FENRA

"You sure you don't want to ride?"

"Walking is good for my leg muscles. Keeps them warm and alive." With my thoughts spinning around and around what we had—and had not—talked about, I forgot to take more care with my limp. Usually, once I am walking, people forget I am lame. Oddly, it is when I am standing still that they most notice the twist that turns my right foot inward.

"So why a horse in the first place?"

"Sometimes people need a practitioner faster than I can walk to them. Besides," I added, looking at him out of the corner of my eye and trying not to smile, "Terith is better company than most people." The silly beast shook his head, making his ears flap. He knew I was talking about him.

Our fifth night on the Road found us without even a Wayfarer's shelter. We had reached the well-used track leading to the small village of Drienz in the latter part of the afternoon, but Arlyn wanted to press on. I did not argue with him, though I was more or less certain we had entered a different Mode two nights before. Without an inn, I had no way to be sure. I seemed to remember that these things were clearer the last time I walked the Road. In any case, without shelter the safer choice for a camp was off the Road completely. The terrain here was well on its way to being mountainous, and we had to find a spot four feet could manage as well as two. Finally I chose a place where spaces between the pines encouraged me to believe some wider, flatter bit of clear ground might be nearby.

I found a clearing far enough off the Road to suit me, with only one low bush, and enough dead leaves and pine needles on the ground to make for softer sleeping, despite the slight slope. I freed the pack that sat on Terith's rump and unbuckled his saddle. While Arlyn gathered rocks to make a fire spot, I began creating wards. I like to use found anchors, stones, twigs, even dried dung if it's big enough and dried enough. For me, using natural objects in their natural state makes the wards stronger. And it saves me from carrying a lot of paraphernalia.

I took my time, placing each anchor carefully after breathing on it. Done right, the warding would make us invisible to almost everyone.

"Fenra?" The sharpness of Arlyn's tone brought my head around. "Where are you?"

Apparently I was concentrating a little too much on invisibility. I moved out of his direct line of sight before restoring myself. "Here, just setting wards."

"You haven't done this before."

"Before we were on the Road." I answered without turning around, continuing to pick up and replace objects in a rough circle that took up almost the whole clearing.

"Do healers generally make good wards?"

I could feel him watching me, and I stifled an eye-roll as I turned around. "Unlike the fire *forran*, I have had recent practice with this, if that's what worries you. Half of healing is warding off sickness. It's only the other half that cures. Now, if you do not mind." I went back to work.

"I'm sure they won't be needed," he said.

"Perhaps so." I did not bother pointing out the pair of eyes just visible on the edge of the light. I waggled my fingers in greeting, and once the fox had had a good look at us, he trotted away.

Hours later my eyes blinked open in the dark. The fire was out, the night too warm to make it necessary for more than cooking. I could see stars through the thin canopy of trees, the outline of Terith, dozing with his nose almost on the ground. If he slept, what woke me?

A shadow passed between Terith and the trees behind him. I sat up.

"Arlyn? You may want to wake up. My wards are about to be tested." It would be some*thing*, not some*one*.

Arlyn surprised me by waking like a soldier, throwing back his blanket, rolling upright, and putting a hand to his knife. "The horse didn't wake up."

I pulled my hair back off my face, re-tying the cord that had loosened while I slept. I stayed cross-legged on my bedroll, my wrists resting on my knees.

"Excuse me." The whisper came out of the darkness on the uphill side. "Can you help me?"

I clapped my hands and a light appeared, hovering over our heads.

"*Now* you can do it." Arlyn thought I could not hear him if he muttered.

A young girl inched forward into the light. She was no more than twelve or thirteen, barefoot, her homespun trousers and tunic wet and dripping, soiled with streaks of slimy green, and torn. Laces, I noticed, not buttons, which told me the Mode. Her light brown hair hung tangled, dripping rivulets of water onto her shoulders, and she had a streak of mud high on her left cheekbone. "Can you help me?" she said, reaching out her hands. "I'm lost." Her face was turned toward me, but her deep blue eyes didn't quite focus.

I looked between her and Arlyn. "And Terith's still asleep," I pointed out. "You would think a helpless little girl like this would wake him up."

"So young to be so cynical."

I shook my head, matching his smile. "Do you recognize her?"

"No, should I? I'd have thought we were still too close to the Road for fetches to manifest."

Interesting that he knew that. "I have heard they have been increasing lately. For a generic fetch this one has great detail. Just look at the dirt in its nail beds, the seaweed and algae in its hair." *A tide pool*, I thought. That's what it smelled like. For a moment the leaves looked like seaweed, but when I looked again they were just leaves. "It must have been someone real. Perhaps the last person who fell for it." I pushed myself to my feet, dusting off my hands on my trousers. As I took a step closer to the wards, the fetch held out its hands to me, looked over its shoulder, squeaked, and ran off into the dark. I found myself rather pleased with it for maintaining form and substance. Most don't have power enough for that level of verisimilitude.

I was not at all pleased by its next attempt. A noise brought us round to the downhill side of our camp. This time we saw a young man, dressed in apprentice practitioner's gray, with the white collar and cuffs and the little black cap everyone had to wear. He was brown-eyed, with dark hair and a very straight, very thin nose. His lips were also thin, but wide for his face. Perfect in every detail, right down to the button hanging crookedly on the front of his jacket. I had sewn that button on myself. I drew in a sharp breath and then pressed my lips tight together.

"Fenra?" The voice was deeper than I remembered.

"Who's it supposed to be?" Arlyn approached the wards, head tilted to one side. "How does it know you?"

"Not so close," I warned him. He stopped, but stayed within touching distance of the invisible barrier. "We studied together."

"Fen, Fen, it's me. It's Hal."

I sighed. I had never liked being called "Fen." The way some people said it you just knew they meant "swamp" and not "marsh." I took a deep breath and flexed my fingers, turning my rings around and rubbing my palms together.

"No, Fen, please. You can't! You have to help me, you don't know what it's been like." The fetch stepped close enough to brush the edge of the ward circle and it tolled like a deep but distant bell.

"Fenra . . ." Arlyn looked between me and the fetch.

I held up my index finger and he nodded, stepping back. I put the tips of my fingers together and drew them apart slowly. Arlyn's eyes narrowed, as if focused on the thin lines of rose light connecting my fingertips.

"No, Fen! Fen, let me in, you can't leave me here. You can't! Fen!" As I paid no attention, keeping my eyes on the light between my hands, the fetch stopped pleading and started to threaten.

"You did this to me, you bitch! Saved yourself and left me to fall. This is all your fault. Help me! Let me in and I'll forgive you."

"Be gone." I made a flicking motion with my fingers, and the rose light jumped like sparks through the wards, showering over the fetch and dissolving it. The voice lasted longer than the body, so the words "help me" and "your fault" echoed in the darkness after the image faded away. I took a final deep breath and dusted my hands off. "Its mistake was to use someone I knew was dead."

"Didn't that bother you?" Arlyn tried to control his tone, but I heard judgment underlying the curiosity.

I waited for my heart to slow down and my breathing to regulate before I turned to face him. "Of course it bothers me." My voice felt rough so I cleared my throat. "That's the point. To bother me enough that I do something stupid and let it in. It certainly wasn't here to make me happy."

"What did it think was your fault?"

I raised my left eyebrow.

"I withdraw the question."

ARLYN

Fenra was quiet the next morning, evidently turning something over in her mind. She'd respond if I spoke to her, but she volunteered nothing. Even the horse looked at her sideways once or twice, nudged her in the back with its long nose as she went around our camp releasing the wards.

So I let her be. If she needed to process what had happened in the night, who was I to interfere?

Late that afternoon, when we'd reached a proper shelter, she stopped and let herself down from the horse's back with the clear intention of going no further.

"There's an inn just ahead," I said, before she could start unpacking. "If we keep going, we'll get there before dark."

She looked at me a little as though she was sorting beans and had come upon a pea. "I had not realized we were so close to the edge of the Mode."

I'd heard that flat tone in my own voice. I swallowed. "*You're* not low, are you?"

She closed her eyes, shook her head, sighed. "Just a little sad," she said. "So you can relax."

Sadness, unhappiness, they go away by themselves when what's causing them is dealt with, is what she meant. Lowness is an illness, not a state of mind. It has to be treated, like any other ailment. Something I know very well. That didn't change the shot of fear that rose into my throat. If Fenra was low, I wouldn't be able to level her, and then who would level me?

Yes, I know. Selfish. But there were bigger things than the two of us at stake.

"So, there's an inn?"

I smiled with relief. "Real food, cooked by someone who knows how. Hot water. Real beds." Her eyes narrowed, but she looked away into the distance, as if she hadn't actually heard what I'd said. The horse snorted. Fenra looked around as if she were waking up, and picked up the pack she'd taken off the beast, settled it again behind the saddle.

We reached the inn with the sun low, but still in the sky. Fenra went with the horse to make sure the stable kids took care of him properly, while I bargained with the innkeeper. I thought about asking for two rooms, but I decided to save our money, even though he changed my letter of credit without questions. When she saw the room, Fenra didn't even blink, she just chose the bed closer to the window. I remembered the fetches and didn't argue with her.

Sometimes you can be more private in a crowd than you can be when there's only you, your companion, and a horse. Something about being alone in a noisy group makes it easier to talk about certain things. As if the presence of others will keep matters from getting out of hand.

"What was his name? Hal?" We sat in a corner table, small enough that we'd be left to ourselves. We had bowls of lamb stew in front of us, full of big chunks of carrot, and potato, and I'm pretty sure some salsify. Two big slabs of bread came along with the stew, dense, heavy with grains and wheatberries. "Beer bread" the landlord had called it. A little heavy for a supper, but then, we hadn't had much in the way of dinner. Fenra swallowed, put down her spoon, picked up her bread, tore a piece off.

"Halkutniarabol, actually," she said. "But no one called him that more than once, not even the instructors. Just as no one called me 'Fen' a second time except him, and then only when he was trying to rile me."

"From the Solni Desert, judging by the name."

She stopped dunking her bread in the stew and looked up at me.

"That's right. I was the only other one from so far away, so they roomed us together."

"Where were you from?"

She bit off the soaked end of the bread, chewed and swallowed. "You have never asked me that before."

I put my spoon down on the tabletop, drummed the fingers of my practitioner's hand. "In the village, our pasts were never important. Now something from mine is taking me—us—to the City, and something from yours tried to kill us last night."

"I am from Ibania."

"Merchant's daughter? Tradespeople?"

"Landowners."

"*Landowners*? How did you manage to escape marriage? Persuade them to let you go to the City?" It's only by going to the City that you find out you're a practitioner. If you never go, you never find out.

She pushed her right leg out from under the table. "It's hard to get someone to marry a cripple. What if it gets passed on? It's not like I was the heir."

"And you never got it fixed . . . ?"

"They *sent* me to get it fixed. They sent me to the City, where I was welcomed into the White Court."

"But your leg—"

"The Court told my escorts that it couldn't be fixed, and they were sent home to my father. Under the circumstances, my family was happy that I had my future provided for by someone else."

"You mean, if you'd had the leg fixed you might have had to go home?" You didn't *have* to become a practitioner, though it meant, among other, less cheerful things, never traveling again.

"And marry someone for the alliance, for the family."

"But you couldn't go back to that, not once you knew you were a practitioner." Not after being on the Road had shown her what she was.

"Exactly." She lowered her eyes to her food, dug in with her spoon.

I changed the subject to the only other one I could think of. "So what happened with Hal, then? How did the fetch know to use him?"

She tilted her head, looked at me sideways. "As I said. It was stupid to use someone I knew to be dead. And to imply that he was somehow in some sort of torment was even more ridiculous."

"How so?"

Fenra turned her spoon over and over in the stew. Finally her fingers stilled, but she didn't let go of the spoon as she sat back. "We were rock climbing. That was another thing that set us apart from the other apprentices. Rock climbing isn't done in the City, not for fun, anyway. But it was where we came from, even though we came from different provinces."

"He fell?"

"Or I dropped him is another way of looking at it."

I almost reached across the table to put my fingers on the back of her hand, but I thought better of it. "Tell me, then I'll know how to look at it."

"We were not supposed to be out that day, the day before an important test. We should have been studying. Fact was, we had done all the studying we needed to do. Throwing a glamour wasn't one of Hal's strong points, and all the studying in the world wasn't going to make him better at it."

"And you?"

"It *was* one of my strong points, so studying wasn't going to make me any better at it either." I waited until she picked out a bit of carrot, looked at it, and set it back in the bowl. "We had climbed this particular face before, so perhaps we were not as careful as we should have been." Here she paused again. "Or rather, there was no 'perhaps' about it. We were in a rush to get down. We were not late, exactly, but if we stayed out much longer we would miss supper, and be caught. We tried to take a shortcut."

"And he fell."

She moved her head up and down. "I was going first, and I remember him laughing, saying I would be there to cushion his fall. About a third of the way down, he slipped somehow, and started to slide past me. I managed to catch hold of his sleeve—our uniforms were made out of this really tough fiber, meant to withstand anything an apprentice could throw at it."

"I remember."

"We managed somehow to link hands—"

"Wrists would have been better. Sorry, sorry, I won't interrupt again." You could have frozen a skating pond with the look she'd given me.

"Linking hands was the best we could do, and lucky to do that much. I had a good solid grip with the fingers of my right hand, but the crevice where I had my toes crammed in wasn't really large enough, and my left foot kept slipping. Enough for a quick hold as I was climbing, but not enough to carry any real weight, for any real time. Then my right leg began to twitch." She looked at it. "I was gathering my strength, but you know, while you are still learning how to practice, you can be easily distracted, and every time my leg twitched I lost focus." She paused, but this time I knew better than to say anything. "When I slipped the third time, he let go of my hand."

I waited again, but clearly she was finished. "So, not your fault."

"Most definitely not my fault." She lifted her eyes to me without moving her head.

"A tragedy," I said. "He sacrificed himself so that you wouldn't fall."

"He did. But the real tragedy is that if he had been a bit more patient, we would both be alive."

"But he couldn't know that."

"No. That's what makes him the hero of this story." She smiled into her mug of ale, as if she saw something there. "To impatience, the leading cause of heroism." She picked up her spoon again and finished eating her stew, even though it had gone cold.

FENRA

I had expected to feel self-conscious the next morning, but luckily Arlyn gave me no sympathetic looks, asked me no solicitous questions. The courtyard was fairly crowded, and a courier rode through while we were waiting for them to bring us Terith. He would change horses here, and eat in the saddle. And sleep there as well.

Arlyn gave me a leg up on Terith's back without any sign of extra care or gentleness. I saw no change in stirrups, saddle, bridle, or reins, but then again, I hardly expected to. These were among the items that didn't change much from Mode to Mode. Medlyn Tierell theorized that once there had been a Mode where horse furniture didn't exist, or at any rate was more primitive, but if he was right, it had to have been long before our time. According to another instructor, there hadn't been any major changes since records were made of such things.

Once we were on our way, it struck me that Arlyn must have lived alone for a long time. He was not much good at keeping his emotions off his face, and not nearly as careful to govern his reactions as most. For example, we left the inn walking on a rough-cobbled Road, where yesterday it had been nothing more than pressed earth with the occasional support of logs set into low areas. Ordinary people—mundanes—do not notice these changes.

It was clear that Arlyn not only noticed, but that he planned to say nothing about it. He looked at the paving with the kind of abstracted half smile you see on someone's face when they return to their home village for the first time in years, pleased to see and recognize even something unremarkable. He really saw it, no doubt of that, yet he said nothing.

I came to the obvious conclusion.

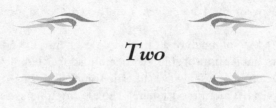

Two

ARLYN

THIS MORNING FENRA'S clothing changed to the White Court's tall black boots, buttercup yellow trousers with two rows of buttons in front, white shirt and cravat, crimson waistcoat, and black frock coat with crimson reverses. Rich green stones in her ears where she'd had silver knots before. The very model of a City practitioner, down to the gray gloves, the silver-headed stick, the black, flat-crowned, curly-brimmed hat.

Even her hair was behaving itself.

I offered her the reins once we were both seated in the barouche. She refused with a polite smile and settled herself into her corner, spreading the lap robe over both of us.

"Have I got something on my face?" I asked her once we were a good piece away from the inn. I know the horse doesn't change, but somehow the beast seemed sleeker, livelier.

"No." Fenra took a deep breath, shifted in the seat until she could face me more easily. She had a look in her eye that promised me nothing good. "I have been watching you react to things."

"It's been a long time since I traveled, I can't enjoy the scenery?"

Her expression didn't change. "Ten days ago you smiled at drainage ditches. Yesterday you smiled at gutters. No one smiles at gutters. You looked at my hat this morning as though at a long lost friend. I think you see what I see. I think you are a practitioner."

Her tone was dry enough I had to lick my lips to answer. "If I am, why don't my clothes change?"

"If you are not, how do you know that clothing changes?" She folded her hands in her lap, tapped her thumbs together, like a teacher waiting for an answer from a backward pupil.

"My cousin might have given me the gift of a practitioner's sight."

She snorted. "Not much of a gift if you never travel. And yes, I did think of that, but it's improbable. As the philosopher Jennock says, the simpler solution is almost always the true solution."

"It's simpler if I'm a practitioner?" I made my tone as sarcastic as possible.

She sighed again, turned her head away from me, spoke as if to the horse's rump.

"You knew there was an inn up ahead and that it had hot water. You knew where the Solni Desert is. You remembered the colors of the apprentice uniform. You knew that there was a time fetches didn't appear close to the Road. You knew that practitioners are discovered when they travel on the Road."

"My cousin could have told me all of that."

"You saw the lines of light between my fingers when I banished the fetch. No matter what your cousin might have told you, you would not have been able to actually see that. And you saw." She looked at me sideways, the brim of her hat casting a shadow over her eyes. "You *could* have been told many things, but you *saw* what you saw."

I had no answer.

"Was there any truth in what you told me? The highwayman and the cousin who rode to the rescue?"

"All of it was true."

"Just not for you."

"Just not for me."

"You are not the highwayman, you are the cousin."

"I'm the cousin."

"Why?"

I decided to misunderstand her. "He was a little older than I, funny, and smart and charming. And he'd been very good to me as a child. He was the only one in the family besides my mother who was pleased for me when they found I was a practitioner. The others were pleased with the advantage they thought it might bring the business." I glanced at her and she nodded. It didn't work that way, but people always thought their own case would be different.

"I loved him, but the others all said he would never amount to anything, that he was born to be hanged."

"And he was hanged?"

I nodded. "I got there too late. I was angry with the family, bitter, most of all because they'd been right." I gathered the reins into my practitioner's hand, rubbed at my face with my right. A gentle touch of her fingertips and Fenra took the reins from me.

"So you pretended you had been on time? That you had saved him?"

"In revenge on them. It sounds childish now."

"Not really. You loved him, and he loved you. The rest of them?" She made a flicking motion with the fingers of her right hand, not unlike the one she'd used against the fetch. "Much sillier things have been done for love."

"I did more. I kept him alive—at least in the eyes of the world. I pretended he'd reformed, set up his own small business in a far-off village, was doing well. I sent letters to him, pretended to get letters back. I even arranged for letters to be sent to the rest of the family, sometimes with money in them. Repayment of loans."

"So when you wanted to disappear, you had a ready-made life to step into."

I spread my hands. "Xandra the practitioner became Arlyn the carpenter."

FENRA

I pursed my lips. I was not so easily distracted. "You still have not answered my question. Why did you hide? Oh." My throat stiffened and made my voice flat. I felt stupid. "The lowness."

"The lowness," he agreed. "At first."

"A practitioner with lowness." I shook my head. "Between apathy, despair, and sudden rages, you could have destroyed so much . . ." I touched the back of his right wrist. "Is that why you tried to kill yourself?"

He sat back against the barouche's green leather upholstery and crossed his right leg over his left. "Turns out we can't."

Again, not an answer. "Practitioners *can* be killed." That, after all, was why I had been preparing to leave the village—in case things got out of hand.

"By others, yes. Apparently we can't kill ourselves."

I had certainly never heard of a practitioner committing suicide. On the other hand, I had never seen a case of the lowness before Arlyn. "You believed you would do less damage in an outer Mode? In a small village?" I glanced at him again but his color was no better and he had a bitter smile on his face. I remembered what he had been like when I first saw him. "You would still have had to be very careful."

"No, actually." He looked ahead, mouth in a straight line, eyes focused on Terith's ears. "I don't have any power." I must have made some protesting sound because he brought his gaze back to me. "The Godstone took it."

Something in his face, his tone, the whiteness of his scarred knuckles made me shiver as though our carriage passed through a mob of ghosts. "Is that the dangerous artifact you have hidden? A 'Godstone'?"

His nod was stiff, as if he also felt cold. "I tried to destroy it. I failed. Sealing it away took all my power."

And maybe a little more. "And if they open the seal, this dangerous thing will be loose?" Another nod. He still looked at me, but as if he did not see me. "Perhaps we will be lucky, perhaps the law of the Red Court will side with you and not the White Court." His eyes snapped into focus. I grinned. "It could happen."

"Well, we do have a secret weapon," he said. I barely felt the pat on the knee he gave me through the thickness of the lap rug. "We have you."

"Somehow I do not feel better."

Two days later I suggested that we stop at Last Inn before the City, rather than press on. We would have arrived before dark, but not much before, a bad time to be looking for lodging. You tend to take the first you find and that's almost always a mistake. Last Inn, while not up to City standards, looked comfortable and inviting, the flagstone area wide enough to allow the largest carriages to turn around. The windows were sizable, though paned with small pieces of glass. The stables were off to one side rather than behind the main building. This Mode is so narrow, I wondered if we would still have the barouche in the morning, or something more elegant still.

"Once we reach the City, *you* can stay in the White Court," Arlyn pointed out. I felt that thinking of him as Arlyn was safer.

"And leave you where? Do you know the City better than I do? Have you ever stayed anywhere other than the Court?"

He frowned, thinking. "Maybe, but it was so long ago . . ."

It must have been. Despite what I had said before, I *should* have heard of Xandra Albainil before Arlyn mentioned him. I had never seen the name on any of the lists, never heard him talked about. His was a story we would have studied, surely? If only for an object lesson on the dangers of overreaching.

How long ago did all this happen? Just how old was Arlyn?

"There we are then," I said aloud. "We will stay here for the night and go in fresh first thing in the morning. I am sure this landlord can suggest a good hotel."

We allowed ourselves to be helped down from the carriage and escorted into the inn while Terith was led away, glancing back at us before he was out of sight around the western corner of the building. We were given a double room, no doubt because of my status, but, oddly, my smiles weren't returned very warmly.

Once in our room, Arlyn swung open the window and stood watching the traffic on the Road. Hooves struck sharp sounds from the hard surface. He rubbed at his face before leaning forward, bracing his hands on the windowsill. Knowing what I knew now, it felt wrong to see him in fawn trousers, low tight-fitting boots, deep blue waistcoat and jacket over a pale lilac shirt and cravat.

"Is there any point in telling them the truth?" I wondered aloud, pulling out the room's one upholstered chair and sitting. "You are not dead, and if they cannot declare you dead, they cannot open your vault."

He looked at me out of the corner of his eye. I squared my shoulders and crossed my arms. "Well?"

At first I thought he would not answer. Finally he turned away from the window, straightening his cuffs. "Logic isn't a great defense when people in power want something. Someone wants to open my vault badly enough that they've gone to the trouble of creating false documents, having a relative tracked down and summoned," he pointed out. "Somehow I feel a small matter like my being alive won't stop them. Alive and powerless, remember. Besides, if there's a way for me to come out of this without having to reveal myself, I'd like to try."

I thought about his intention to return to the village. Could he do that if everyone knew he had been a practitioner? Would the White Court even allow it? "Wheels within wheels," I said. "They might very well decide that if you are not a practitioner anymore, your vault is no longer your property."

"I hope that wasn't meant to make me feel better."

The Road expanded as it entered the City, becoming a smoothly paved thoroughfare wide enough for three or possibly four carriages to move abreast of each other. Though there were now buildings all around us, the openness of the Road gave the illusion of space. The whole inner City was open squares surrounded by narrow, elegant houses, spacious thoroughfares, well-manicured parks. Before we were sent out on the Road, part of our training as apprentices involved patrolling and policing the City's darker places, where the poorer and less honest segments of society lived. Since the White Court took charge of it, everyone else—including the Red Court—could pretend these areas didn't exist at all.

Hal used to say that if we really wanted to deal with the less honest among our inhabitants, there were other, richer places to look.

"Were there gates in the City walls in your day?" I asked Arlyn as we clipped along the main street. Terith at least was enjoying himself, nodding at the other horses that passed. This early in the day, the few carriages were outnumbered by carts and wagons making deliveries. These vehicles were banned from the city center after midmorning. We were looking for Bridge Square, and the hotel recommended to us.

Arlyn looked sideways at me. "How old do you think I am?" His smile did not reach his eyes. Of course, it rarely did.

"I think you will tell me when you are ready." I gave him the same kind of smile back. "When I studied under my mentor, the Lorist Medlyn Tierell —"

"You never said your mentor was a lorist." Arlyn's tone was speculative.

"Senior Lorist, in fact. His books say that the City walls once had gates, with armed guards." Medlyn kept these books apart, in his own private library, for safety, though mundanes could not read them, any more than they could see the changes as they traveled the Road.

"Your earrings just flashed. We must have passed the perimeter."

I raised a hand to my ears but let it fall. After so many years, you can forget you are wearing them. The earrings mark a practitioner's class and position. We are given our first pair the day we are accepted to the White Court, our final pair on the day we graduate. The Court could keep track of us through them, and it's said that only death will allow them to be removed. When I apprenticed here, observation wasn't carefully kept, and Medlyn had told me this was still so. Now it was far more likely that off in some dusty old room in the least comfortable tower some slow or careless apprentice deserving of punishment was sitting, attending to the device that told the White Court who had entered the City and when. Hal and I had never been able to prove it, but we believed the reports were rarely if ever read.

I glanced at Arlyn's unadorned ears. *How had he managed to get them off?* Was the loss of his power a kind of death?

I have been a graduate long enough that I do not have to ask for permission to leave or enter the City, though there are quite a few old stick-in-the-muds who believe I should be required to report in. Technically, I suppose those rules still existed, but they hadn't been enforced in anyone's lifetime. Unless . . . I looked sideways at Arlyn again.

Ginglen Hotel stood exactly where we had been told to look for it, one of a row of three-story houses, facing into the narrow but sunlit Bridge Square, one of the smaller, less fashionable squares, in an area populated more by merchants and professionals, less by the monied and the noble. The hotel had a façade of dressed stone on the ground story, and stuccoed brick on the upper two. Traveling alone, I could have gone to a more fashionable address, or straight to the White Court itself, but Arlyn, while clearly a gentleman from his dress, was more limited in where he would be welcomed—and in what he could pay for. I remember my mother once saying that it was better to be slightly too good for your surroundings than the other way around.

As we drew up in front of the hotel an hostler's boy immediately ran out and took charge of Terith and the barouche, ready to lead them off around the corner to the mews that no doubt ran behind the houses. Terith gave me a snort and a roll of his eyes, but he knew what it meant to be in the City as well as I did.

The host was a small man so thin as to be skeletal. He was

scrupulously dressed, down to the contrasting buttons on jacket and waistcoat. Even the footman standing at the ready was carefully and neatly fitted out in shirt, waistcoat, breeches, hose, and buckled shoes. The hotel might not be on the best street, but the staff clearly wanted to attract the best clientele they could.

"I have a two-room suite available for immediate occupancy, Practitioner, or will you be going to the White Court?" The stiffness in his smile equaled that of his posture.

"We will take the suite, thank you, Dom . . . ?"

"Ginglen, Practitioner. Owner and host."

"Dom Ginglen. I am Practitioner Fenra Lowens and this is my client, Arlyn Albainil."

"Albainil? The cabinet maker?" This time the smile seemed more genuine. "Or no, apologies, that would have to have been your grandfather. I have an Albainil piece in my private drawing room. It's our family's pride and joy."

"I'd love to see it, when time allows." Arlyn bowed. *His grandfather, was it?*

"Of course, of course. In the meantime, Itzen here will take you to your suite. Any luggage?" He would know, of course, that practitioners traveled light.

"Just two cases in our carriage, if you could have them sent up?" Arlyn said. "Oh, and when it is convenient, I would like to have a message sent, if I may."

"Give your message to Itzen, he'll take it himself. And for now, if you will." He opened the ledger to another page. "If you would sign here, Practitioner, I will send the bill to the White Court." *Was I imagining it, or did his face stiffen on those last words?*

I drew a leather wallet no thicker than an envelope from my breast pocket. "That won't be necessary. While he is out on his errands, perhaps Dom Itzen would be kind enough to have this changed for me." I extracted my last letter of credit from the wallet and handed it over. Smiling, Ginglen passed it to Itzen without looking at it.

"Two bedrooms?" Arlyn murmured in my ear as we followed the footman up the stairs.

"More comfortable, don't you think?" I murmured back. After three weeks on the Road, he had to be as tired of my face as I was of his.

ARLYN

"What message must you send?" Fenra stabbed her walking stick into the stand provided for it, placed hat and gloves on the nearby table, untied her cravat. As I remembered them, a practitioner's clothes were always comfortable, but it had been a long time since Fenra had worn so many layers. Like me, she was more used to the gowns and tunics of the outer Mode. My hat joined hers, and I had my jacket off before she had even started on her buttons.

"I'd like to delay as much as we can," I said. Fenra carefully hung her jacket in the closet provided for it, and just as carefully left mine where it was. "I'll tell the advocate of the Red Court I've arrived, and ask for an appointment."

"You still have the letter?" she asked. She sat down on one of the narrow-legged chairs and stretched her legs out in front of her. At first I thought she was contemplating removing her boots, then I realized she was admiring them. This was a side of her I had never seen.

I had to pat all my pockets to find the one that held the letter. I almost missed it. It was paper now, not parchment, and it folded more closely than when I first received it. I placed it in Fenra's outstretched hand.

"You'll notice it lacks the White Court letterhead," I said.

"Then I agree." She tapped the folded letter on the leather-topped table. "Our message should be to the advocate of the Red and not the White. Let him—" She flicked the letter open again. "Yes, it's a him, let him explain things to us."

I leaned back against the table, considering whether to take off my own boots. "You can't come. Simple cabinet makers don't travel with practitioners, no matter what you told the host."

She drummed her fingers on the arm of her chair, frowning. "That may have been a mistake, though I cannot see what else I could have done if we wanted to stay together." She looked at me with raised brows. "Can we think of a reason for you to have a practitioner with you?"

I took the letter from her, tucked it away again. "I'll hold you in reserve," I said. "When the advocate tells me what this is about, I can call you in as my advisor. Until then, you're just a friend who accompanied me to the City, as the only person in the village who's been here."

"It would mean your going alone to the first meeting."

"I'm not without my own resources," I said.

"Welcome, Dom Albainil, welcome. I'm Advocate Lossingter, as I'm sure you must realize. Come in, come in. Can I get you a beverage? Jossen! A lemon drink for Dom Albainil. Ice?" he said, finally turning to look me in the eye.

"Certainly." I sat down in what was obviously the client's chair and let the fussy little man—evidently the advocate who'd written the letter—sit down on the other side of a hideously carved and paper-strewn desk. Much the same description could be used for the advocate himself, at least the hideous part. An uglier, more overdressed man I'd never seen. Even his hair wasn't natural, neither in color nor in abundance.

"It was very good of you to come so promptly, Dom, very good indeed." This opening was interrupted by the appearance of the clerk with a tray carrying a beautiful blue glass pitcher, two matching glasses, three lemons, and a bowl of ice. "Was the trip very trying?" He waved away the clerk, preparing the drinks himself.

"As comfortable as you can be when you're not at home." I used the broadest country accent I could, considering where I was supposed to have been living my whole life. Travel to the City wouldn't entirely remove all signs of a person's class, profession, or area of origin.

Lossingter stopped fussing with the lemons, added ice, and handed one of the glasses to me. "All the more reason to thank you for coming so promptly, and to move forward quickly so that you can return home."

He'd stopped being so twitchy once he had the lemon drink in his hand—meaning there was something unmentioned either in his drink, or in mine. We saluted each other, raising our glasses to eye level. I put the edge of my glass to my lips, but lowered it, drink untasted. Once I would have known immediately if the drink had been doctored. I began to regret not bringing Fenra with me.

"Tell me, how much do you know about your late cousin, Xandra Albainil?"

I rested my lemon drink on the small stand to the left of my chair.

The sweating glass would stain the dark wood if something wasn't done, though I wasn't offered a coaster. I pressed my lips together, reminding myself I was not at home. "'Cousin' is a nice loose word, isn't it?" I said. "I've heard the name, of course, but I never knew him personally." I shrugged. "I don't know of anyone in the family who did. May I see his testament? Did he name me specifically?"

"The testament itself is with the Red Court processor." Lossinger folded his hands on the desk in front of him. His own glass rested on the tray, half empty. "I did read it myself, however, and as I believe I told you in my letter, you are cited by name. Records were consulted. Under the circumstances, the White Court was happy to help with location difficulties, and here you are." His smile was as full as if he'd done all the work himself.

"I see. And what do I do now? There are instructions, I suppose? I'll have to see the testament myself eventually, won't I?"

"Oh yes." This time the smile didn't reach his eyes. "We'll give you every assistance, naturally. However—" Here he paused, pulling down his brows and tilting his head to show me how serious he was. "Things are more complicated than you might suppose."

No, they're more complicated than you *might suppose.* "How? Exactly?"

"Did you know that your distant kinsman was a practitioner?"

I trusted that my hesitation was so small as to go unnoticed. "Was *he* the one, then? I knew there'd been one in the family somewhere. At least, that's what my father always said."

"Your father was quite right. What most people don't realize is that the testaments of practitioners are handled a little differently from those of ordinary people."

I nodded to show I was following.

"Yes," he cleared his throat, "in the case of practitioners the White Court is always considered as a party in the testament, whether they are specifically named or not."

"Why would that be? Is it like a tax or something?" I did my best to look suspicious, and hard done by.

"Ah no, no, no. Nothing like that, I assure you. It's simply that any practitioner will have left scientific artifacts and appliances—machines, theorems—that can't be used by ordinary people. The White Court reserves to itself the right to take such property and examine, distribute, or even dispose of it as they find necessary. Everything else named in your kinsman's testament, any monies, real property, jewelry, and so forth, is yours to deal with as specified in the document." Now he leaned forward, his elbows on the desk, and spoke to me in a more intimate voice. "The fact is, my dear Dom Albainil, the residue of Xandra Albainil's estate will come to you." He raised his eyebrows in a suggestive manner. I looked as innocent as possible.

"Well, that's all right, then," I said. "When does this all happen?"

"I notified the White Court when I got your message yesterday and they will see you in their receiving hall tomorrow, at midmorning."

"Tomorrow? So quickly? Did Xandra leave lots of important scientific stuff, then?" My stomach knotted, my mouth dried. I took a tentative sip of my lemon drink.

"Oh no, you needn't be concerned about that. You must understand that Practitioner Albainil has been missing, and presumed dead of course, for quite some time. Fifty years at least, perhaps more. It's hard to tell with practitioners, isn't it?" He waited for me to nod again. "Whatever items he left behind would be so out-of-date that they'd only be of historical or academic interest to the White Court. Still, they have their rights."

"That makes sense." I picked up my gloves and stood. "I'll see you at midmorning tomorrow, then? Will we meet at the White Court gate?"

"Best you come here first, Dom. Best for you to appear with your advocate at your side. If you are here just before midmorning, I'll have a carriage waiting to take us over. It's not far."

Meaning you think they'll try to rob you of your fee if you don't deliver me yourself.

"Until tomorrow, then." The clerk showed me out.

⌒

FENRA

The moment Arlyn was out of sight I began to think I should have gone with him after all. I could not know what, or who, was waiting for him at the advocate's offices. But following him unnoticed would now be difficult. In my own clothing, any passerby would know me for a practitioner, and their interest might alert Arlyn. On the other hand, even if I could persuade the staff of the hotel to lend or sell me ordinary clothing, practitioners were required to wear their colors while in the City, and so my clothing would change, whether I liked it or not. Ordinarily I could see why, and ordinarily I would agree. But ordinary was not a good description of our circumstances.

Rather than sit and worry, I decided to attend the White Court and see my mentor, the Senior Lorist Medlyn Tierell. He would be surprised to see me in the City—usually he came to me—but I felt his knowledge and advice would be more than useful to us.

Much as I would have liked to have Terith with me, he had earned his rest, and I asked the footman who had brought up our baggage to send for a public conveyance. I did not want to walk all the way and arrive dusty and worn out.

As instructed, the carriage driver took me as close as he could get to the West Bridge. The White Court, or the Practitioners' Court as it's sometimes called, is made up of a number of interconnecting buildings, towers, laneways, arcades, courtyards, gardens, and patios that stand on a long, narrow promontory cut off from the rest of the City by the rivers Garro and Denil, accessible only by the West and East Bridges. Once or twice in the past there has been animosity between the two Courts, and the White's location made it easy to cut itself off from the rest of the City. Stories say that once the then-Headmaster had diverted the flow of the Garro to irrigate new gardens and fountains. When the City protested, both bridges were closed, and the City forgot, for a time, that the White Court even existed.

A mild drizzle had started while I was in the carriage, but as I walked across the West Bridge the sky cleared and the rain faded away. Except on very rare occasions, it only rains on the White Court at night.

I passed by dormitories and through study halls, threading my way along the narrow alleys, wider promenades, and open squares that made the White Court into a maze it can take months to learn. Almost everyone I passed wore apprentice gray, and almost all of them peered at me from the corners of their eyes, as if to see a practitioner was a rare thing. I expected at any moment to be approached, as any unrecognized practitioner would have been in my day. It seemed that apprentices were not as curious today as we were then.

Finally I arrived at the bulky five-story red stone tower that held the teaching staff's offices and, for some, their personal workrooms. I found Medlyn at his desk with a thick book open in front of him, exactly as he had been when I had last seen him in this room. Perhaps because it had been so long since I had seen him in his office, I saw for the first time how old he had become. Not all practitioners can stop their own aging, though most can slow it, and my professor was obviously now in the second category. In the past, many people at the Red Court have approached the White to have their aging stopped, or at least slowed. Official policy was that it wouldn't be done. Which didn't mean it *hadn't* been done once or twice when the two Courts had been on exceptionally good terms.

Medlyn's wards flashed as I stepped through them, and he looked up, his face glowing with delight when he recognized me. By the time he pushed himself to his feet, however, the look became troubled. Hands outstretched, he returned my kiss on his cheek with an embrace of his own. He waved me to his visitor's chair and resumed his own seat, taking his weight on his wrists as he lowered himself.

"You didn't say you were coming to court when I saw you last—not that I'm not pleased to see you—but this isn't the best time for visits to the City. Had you sent me a note, I would have been happy to come to you."

Though he held the position of Senior Lorist, Medlyn's passion was transportation and movement. I knew that he must have perfected at least one of the *forrans* he had been working on when I was still his apprentice, but unfortunately etiquette prevented me from asking for specific information. A practitioner's *forrans* were private until they were ready to be revealed to the Court and shared.

Behind him the door to his laboratory stood open, and I could see a number of familiar olive wood models of bridges and fountains sitting on the shelves. "Looks like you have even more models now—though they are a bit dusty."

Medlyn did not so much turn his head as his whole upper body. "Well, they come in handy."

I wondered if Medlyn was actually ill, rather than just getting old. I would watch for a chance to ask, though again it would go against etiquette.

"The problem that brings me didn't surface until after you were gone," I said. He smiled and nodded for me to continue. "I am not sure where to begin, but . . . what do you know of Xandra Albainil?"

He pulled his head back into the deep collar of his blazing white shirt. "Xandra? Xandra Albainil? That's a name I haven't heard since my own student days. How on earth did you come across it?" His hands trembled as he shut his book and set it to one side. His smile was exactly the same as it had ever been. "He was a notable scholar." Medlyn frowned without looking up. "It's true that when I've come across his name before—and that's rarely—it's always been in connection with some notable *forran* that's either known to be his or at least is attributed to him."

"Such as?"

"Well, he created the earrings, for one thing, that's known."

Which proved Arlyn was *older than he claimed to be.* I lowered my practitioner's hand from my earlobe.

"And the *forran* that makes the vaults possible, that's always been considered one of his—at least by those who bother to look into these things, which frankly, my dear, generally means me." Medlyn tapped on the desk top with the index finger of his right hand. "The reason I'd heard of him is that he'd spent quite a bit of time investigating alternative ways to travel, and I had a look at his work for a project of my own." He tilted his head in the direction of the open workshop door behind him. "I read once that he'd developed a *forran* that found another world, but I think that was just rumor. More important, for his time at least, Xandra Albainil was probably the White Court's greatest student of the nature of the world itself, its origins, and how the Modes function. He was one of the early supporters of the theory that it was practitioners who created the Modes—not the other way around—but I couldn't find that he'd proved it one way or another."

I remembered Arlyn's casual reference to a Maker and frowned.

"Now, are you ready to tell me why you're here, now, asking about a practitioner whose name has crossed no one else's lips in all of my lifetime?"

"It appears that a friend of mine is a distant relative—very distant, from what you are saying—and he has been named executor of Xandra's testament, as the practitioner has been declared dead."

Medlyn folded his hands across his stomach. Veins stood out on their backs, and his bony wrists poked out of cuffs that were suddenly too large. How had I not seen this before? "I've heard nothing of this," he said finally. "Though I must tell you that I've been neglecting my chair on the council lately, and focusing on my own concerns. I'll say one thing, however. Having a practitioner declared dead is not something entered into lightly. The mandatory wait from the last known contact is very long." He frowned. "Very long." He raised his eyes to me. "Even so, there is usually some highly significant reason prompting such an action."

"I suppose so," I said. He knew there was something I was not telling him, but he was honorable enough not to ask. Or he knew me well enough to know that, left to myself, I would tell him freely. Eventually. "But as I said, my friend, Arlyn Albainil, is summoned to attend to his relative's affairs."

"I would have given odds against their finding a kinsman, after so long a time."

"Well, he is not exactly—"

We were interrupted by a young woman in apprentice gray pushing a familiar high-wheeled, rosewood cart carrying two etched glass goblets and a tall crystal pitcher from which the most delicious smell escaped. I noticed, again as if no time had passed, that several leather-bound volumes of what Medlyn had always called the Practical Biographies were stacked on the lower shelf. I had not seen any signal from him, but I did not need to. Once upon a time that apprentice would have been me. I waited until the girl left, trying hard to disguise her interest in me, before taking up the pitcher. I noted with some amusement that what I poured into Medlyn's glass was different in scent and color from what I poured into mine. Apparently he had perfected another of the *forrans* he had been working on when I was a student. Medlyn waited until we both had our drinks in front of us before resuming our discussion.

"Of course they'd need a blood relation if they want to open Xandra's vault. I must warn you, I've seen the *forran* for this procedure, and it's very dangerous for the person involved. In fact, if that person isn't a practitioner, it's not advised."

I fidgeted in my seat. Suddenly it wasn't as comfortable as it had been a moment before. "I thought it might be something like that," I admitted. "How should I advise my friend?"

Medlyn took a careful sip and sat back. His eyes narrowed and I knew he was scanning his brain for all possibilities and consequences before answering my question. Including the consequence of answering the question.

His words when he finally spoke were exactly what I would have expected. "If it appears that something of great value could be in the vault, that would be one thing . . . Let us see what the archives tell us."

Obedient to his gesture, I set down my glass and fetched out the top book from the shelf under the cart—as always, it was the one needed at the moment. For years Medlyn had kept an archive of all practitioners, researching and adding new information whenever any was found. Though the lists were alphabetical, the size of each individual volume never changed, a feat maintained by a *forran* designed by Medlyn himself. He always said he would pass along the knowledge before he died.

I wondered if he had recently done so.

Medlyn held the book up between his two hands and let it fall open, as it always did, to the page he wanted. "Albainil's dates are given in the usual way," Medlyn said after reading for a while. "That is, in relation to other events of note. Without looking up the events themselves, all I can tell you quickly is that we are speaking of many generations of mundanes. Hmmm. *Forrans* of location." Medlyn looked up. "That would be the earrings."

As the movement of his finger reached the end of the entry, Medlyn grew even paler, and his hands trembled again. "There is a different hand here," he said, tapping the page with his index finger. "And yet it isn't different." He drew in his brows. "More as though it were added after many years."

"What does it say?" I asked, though I had a feeling I knew already.

"Apparently that he planned to locate a Godstone." I heard the capital letter in his tone. "In those days the White Court was more democratic than we are at present. If he'd wanted to do such a search now, he would have to ask permission of the High Council. In his time, the council was really just an informal group sharing news of their individual projects."

"Why? I mean, does he say why he wanted to find one?"

"His stated purpose was to take control of the Modes."

My entire body felt numb, and his next words seemed to come from far away.

"Unfortunately, it isn't clear whether he succeeded. This last entry indicates that he entered his vault one day and was never heard from again." He sat back and pressed the palms of his hands together, his frown creating a pattern of wrinkles around his eyes.

"Would it be dangerous?"

"My dear. An artifact that could control the Modes? Anything could be possible. He could have changed or even destroyed the world itself."

And the thing was still in his vault, I thought.

"There could be no good reason for anyone to want such a thing," I said aloud.

"If it exists," Medlyn said with a pointed change in tone. "If this is more than theory and Xandra Albainil *was* experimenting with the Modes, and if he decided of his own free will to seal the thing away and disappear, it isn't because the thing is harmless."

My old mentor's words hung between us as if suspended in the suddenly cold air. I folded my arms and tucked my frozen fingers into my armpits. "And my friend?" I said. "What should we do?"

"If he is indeed a friend," he said finally, still with that odd tone. "Tell him to go to the Red Court now. Immediately."

Strange advice from a practitioner. "Why the Red Court?"

"As things are between the Courts at the moment, the Red may be his best friend. In the event of a hearing, they would insist that the White Court be required to state exactly why they needed your friend—what they hoped to gain. If the White refused to cooperate, your friend would be excused outright. If they reveal their purpose, they would need to convince the Red Court that the technology—as they call it now—hidden away is worth the risk to your friend's life. In effect, they must agree to force him to take part. And that, given the current political climate, is extremely unlikely to happen."

Moving automatically, I took the archive back from Medlyn and returned it to the cart. I picked up my half-empty glass and found my drink still cold. "I wish I did not know any of this," I said.

"The first time you said that was the last time you sat where you're sitting now. When you told me you were going to leave before your exams for second class." For a moment Medlyn's smile took away all his years.

"You did not argue with me, as I recall."

"No," he said, his smile fading. "But the Court was becoming an unpleasant place, even then—fewer scholars, more politicians. You wouldn't have wanted to use your strength in the way you would have had to, to survive. And I wouldn't have wanted to watch you doing it. Better that you, and the others like you, should ride things out in safer, saner places."

"I'm not sure I have the strength to face what might be coming now."

"You have it, trust me." Medlyn took a small sip from his own glass.

ARLYN

When I stepped into our suite and found the rooms empty, I realized I'd expected Fenra to be waiting for me. Suddenly I felt like just leaving, walking out and never coming back. Fenra, for one, would be safer without me. But the reasons I couldn't do that hadn't changed. Even setting aside the lowness, I'd had to bring her. I needed her for more than keeping me level. I was almost certain that if I'd explained she would have agreed to help me anyway—she was a better person than I had ever been—but almost wasn't good enough.

I heard what were unmistakably her feet on the stairs. I stood to one side of the windows so as not to have the light behind me. She would have no trouble seeing my face. As the door swung open we both spoke at once.

"What's happened?"

"You first," she said, closing the door behind her and tossing her gloves and hat onto the table. Her face had a look I didn't remember seeing before, not even when the Ullios had brought her their mostly dead child. She turned the second chair at the dining table around and sat down facing me. She began to work off her left boot with the toe of her right, face squinching with the effort. Once it was off, Fenra held still, boot in hand, eyes lowered, listening to my summary of what I'd learned at the advocate's, finally nodding. Obviously something besides our present difficulties worried her. She sat up straighter and focused her eyes on me.

"Just what is the procedure to open another's vault? Why might it kill you if you were not a practitioner?"

"You don't have a vault of your own." It wasn't phrased as a question, but I waited for her to respond nonetheless.

"Never saw the need for one," she agreed. "My work has always been healing, and there isn't much secrecy or danger to that. Except for classwork, I have created no artifacts, accumulated no potions other than what I make or carry with me as needed. A vault would have been an unnecessary use of power." She pulled off her right boot, set it down next to the left, stretched her toes, flexed her feet. In the village she'd worn mostly sandals, like everyone else.

"You know the theory, however?"

"A little more than that. For our exams we had to create a vault, even if it was just a small one. Whether we could make one or not had some bearing on what class of practitioner we would join. Of course, it was a long time ago. I am not sure I could make one now."

Which told me that Fenra was not the class three practitioner she pretended to be, something I'd long suspected. It wasn't until class two that you needed to be able to create small vaults of the kind Fenra referred to. I wondered, not for the first time, just how old Fenra was. True, one of her professors was still living, but considering that the average lifespan of practitioners was indefinite, the old man's continued existence proved very little.

"So you remember then that there's no artifact, no external object used to seal the vault when it's created, though one can be made later as a key. The pattern comes out of the practitioner herself, created from the practitioner's own blood and bone, and only that same blood and bone can access the pattern. Without a key, there is no *forran*, no other object that anyone else can use to open the seal."

"And your vault has no key. So without the blood and bone of the practitioner himself, you can use the next best thing, a blood kinsman?" She rose, carrying her boots to place them outside the door, where the boot boy would pick them up for cleaning.

"A powerful enough practitioner can. I've seen it done successfully ... once." When I didn't elaborate, Fenra stopped, leaned against the closed door.

"Successfully in that the vault was opened but the kinsman did not survive? That's what Medlyn told me might happen," she added. "But it won't kill *you*, will it? Can they force you to take part? I mean, seeing as you no longer have power to protect yourself."

I didn't answer. This was the time for me to tell her what I needed her for. My teeth clenched. Fenra stayed leaning against the door of the suite, arms crossed, looking at me. From her face, I didn't have to tell her.

"Arlyn. I know what you said before, but I think we should leave. Medlyn said that if you go to the Red Court you can question the summons, make the White Court show their hand." She explained in more detail what her professor had advised.

"Perhaps we shouldn't have come," I said finally.

"It's not too late," she said. "We can still leave."

"You might be allowed to come and go as you please, but it may be too late for me. The advocate has already told the White Court I'm here, remember, made an appointment for tomorrow."

Fenra opened her mouth to protest, but she must have remembered that she was talking to another practitioner, and not an ordinary citizen. It's long been the White Court's policy not to let the people know exactly how much they can do. They didn't keep track of every citizen all the time, though that was due to lack of resources, not desire. They could, however, pinpoint any individual in the City and its immediate environs, and someone who didn't arrive for an appointment was definitely someone they would want to locate.

As for going to the Red Court, that could be a problem as well. A practitioner could easily find someone to do him the favor of handing us over, in exchange for some favor in return.

"I am coming with you tomorrow."

Now it was my turn to open my mouth to protest, and close it again without speaking. That was, after all, exactly why she was with me. "How do we explain your presence?"

"Simple." She went into her bedroom, emerged slippers in hand. "When you thought it was just the advocate you needed to see, your friend who accompanied you to the City minded her own business. Once you learned that the White Court was involved, you got nervous, and your friend assured you there was nothing to worry about, and came with you to prove it." She waited until I nodded before continuing. "I would not be expected to know what unsealing a vault involves," she pointed out. "I am only a third-class practitioner, remember." She kept her eyes fixed on mine until I nodded my agreement.

"So," she said. "Tell me about the Godstone."

Three

FENRA

THE WAY ADVOCATE Lossinger's mouth tightened made his unhappiness with my presence obvious. Unfortunately for him, no mundane had any say over the comings and goings of a practitioner, and besides, Arlyn wanted me there.

After as short a carriage ride as the advocate had promised we entered the Court by the East Bridge, the one set aside for mundanes. Unlike the West Bridge I had used the day before, this entrance had guards checking the business of everyone entering. Their blue tabards, with the old-fashioned cut, were reassuringly familiar.

For me they had no questions, only informal salutes. The others had to explain themselves.

"Advocate Lossinger and Dom Arlyn Albainil to meet with Practitioner Metenari."

When I heard the name, I wished I had skipped breakfast. Of all the people who had been apprentices with me, Santaron Metenari was my least favorite. I hovered in the background when we entered the conference room in the North Tower. I noticed his jacket was too tight, something you rarely see among practitioners. Like everyone else our clothes conform to our bodies as we travel from Mode to Mode—another element that mundanes do not notice. When not traveling, we buy new clothes or have ours altered like any mundane. From the tightness of his clothing, Metenari had not been out of the City since his weight gain and had not bothered to buy new. Now that I thought about it, his clothing had always been on the tight side. Even when we were apprentices, he somehow arranged to be courier for only one mouth instead of the usual six. His eyes flicked over me, surprise flashing over his face for an instant as he took in my clothing, before he dismissed me with a charming smile that didn't reach his eyes. I could tell he did not recognize me.

He should have. I had entered the White Court several years after Metenari, but I caught up to his level quickly, as he was among the stodgier of my fellow apprentices. As I traveled more, my own progression slowed, however, and he lost interest in me. Santaron had a keen political sense, and took little notice of classes inferior to his own. Hal used to say that no one knew better where the potions were poured.

"Welcome, welcome. You're Arlyn Albainil? Executor of the testament of the practitioner Xandra Albainil?"

Arlyn inclined his head in a shallow bow. "So they tell me."

"It's wonderful that you were able to come so promptly, Dom Albainil. I take it your advocate has explained the White Court's interest in this matter?" Metenari turned to me. "You're more than welcome to sit in the observation area, Practitioner."

"I am here as a friend of Dom Albainil's," I said, putting on as strong a country accent as I could. Instinct told me that if he actually did not remember me, I should keep it that way.

"Wonderful." He widened his eyes at me, but there was still no recognition in them. I kept my face straight.

"I knew nothing of this," Lossinger was quick to chime in.

"Practitioner Lowens is from my village," Arlyn said, as we had planned. "She was good enough to come with me when I told her I was coming to the City. I've never traveled so far before," he added, making himself sound even more of a bumpkin than I was pretending to be. He'd even pronounced my name as strangely as he could, LoW-ENSS, and Metenari still did not react. Though it suited my purpose, I had to admit it annoyed me to be so easily forgotten.

When we were finally all sitting at the round table in the center of the room, Metenari offered refreshments, but in such an abstracted way that no one accepted, though obviously the advocate wanted to. His face bright with enthusiasm, Metenari opened a folder, much stained, full of papers which crackled with age. I sat up straight. These were clearly not documents known to Medlyn Tierell, and I studied them as carefully as I could while feigning disinterest. My old mentor would be interested in as much detail as I could get him. I could see a faint haze of purple, as if someone had shaken open lilac blooms over the paper. So the papers were so old they required preservation. I resisted the urge to look at Arlyn. Just how old was *he*?

"Advocate Lossinger has explained that we need your assistance to open your kinsman's vault, though we do not need your permission." Metenari's musical tones made this a statement, not a question, but Arlyn answered it anyway.

"Yes, that's right, but my friend here tells me that it's dangerous for me, what you're suggesting. I know you've your rights and all, but does that mean I should put myself at risk? I mean, I have rights too, you know."

Two little red spots appeared in Metenari's cheeks, and he took a deep, calming breath, his lips quirking into a stiff smile.

"Let me assure you, Dom Albainil, that with all due respect to your friend's learning—" I could almost see the words "village practitioner" passing through his mind, "—there is no longer any danger associated with aiding the White Court in the opening of a practitioner relative's vault." Now his tone was one of a patient teacher instructing a politically important but backward student. "In the past that may well have been so, but our techniques and knowledge have moved quite a fair piece along the Road from those primitive days."

It was on the tip of my tongue to ask him how many such unsealings had been done with the "new" techniques, but caution prevailed.

"And let me assure you further of the utmost importance of what we are doing. You would be assisting in a project of the highest level of value to the entire world."

Arlyn was nodding, but his frown and wrinkled brow showed him perplexed. "I see," he said finally. "Though I still don't know how you think I can help you. I don't know anything about the practice."

"Ah." Metenari's tone was more than condescending. "You let me worry about that, Dom. You may know something you're not even aware you know."

"If you say so. Still . . ." Arlyn twisted his mouth to one side and tilted his head. "It's all so new—I'd like to think it over some."

"If you think that's wise, of course." Again his tone was gentle, but I had seen Metenari's lips crimp in the corners, and thought that what he really wanted was to take this country lad by the neck and shake him into immediate agreement. "I must advise you, as a friend, not to take too long, however, Dom Albainil. As you pointed out, we all have our rights here, and I would be ready to take this matter to the Red Court if it should become necessary."

Advocate Lossingter perked up at the thought of more fees. I do not know who he thought was going to pay them.

"Ah, well, no," Arlyn said. "I'd just like a day or two to get used to the idea that I'll be participating in some scientific project. This is all so new to me."

"Naturally so. I look forward to hearing from you." He turned and nodded just enough to be polite. "Advocate. Practitioner."

I stood when Arlyn did, returned Metenari's nod without meeting his eye. Lossingter scrambled to follow us out.

ARLYN

It took us longer than it should have to get rid of the advocate. I could see signs of impatience in the tight muscle of Fenra's jaw. Finally we agreed to meet with him again the following day to "review our options." At that, he only left us at the East Gate because it was clear we weren't returning to our hotel, after all. Instead, Fenra linked her arm through mine and led me back into the maze of the White Court's buildings.

"You knew him," I said as we crossed through a corner of the rose gardens. "The plump practitioner." The corner of her mouth twitched, but she didn't smile.

"I hope it wasn't that obvious to him, though come to think of it, Metenari probably takes it for granted that other people recognize him." She slowed down and I matched my pace to hers. "From what I remember of him," Fenra said, knocking a loose piece of gravel to one side of the flagstone path with her toe, "and from his attitude this morning, this is his project." She waited, hand outstretched to the nose of a passing dog, until the elderly practitioner on the other end of the leash pulled him away with a sidewise look at our linked arms. "Do you believe what he had to say about the unsealing ceremony?"

"Not for a minute. I'm with your old mentor on this one. I don't doubt the White Court has made some advances since my time, but the fundamental rules of the practice don't change no matter how hard you work on them." I stuck my hands into the pockets of my trousers.

"So why did you agree? Are we not just delaying the inevitable?"

We entered a large rectangular courtyard almost filled by a long pool full of water lilies. The last time I'd seen this place, the water had been kept clear—this was called the Mirror Patio because it reflected the buildings around it. I thought she meant to walk through the patio, but instead she led me to one of the stone benches under the arcade, sitting back with a sigh and taking a deep breath.

"It's not just a delay," I told her, keeping my voice low. "I have a different plan entirely. We, or rather you, are going to reseal the vault. Then I can't be used to unlock it."

She went very still, paled as much as someone of her complexion could. Her nostrils flared, her right hand began to close. Her eyes fixed on the edge of the path.

"Reseal," she said finally, without looking at me. She rose to her feet. "Come with me." She started down another path without checking to see if I followed. I did, of course, but not without asking where we were going.

"To Medlyn Tierell. He can help us."

Naturally I recognized the building she led me to, the barred windows on the first floor, the wide treads of the staircase, even the worn spot on the tiled corner where over time the touch of fingers had worn away the glaze. Medlyn Tierell's office and workroom were on the third floor. My office had been on the sixth, the top floor. I liked to be up high.

I'd always avoided him when he made his visits to the village, and I admit to being a little shocked at how old Fenra's mentor looked. I'd never known any practitioner to show that much age. He sat looking at the door as we arrived, as if he'd been expecting us, the eyes bright and sparkling in his wrinkled face. His hand—his practitioner's hand—shook as he held it out to Fenra. His head, when he turned his smile to me, trembled only slightly less.

Once we were sitting down—me in the only visitor's chair, Fenra with one hip propped on the edge of the desk—Fenra spoke.

"Now, tell Medlyn what you told me."

"I want Fenra to reseal my vault."

"So, you found the Godstone after all, though the records don't say so?" I could tell from the surprised look Fenra gave him that she hadn't told Tierell who I really was. The old man was hard to fool. "Fenra is more powerful than she pretends, but is she powerful enough to lock the Godstone away?" Medlyn's voice had an undertone I didn't understand, at first.

"It took all of his own power to contain it," Fenra told him.

I shut my eyes, called myself every kind of fool. No wonder she'd brought me here—where she had support—before letting me go any further. I used to be able to explain things clearly. "Fenra, I'm so sorry." Which was worse, I wondered, that she believed I would happily use her—or anyone—to undo my mistake? Or that she was right?

She let her gray eyes focus on the top of the desk before lifting them again to mine. "What exactly are you sorry for?"

"This misunderstanding." I looked at them both, made sure I had their attention. "I never intended to replace my seal with yours, but to intertwine yours with mine, changing it enough that anyone trying to use only my pattern would fail."

Fenra turned to her old mentor. "Would that work?"

Tierell's eyes looked into the middle distance for a moment, and then he smiled. "Not only should it work, but it should be permanent." Fenra pulled her right leg up, propping her heel on the desk. "The vault can never be opened, since no one else will know there are two seals," the old man continued. "And all with no danger to you, since you will just be adding to the existing seal, not making it yourself."

I was relieved Tierell saw so quickly. "Once we have Fenra's seal in place as well as mine, nothing Metenari tries will work."

"So. To the practical," Fenra said, straightened to her feet. "I have only made one such seal in my life. All theory, no practice."

"Have no worries," Tierell said. "I remember examining you on that occasion, and I assure you you'll have no trouble when the time comes."

I thought so. Lack of practice might affect execution, but no practitioner above third class had a poor enough memory to forget how to do something once they'd learned how. And Fenra was not third class, or even second, whatever she pretended.

"Good." Fenra nodded again, and picked up her stick from where she'd laid it down on the desk. Then her brow furrowed. "Can we do this intertwining without undoing your seal?"

"Yes." I didn't voice my uncertainty aloud. If I was wrong, and our attempt to strengthen it opened my seal—my breathing speeded up, and I did my best to control it. I didn't want to face the Godstone again. I had a back-up plan, but I hoped I wouldn't have to try it.

~

FENRA

As we were preparing to leave, Medlyn came out from behind his desk and asked Arlyn to give us a moment in private. Arlyn eyed the closed office door skeptically.

"I'd rather not stand in the corridor. Do you mind if I . . ." He gestured toward Medlyn's workroom.

"Not at all, please."

Once the workroom door was closed behind him, Medlyn turned to me, gesturing for me to take the chair. He leaned back against the front of his desk and smiled at me, ankles and wrists crossed. "Tell me," he said. "What's troubling you?"

"You mean, besides the idea of the Godstone?"

"You knew about that before you came in. There's more now."

I took a deep breath and let it out slowly. "I am facing a harsh truth," I said. "Any rational being, most especially any practitioner, knows that a single person has no importance compared to the fate of the world . . . you just never really think that's going to apply to you."

Medlyn reached over and patted me on the shoulder. "When I was younger than you are now, I might have thought of this as a great adventure. Now? I have to say I'm more than a little relieved that I'm too old to do anything but give advice."

I took his hand in mine. I felt bone, and skin, and very little else. "Is there something I could do?"

His laugh brought a smile to my face. "Here when I've just told you I'd be relieved to miss this particular experience? Thank you, my dear, it's kind of you to offer. But I've reached the point where not even someone of your talent can do very much. It's come a little faster than I expected, but I've seen my days, and I'm ready." He patted my hand. "I'm glad you came, though I'm sorry for what brought you. I'm sorry . . ."

"At least come outside with us, just for a bit, for fresh air if nothing else. It's a month at least since you came to see me, have you left your rooms since then?"

His face broke into a warm smile. "Oh my dear, you'd be surprised where I've been lately, and how I got there." He held up his hand. I thought I saw it trembling. "But not today," he added. "Though I thank you for asking."

"In that case." I stood. "I will take my final leave now, if I may. I do not know where my next days will take me."

He took my face in his hands. His fingers were like ice. "You were not my best pupil—though you came close! Very close. But you were my favorite." His eyes twinkled. "Be well, Practitioner Fenra Lowens, craft-daughter."

"Farewell, Lorist Medlyn Tierell, craft-father. I will." This would be the first and last time we used those words to each other. A relationship always known, and never before acknowledged.

I was at the workroom door, my hand on the knob, when his soft voice called me back.

"I almost forgot. I must have known you were coming today, my dear. I brought this with me. It would have found its way to you, after, but since you *are* here, take it with you now." He pulled at a chain around his neck, freeing a gold locket with a blue enameled design on the cover from under his shirt. He held it out to me. I stepped toward him, hand outstretched, and then I hesitated.

"I have no blood kin left, Fenra, no kinsman of any kind," he said. "With this, you will not need one."

The locket was still warm from his skin.

Luckily for me, Arlyn did not want to talk on the way back to the hotel. He must have seen that Medlyn's private words had affected me greatly. We both pretended to find the buildings we passed of overwhelming interest. Once back at Ginglen's, I let Arlyn go up to our rooms alone. I needed to see Terith. He would be missing me as much as I was missing him. I told him my news while I stroked his face, and he told me his, puffing air into my ears. He had been well looked after, I was happy to hear, though he suspected Dom Ginglen paid for first-class oats and received second-class. Suddenly Terith raised his head, and flicked his ears in the direction of the far end of the mews. Lifting my eyebrows, I followed his suggestion and went to find the source of the noise.

The mews was narrow, but long, and probably ran behind both the hotel and the house next to it. The cobbled yard was surrounded by a dressed stone wall, with a gate in the end nearest Terith's stall. I followed the narrow yard to the far end, where I pushed open the stall door. Three faces looked up at me, two startled adults with mouths open, and one young girl with lips pressed tight together to keep in the sound of pain. Pain I could do something about.

"Let me help," I said, stepping into the stall and crouching down next to where Dom Ginglen knelt at the girl's side, a footman I had never seen before on the other. Both adult faces were now impassive, though the footman was noticeably pale.

"Thank you, Practitioner, but it isn't necessary. We can manage."

"You are not managing, that is the point. Let me help."

"We can't pay," the footman blurted out.

"Did I ask for payment?" As gently as I could, considering my feeling of urgency, I took him by the shoulders and set him aside.

"Practitioners can't work without a permit," Ginglen said. "And we can't afford one."

I looked up from the injured girl's sweating face. "What? Since when?"

"Since four years ago come the equinox," he said. "We tried to make an appointment, but even the surety they asked for was more than we had. We tried to take care of her ourselves, but she's getting worse."

"Of course she is," I said. "An infection has started." I sighed and rubbed the palms of my hands together.

"They said," the footman offered tentatively. "They told us that in any case a new theorem would have to be written, and that would cost extra." I noted the new City term for a *forran*. Things had been heading in this direction even while I was still an apprentice.

"We tried to sell our Albainil, but the only one who offered to buy it wouldn't give enough."

"What next?" I shook my head. "This is why I stay away from the City." When I felt my hands had reached the temperature of the girl's skin, I set them carefully around her face. I felt the source of infection as a cold spot just under the skin, below her left armpit. A rib had cracked and a small sliver of bone had broken free and punctured the lung. Just a tiny puncture, but made slightly worse every time the girl took a breath.

"Sleep," I told her, and she did.

This was a garden, a little overgrown, as if the gardening staff had been on holiday since the start of the season. I walked quickly past what would normally be pretty borders, beds of flowers, and topiary animals of the kind that would delight a child, ponies, puppies, kittens, and rabbits. Then I found a place where a spiky-leaved ivy had overgrown its ornamental urn, poking a new tendril into a nearby rosebush.

Very subtle, I told myself.

As I was restoring the errant tendril to its own place and rearranging the disturbed rosebush, I heard something rustling behind me, but by the time I turned around I saw nothing moving. I smiled. Had the topiary bunny twitched its nose? I gave it a pat on the head as I fell back on my rump, grazing the palms of my hands on the cobblestones of the stall. My knees were jelly, and though my arms propped me up, I could feel my elbows shaking.

"Her color is better," the footman said, as he took the girl's hand. "She's my niece," he added, turning to me. "She just came to work with us last month. I don't know what I would have told my sister." He cleared his throat. "Thank you, Practitioner. Are you going to get into trouble for this?"

"Not if no one speaks of it."

"We were told the White Court would know if someone used the practice without permission. Some machine tells them."

If I had been stronger, I would have laughed out loud. "That they cannot do, I assure you. It's only something they say when they wish to frighten people into obeying. There's no one left who can create so complicated a machine as that." Maybe Arlyn could have written such a *forran*, when he was Xandra. "I am sorry things in the City have reached this state. It was different in my day."

"You say we don't have to pay you, but surely—we won't charge you for the room."

"It's not me you are charging, it is my companion."

"There must be something—"

"There is." I accepted his help in rising to my feet. My legs still felt rubbery, but a meal should take care of that. "Look after Terith." I waved at where he looked with interest through the opening in his own stall. "If something should happen to me, take care of him." I turned back with my hand on the stall door. "Oh, and by the way, you should check the quality of your oats. Your grain merchant may be cheating you."

We had only been in the City a little more than two days, and the change in Arlyn was noticeable. On the Road, he had been ready to follow my lead, relying on my more recent experience. This City version of Arlyn was more confident, assertive, even decisive. In the village, I had only seen him like this when he was beginning a new piece of furniture.

I felt I was starting to understand what might have caused Arlyn's lowness. A practitioner's power began with an unconscious inclination to travel, and grew, through study and practice, to a fully conscious awareness of internal power—an awareness that never faded. Having the awareness without the power would be more than enough to loosen Arlyn's grip on his self, and on the world around him. It had taken him a long time, exactly how long I was still afraid to ask, to reach the state of catatonia in which I had found him. I wondered how long it would have taken him to die of it. If, indeed, that was possible.

Back in our suite, I found Arlyn leaning over the dining table, braced on his hands. He must have begged some paper from one of the servants. The suite had come furnished with writing instruments and paper, but there hadn't been so large a piece as this. Arlyn straightened as I joined him at the table, rubbing his upper lip before resting his chin in his hand.

"I couldn't get any colored inks," he said. "Can you color these lines for me?"

I frowned at the paper. "Is this ink made from cuttlefish or charcoal?"

"Cuttlefish, why?"

"I do better with things that once lived—and do not talk to me about charcoal, the cuttlefish lived more recently."

I forgot how tired I was when I examined Arlyn's design more closely. Asymmetrical, and yet pleasing to the eye. I knew what it was, of course, I just hadn't thought Arlyn still capable of drawing one.

"What colors do you need?"

"This line here." He pointed to the line that, while weaving into and through the rest of the pattern, also formed an outer edge. "This should be gold. This one, lavender, this one sage, and this last periwinkle."

I raised my eyes to his face. "Rather sprightly, aren't they?"

He shrugged. "My focus is crystal," he said. "Only the prism colors come to me."

I nodded, took a deep breath, touched the tip of my left index finger to a spot on the gold line and released the air in my lungs as slowly as I could. There was linen in the paper, glue from bones, and that was enough to help me. For a moment I thought my focus hadn't been tight enough, but suddenly the line under my finger changed to the warm orange color practitioners call gold. The other colors came more easily as my confidence grew. I remembered, touching the locket under my shirt, that Medlyn used to say there's a reason what we do is called "practice." When the colors were all in place I took a step back, brushing the ends of my fingers together to clear the residue of power.

"Well." I coughed and cleared my throat. Arlyn handed me a glass of water from the pitcher on the sideboard. "How does it look?"

A shadow crossed over Arlyn's face. "I can't see the colors, you'll have to tell me."

I opened my mouth to respond—and got to a chair just fast enough to save myself from hitting the floor. Arlyn pushed a plate of biscuits within my reach and poured me out a cup of chocolate from a pot covered by an insulating hat. As I crammed a biscuit into my mouth, he took my practitioner's hand and began massaging it. I could feel the roughness and the calluses on his carpenter's fingers.

"What will this do?" I asked, indicating the paper with a nod.

"When we reach the vault, you'll use it to call up my seal without having to touch it," he said, giving the heel of my hand a final rub with his thumbs. "This is the pattern you need to look for."

My breath stopped in my throat. His pattern. Arlyn's pattern. It was as though he put his beating heart into my hand.

"How do you know I can use it?" I had a sudden image of what would happen if the pattern failed just at the wrong moment.

For an answer, Arlyn indicated the sheet of paper. "You were able to add the necessary colors to this version, so obviously you can interact with it. What?" he added when I did not respond.

"I am thinking how much I could have learned from you over the years of our acquaintance if I had known who and what you were."

He went still as a statue for a long moment before answering. "You might have learned how to be so sure of yourself that you endangered the whole world."

⟡

ARLYN

We passed through the arched stone opening leading to the enclosed space known as the Headmaster's gardens. I wondered who was Head now, and whether the position was the same as when I had turned it down. I looked around with interest, recognizing very little except the pattern of the original flagstone paths, grown uneven and mossy under trees and shady areas. The oaks I remembered had been replaced with fruit trees, apples, peaches, oranges, medlars, pomegranates, and mulberries—most of which wouldn't bear in the same season outside the White Court. New gravel walkways led off toward the distant buildings. Long shadows caused by the setting sun.

We mingled easily with the students still using the gardens. I walked with my hands clasped behind my back, while Fenra used her silver-headed stick to point out plants and fruits of interest. There were several other people, including one group of seven recently arrived students on a guided tour, doing much the same thing. I got a couple of sideways glances, but otherwise no one took any notice of us.

We'd agreed that Medlyn Tierell's office would make the perfect place to wait until dark. It was obvious Fenra wanted to check on the old man, and I wouldn't mind a chance to ask about some of his wooden models, especially the ones I didn't recognize.

We entered a familiar courtyard through a tiled arch I'd never seen before. "Is the Singing Tower much changed?" I looked up and around until I saw the familiar façade against the darkening blue of the sky.

"Not when I was here last," Fenra said. "Too many offices and work-spaces in use for much renovation to be done." We walked by three splashing fountains, along another arcade, and through another arched opening, this time one I remembered, and finally to the ancient red stone building we had visited that morning.

When we reached the third-floor landing I took hold of Fenra's sleeve cuff with the fingers of my right hand. With my left I indicated the office's open door.

"No worries," she said, her teeth flashing white against her skin. "He always leaves it that way if he's alone." She rapped on the door's middle panel with one knuckle, pushed the door wider open at the same time.

"Medlyn . . ." Her rough voice died away. The room was emptier than empty. No carpets, no books, no artifacts, no lorist. The same furniture, but all polished and clean. And two boxes, I now saw, full of scrolls and books. I had hold of Fenra just above her elbow; I could tell she felt no touch at all. I increased the pressure of my fingers, gave her a little shake.

"Fenra."

"No, it's all right." She touched her cravat as if she was feeling for something under it. "I think I knew when he gave me—but I hoped I was wrong." She patted my hand and I let go of her.

"Do we have a new plan?"

"I think so." She pulled a locket out of her shirt by its chain and a light dawned. Tierell must have given it to her—I'd never seen it before. "If you agree," she said, "there is somewhere even safer for us to wait."

Fenra stood in the cleared space in front of the desk, and reached for my practitioner's hand with her right. She held the locket up to eye height in her practitioner's hand, rubbed it between her fingers. The enameled side would feel warm, the gold side cool. Finally she held it to her forehead, lowered it, manipulated the tiny release to open it, enameled side up.

At first nothing happened, but just as a frown began to wrinkle her brow, fog spilled slowly out of the open locket. I stood closer to her, careful not to squeeze her hand.

I saw the fog, I saw the pattern as it began to emerge. I saw no colors. None.

The pattern moved, expanded, settled around us. The fog disappeared. We were standing in an oval room paneled in a dark streaky wood. Wood paneling in an oval room? I went immediately to the wall to examine the joints. As I suspected, there were none.

"It feels like him." Fenra spun slowly around, chin lifted.

"What do you see?" I asked her. I could tell she was looking at more than bare walls.

"Books," she said. Her smile was sad as she stroked the air between two shelves. "You?"

"Oval room. Zebrawood paneling. Could use a little oil." I pointed. "Oak bookshelves, empty, section of paneling, bare, a couple of cupboards, and over there some chests."

She spun around on her heel. "Nothing more?" Her brow wrinkled, then her expression cleared. She took my practitioner's hand in hers. The room leaped into life around me.

She was right. Books, and not just books. Scrolls, tablets, loose papers covered with elegant writing, mechanisms I didn't understand but that my fingers itched to touch. Since she had me by the hand, I followed Fenra over to a shelf holding a series of wooden models of bridges, fountains, and other public structures. Some of these I'd seen before. Tierell must have moved them from his workroom before he faded. They must have been very precious to him. I wondered if he'd carved them himself, but I didn't want to ask. Fenra touched a bridge model with a strange smile on her face, and I knew she was lost in some memory. I pulled a book off the shelves, and when I let go of Fenra's hand, I could still see it.

Two hours later I put the book back on the shelf, and it and every other artifact in the room disappeared.

"It's time, I think."

She pulled the locket out.

FENRA

I tucked the locket back into my shirt and shifted the chain until it settled in place. My heart felt unexpectedly light. I thought I had lost him, but it seemed I could visit Medlyn whenever I wanted. As long as the world didn't end first.

"So, how does it work? Can we access the seal from here?" I found myself reluctant to leave; Medlyn's old office, even without him in it, felt so much like home.

"There's too much of your old professor left in this room." Arlyn frowned, the muscles in his face standing out under the skin. "Too much of him, and too many layers," he said finally. "The pattern is designed for me, or rather, as it's my pattern, it *is* me. I can—I could— access my seal from anywhere. But from my mind to the paper, from the paper filtered through you—"

"Too many layers," I finished for him. "Too much between me and the workshop. As though we were searching for a room on the wrong floor," I added, remembering an analogy Medlyn had used once in class. "So what now?"

"Now we remove a layer," he said. I raised my left eyebrow. I knew that irritated him. "We go to the right floor—or rather, dropping the metaphor, the right tower. Follow me."

More decisive, I thought, which had to be a good thing. It had been a while since I had leveled him, and indecision is a symptom of lowness.

Arlyn led me out of the building, heading down the narrow alley that ran along the eastern wall. I could hear the soft sound of the Denil River on the other side of the stone. Following the wall brought us to the rear entrance of the Singing Tower, the oldest building on White Court grounds. The wall in this rear section of the building was made up of blocks of stone as large as a riding horse. According to what we had been told as apprentices, this was all that remained of the outer wall of a watchtower, used as part of the original White Court buildings. There were other such remnants throughout the City, but none so large as this.

The stone staircase, solid despite showing the wear of centuries of passing feet, climbed up the inner side of this outer wall. Halfway to the next floor, between landings, Arlyn stopped, placing the palms of his hands flat against the stone.

"Lost?" I said.

"There's a door here." Arlyn frowned at the mortar between the fitted stones as if he saw mold.

"Arlyn." I heard the rise in the pitch of my voice and cleared my throat. "This is an exterior wall. There's nothing on the other side but air, and thick as the wall is, there's no room for a hidden passage."

"I tell you there is a door here." Arlyn's tone was gentler than his words. "See for yourself." He gestured with an open hand.

A door appeared in the wall. Not as though it had just been conjured into existence, but as though it had always been there. Not a secret door, made up of the stones themselves, but an ordinary, sturdy, raised-paneled door with enormous hammered hinges and decorative brass bosses gone dark with age. I touched it with the tips of my fingers and felt wood, not stone. I shut my mouth.

"It's not there unless you know it's there," Arlyn said.

"So I see." I swallowed and squared my shoulders. "Is it locked?"

"It doesn't need to be. It's not there unless—"

"You know it's there. Yes, I understood you. But once you had told someone, did you not have to lock it?"

"I've only—I've never told anyone before," he said. I pretended not to notice the hesitation.

Weeks ago even this attempt to compliment me would have made me smile with pleasure, but now the cold feeling in the pit of my stomach only spread a little more. The latch felt as real as the door, cold, heavy iron, turned, bent, and smoothed by an expert hand. When I lifted it, the door swung inward of its own accord, as though counterweighted. I gasped before I could stop myself. I had thought this workroom would be like Arlyn's woodworking shop. Sawdust. Tools. Signs of work abandoned half finished.

Instead, this room was more like a library. Polished wooden floors, paneled walls, even a multi-colored, inlaid, coffered ceiling. Chests, cabinets, tables, chairs, benches to sit on and to work at. There were even bare bookshelves along one wall. I should not have been surprised; the Albainil family were woodworkers and furniture builders, and Arlyn certainly had the skill.

Astonishing as the look of the room was, the first thing I noticed on entering was how it felt. Medlyn Tierell's office, his workroom, even his vault had felt like him, his essence of calm and patience. His steadiness and warmth welcomed me from the moment of entry. This place felt harsh and cold, for all that it was so beautiful. I could not imagine anyone feeling welcome here.

"Do you feel it?" Arlyn was looking around the room with narrowed eyes. He stood stiff and upright, hands clenched into fists as though ready to defend himself against something only he could see. Finally he relaxed and turned to me. "What a terrible man Xandra Albainil must have been." I had never heard his voice so flat. "Do you feel it?"

"I feel a chill, and . . . impatience. That is you?" I also felt a sterile emptiness, but I could not say that to my friend.

"It's the man I was." Arlyn did not look at me when he said this. I am not sure what he would have seen on my face. True, I had never felt this from Arlyn himself, this impatience with lesser minds, this feeling that he could not be troubled to explain. The Arlyn I knew might have been withdrawn on occasion, but that was the lowness more than any-thing else. Children liked him, dogs liked him, even cats liked him. Could a person change so much as this?

Still, I felt nothing evil in this room, unless the coldness itself was some form of it.

⬷

"My dear boy, what are you doing here? You're supposed to be follow-ing Practitioner Lowens."

It wasn't often that Metenari felt like a fool. In fact, the last time had happened so long ago he wasn't sure he could date it. She hadn't been in his class long, he told himself. They'd both been apprenticed early. Too bad her promise had faded so quickly. Made it to second class by the skin of her teeth. Metenari himself always thought Lorist Tierell had pulled some strings. Not that it mattered in the larger scheme of things. She'd taken herself off to work as a healer. There were always a few apprentices who preferred the Road and the outer Modes to work-ing in the City. Metenari shivered at the thought. Necessary, of course, and if she had no better ambition, perhaps it was a good thing her power leveled off so early.

As soon as he remembered who she was, he'd sent out an alert. She wasn't so powerful in herself, perhaps, but for her to show up with Arlyn Albainil at such a crucial juncture could be more than coinci-dence. She knew *something*, that much was clear, or she would have introduced herself. He'd had them followed back to their inn, and his caution was doubly justified when they were observed re-entering White Court grounds. They'd paid a second visit to Medlyn Tierell—they obviously didn't know the old man had faded—but there was no telling what they'd learned from the Lorist in the first place.

And now here was Noxyn standing in front of him, panting from his run up the stairs.

Metenari set his cup of orange ginger tea on the purple hexagonal tile meant to keep his drinks from marking the top of the desk. "Well, explain yourself, man, why aren't you following them as ordered?"

"I've left Predax watching. I thought you'd need to know . . . they went through a door in the south stairwell of the Singing Tower."

Metenari sat up and leaned forward, feeling his spirits lift. "There is no door in the south stairwell." He spoke as evenly as if he were merely lecturing to a class of apprentices.

Noxyn looked triumphant. "No sir, there isn't. But the Albainil man told her there was, and then there was. I could see it, clear as sunlight."

Metenari fell back in his chair. He'd heard of this *forran*, he'd read about it in the same collection of old journals where he'd first found mention of the Godstone. "Tell me exactly what he said. As closely as you can to his own words."

Noxyn frowned. "He told her to see it, or asked her if she saw it, something like that. And then the door was there."

"And you saw it also. No, you needn't answer." Of course the boy had seen it. He'd been close enough to hear the words, so he was close enough to be within the influence of the *forran* triggered by those words. Just as Metenari had thought, the knowledge of the door and its secret must have been passed down through the Albainil family for generations, until it reached this Arlyn person. Some resonance, some remnant of the pattern common to the whole family would be enough to set the *forran* in motion, once Albainil was in the right place, and the right words were spoken. Metenari shivered. Imagine if the man had died without passing it along. Where would the world be then? As it was, there might now be a chance to save everything.

"What is it, Practitioner?"

"It's ingenious, is what it is," he said now. No use in frightening the boy. "Think of it, Noxyn. The ultimate security. No one can even see the door unless they are told it's there. No need for special locking *forrans*, no guards physical or practical. Just never tell anyone about the door and no one will ever know to look for it." *Genius, misguided certainly, but the man had been a real genius.*

"I've never heard of such a *forran*." Noxyn was visibly searching through his memories.

"You wouldn't have, my boy. This particular *forran* would be older than all of us. Perhaps it's in a book of deep lore we haven't found yet, though come to think of it . . ." Medlyn Tierell might have found it, and explained it to his favorite pupil when she came to ask him about it. When the summons had come to him, Arlyn Albainil must have confided in his friend the practitioner, and Fenra Lowens had come to the City to find out what she could get for herself. *Maybe Fenra's tired of the outer Modes and grown ambitious.*

"Those two know much more than they're telling." Metenari drummed his fingers on the tabletop, slowing as a chill cloud of horror fell over him. "Did you still see the door once they'd gone through?"

"Yes, Practitioner."

He closed his eyes in relief, pushed back his chair and got to his feet. "We must hurry; it may not be too late. They've no idea of the danger they're in." Fenra Lowens was barely a second-class practitioner. What had the lorist been thinking, sending her in there alone? The God-stone would eat her alive if he didn't get there in time.

—⌁—

ARLYN

"The vault is like the door," I told her. "You have to be told where it is."

"How on earth did Santy think he was going to open it?"

"Santy?"

"Santaron." Fenra lifted both eyebrows and shoulders. "It's what we used to call Metenari. He was such a fussy man. He reminded me so much of my older sister . . ."

"Even fussy men know valuable objects when they see them. Perhaps especially them." I couldn't stop looking around me. Everything was familiar, yet everything felt as though it belonged to someone else. Which was true, in a way. "If he discovered enough to know that I—that Xandra had left something valuable in his vault, then he might have deduced why no one knew where the vault was."

"If he doesn't know where, how could he use a relative?"

I looked at her out of the corner of my eye, but she wasn't looking at me, so my piercing glance was wasted. "He must have thought that I knew, or, failing that, that he could extract the location from the pattern."

"The pattern he assumes you have in common with your supposed cousin because of your pretended kinship?"

"Exactly."

"Uh-huh." She turned slowly, hands on hips, scanning the walls. "Where's this vault, then? The sooner we start, the sooner we can leave this place."

I turned slowly around, completely at a loss. "I don't know," I admitted finally. "I can't see it anymore. Here, take the pattern." I handed it to her. "Hold it up in front of your eyes and turn until you see it appear on one of the walls."

Fenra went to the section of wall closest to her, loosened the muscles in her shoulders, and held up my pattern. I realized watching her that she was bareheaded, as well as empty-handed. She must have left her hat, stick, and gloves in her old mentor's vault. I hadn't noticed until now. There was dust on the sleeve of her jacket, though I knew there was none in the room. Her boots were scuffed, her hair beginning to escape its braiding.

She had examined three-quarters of the room when she finally stopped.

"I see it," she said. A larger version of my pattern had appeared on the wall, as though projected. I knew Fenra would see the same colors she'd added to the paper version; again, all I saw were fuzzy lines like smeared charcoal.

"Odd," I said. "That's not where I thought it was."

"Good thing you did not tell me, then." She didn't take her eyes away from the pattern, but her voice hardened.

"Do you know you get sarcastic when you're angry?"

"I am not angry, I am afraid."

That silenced me for a minute. "I'm sorry, there's no other way."

"Which doesn't stop me from being frightened." She breathed out slowly through her nose, spread her feet a more comfortable distance apart. "Now what?"

"Reach through the pattern, use it as a template, touch the pattern on the wall, but as lightly as you possibly can. Careful that you don't push through it. Once you feel comfortable, draw a section of your own pattern on mine. Start in one of the spots where the lavender line crosses over the gold. It doesn't have to be complicated, just enough to change what's there."

She swallowed, loosened her cravat with her free hand. "I wish I had thought to take my jacket off."

"Do you need to? I could cut or perhaps tear—"

"No, I can manage. It's just now I feel too warm." The arm holding up the paper trembled slightly, but the hand Fenra reached through did not. I felt my muscles tense as the tip of her left index finger came closer and closer to the gray smudges that were all I could see. Fenra leaned forward, just a few degrees, and I caught myself just before speaking to her. She wasn't overbalancing, it was the sheet of paper that pulled her forward, as if it reached toward the larger pattern on the wall. There was nothing I could do now to help her, and plenty that could cause her harm.

I found I was leaning forward in sympathy. I straightened only to find I was leaning once again, my hand reaching out as if to take the paper. I could see her moving her finger ever so delicately, as if she lightly tickled the eyebrows of a cat. Her tongue came out of her mouth, pressed against her upper lip. Suddenly I saw the little girl Fenra had once been. She shifted her feet. I just had time to wonder whether her leg hurt her when the door to my workroom slammed open.

Before I could react, Metenari was in the room, calling to Fenra to stop, to be careful, that she didn't know what opening the seal would do. I jumped to intercept him just as he reached her and set his hands on her shoulders, inadvertently giving her the smallest and shallowest of pushes.

But it was enough. Pushed from her weaker side, Fenra stumbled forward, her outstretched practitioner's hand passed through the pattern she had been drawing on, the paper in her right hand swept into the larger version of itself, and the lock unsealed.

"Run!" Metenari called to someone over his shoulder, his hands wrapped around Fenra's upper arm, holding her back against the force that threatened to pull her through the opening. Two men I hadn't noticed were out the door and running before I could react. I reached Fenra just as she passed the length of her arm into the opening.

"Hold her! Tight as you can." Metenari freed his practitioner's hand and began making passes in the air. At first the forward pressure lessened, but instead of continuing to fade, we started being pulled to one side.

"No!" I shouted. "Back!"

"Can't." Metenari spoke through clenched teeth, sweat breaking out on his forehead. "Not all of us."

Of course not. "Gate," I shouted, jerking my head toward the right-hand side of the opening.

He looked at me startled, glanced above my shoulder, nodded, and changed the symbols he was drawing. Suddenly all sound stopped, the wind died down, and Fenra and I were standing on a small rise, overlooking a rocky plain. Fenra fell to her knees in front of me, gasping for air, coughing. I helped her sit up.

"He didn't see the gate until I told him it was there." I think I was giggling.

Fenra panted, her face gray, lips trembling. "You know where we are?"

I looked around me with a sinking heart. "I've been here before."

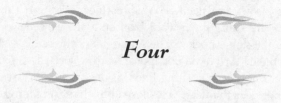

Four

FENRA

"I DID NOT SENSE him at all until he grabbed me." I rubbed the palms of my hands dry on my trousers before Arlyn could see them trembling. No one had ever been able to take me unaware, not even as a child. I sat on hot dry ground, left foot tucked under my right knee. Arlyn had his arm around my shoulders.

"You were focused on what you were doing, and rightly so." Arlyn withdrew his arm, moving his hand to my shoulder. "A mistake could have killed everyone—as you saw. I didn't notice him myself until it was too late."

"You could not expect to sense him," I said. His mouth twisted and he looked away. I regretted speaking. I shifted to get my legs under me and Arlyn stood up to offer me a hand. A slight dizziness, a sparkle to the air, and I felt balanced again, and able to look around me.

"Where are we?"

I had never seen country like this. Boulders as tall as two people, dry, sunbaked ground, little more than rocks and hard earth interspersed with the occasional stiff growth that might have been heather in a damper climate. This place hadn't seen rain in months, possibly years. I shrugged off my jacket, following it with my vest. My cravat I untied, using it to wrap up my hair in a loose turban—anything to get it off my neck. *Scrub desert*, I thought. I had read of it, though reading couldn't give you the feeling of air so dry your body's sweat evaporated as it formed. Still, there was life here. I could feel the presence of tiny and sensible beings sleeping until evening, when cooler air would bring them out.

Arlyn made no effort to remove his own jacket, though I could see a sheen of dampness disappearing from his upper lip.

"A parasol is what we need—" My breath caught in my throat. I patted myself. Silk shirt, linen-wool trousers with carved wooden buttons. Cuffs the same, pockets the same, boots the same. "My clothes haven't changed." Arlyn did not appear at all surprised. "We are in the same Mode. But there's nothing like this near the City."

"We're not near the City." He clasped his hands, tapping his upper lip with his extended index fingers, something I had never seen him do. A crease between his eyebrows made my heart beat faster. What worried him that much?

"Arlyn." I weighted my voice with all the authority I had in me.

He lowered his hands.

"When I first found this lovely spot, I thought it was a new Mode. In fact, I thought I'd found the secret of how the Modes come into being."

"What do you mean? The Modes have always been . . ." I remembered what Medlyn had told me; I remembered that I did not know Arlyn's age. "You are saying they haven't always been."

"No, there used to be more. Practitioners traveling on the Road used to find themselves in a brand-new Mode. Not all the time, but sometimes. I wanted to know how it happened, so I created a *forran* I thought would work." He waved around him. "This is what I found instead."

I checked the knots in my cravat with the tips of my fingers. "This isn't our world at all." I could not believe how normal my voice sounded. "What world is it, then?"

"They have their own name for it, of course, but we—I always called it the New Zone."

"That's it?" I licked my lips, feeling the moisture evaporate almost before it formed. "The mind that came up with the door that wasn't there couldn't think of a better name than 'New Zone'?"

Arlyn started to speak, stopped, and pressed both lips and eyes closed. Finally he turned to stare at the western horizon, as if he expected something.

I squinted. The same landscape as far as the eye could see. I could feel my shirt wet against my back. "So what now? You teach me the *forran* and we go back through the gate, or portal, or maybe you call it a 'door'?"

Arlyn shook his head, but clearly not in response to me. He looked around again, apparently did not see what he was looking for, and sat down cross-legged on the ground. He indicated the space in front of him. "You might want to sit."

At first the ground felt too hot to sit on, but after removing three sharp rocks from under me, I became reasonably comfortable.

Arlyn, wrists resting on his knees, looked as if the heat no longer affected him. "Don't roll up your sleeves," he warned me as I took hold of my right cuff. "The sun is strong enough to burn even your skin." He waited until I'd re-buttoned my cuffs before continuing. "We need to find another gate." Arlyn waved a hand in the air over his head. "One that you can use. This particular *forran* is one-use only."

"Wonderful. What genius thought that was a good idea?" Arlyn kept his eyes fixed on mine, but did not respond. "Ah, sorry." Not that I felt apologetic.

"Yes, well. It seemed like the right thing to do at the time." His tone was flatter than his words suggested. "That gate, and a few others, were meant to be used as bolt holes. I didn't want anyone and everyone to be able to follow me through."

"So Metenari can't follow us."

Arlyn shrugged. "Does he want to? He wanted the vault unsealed—"

"And now it's unsealed." *By accident.* I squeezed my eyes shut. "I do not think he meant to push us through the gate."

"Not at all. Though you have to admit we're neatly out of the way. With Lorist Tierell faded, there isn't even anyone to ask questions about us."

"Ginglen, at the hotel—" I began. *He would remember me*, I thought.

"If he does anything at all, he'll apply to the White Court because a practitioner might be missing, leaving a horse in his stable. I don't expect anything will come of that, do you?"

Except I had asked *them* to look after Terith. I began to speak, and stopped with the words unsaid. I examined Arlyn's face more closely. He looked at a rock to the left of his knee, focused, but not seeing it. I had seen that look before. The face of someone at a loss.

"Arlyn, give me your hands." I reached mine out to him. He flinched away from me and caught himself.

"You might need all your strength," he said, still withholding his hands. "When we find another gate, you'll have to get us out of here."

"My strength isn't all we will need. Without your knowledge, and

the focus that being level gives you, we might be trapped here forever."
He looked at me a long time before he finally put his hands in mine. I
closed my eyes.

Most often Arlyn is lost in a deep fog. A wind comes with me, or
brings me, I do not know which, and blows the fog away. At first just
tiny wisps of it, then more and more until I see Arlyn completely.
There are dark areas in the fog, and I wonder if it contains other lives,
other beings, lost perhaps, or simply living their lives in their own
time and place. I do not know what the fog is, or where it comes from,
or whose mind contains it, mine or his.

But this time the fog is dense, almost solid, as it is when Arlyn waits
too long to be leveled. The light, brisk breeze I expected to use would
do no good now. It would have to be closer to a gale. I struggle to call
the wind to me, but nothing happens. Just as I think I feel a slight movement of cool air, a dark rivulet in the fog reaches out for me, touching
me on the ankle the way a strange but friendly dog will sniff at you. I
hold still, as I would for the dog, but my heart pounds and my skin is
cold with sweat. A black terror washes over me. For a moment I am
frozen in place, then my lungs heave, and I suck in a great breath of air.
I can smell brine, and the sea. And then it is gone, and I am alone again.

This has never happened before. Could it be because we are in a new
world? I concentrate again, calling the wind, but cautiously, as delicately as I can. Finally it comes, bringing back with it the smell of the
sea, and my heart beats faster. But this time the wind behaves itself,
blowing away more and more of the fog, until I can see Arlyn sitting on
a rock jutting out of a sandy beach, the water just lapping at his toes.

The fog has gone.

When I opened my eyes Arlyn was looking at me. Whatever the
lowness is, it doesn't affect his appearance—at least, not until the point
where he does not care whether he bathes, or if his clothes are clean,
or his hair needs cutting. But when he is level, he carries himself differently, there is a spark of light in his eyes. I thought I would be able
to get a straight answer out of him now.

"How do we get out of here?"

"You know how you call animals to you? Do the same, but call for
help to come."

I blinked in surprise. Definitely not the answer I expected. "Help isn't an animal, Arlyn. It's more of an abstract idea. How do you suggest I call it?"

He shrugged. "How do you call animals? I mean an animal you can't see, but you know must be out there? Not every practitioner can do it."

That was true. Annoying, but true. "I visualize it."

"Well, help is out there. Visualize it."

I decided I almost preferred his listlessness to this know-it-all attitude. "I take it back. I could not have learned anything from you in the past few years."

"Humor me."

ARLYN

Fenra shut her eyes, laid her hands palms up on her thighs. It's a good thing her skin is so dark, or this sun would have been burning her already. She slowed her breathing down to where I couldn't be sure her lungs were still working. When I saw her shoulders lower, the muscles of her face relax, I asked her:

"Which way is help? Point to it."

I knew it was coming, but I still jumped when Fenra's arm shot out, pointed to her left and slightly behind her. Not the way I expected, but I trusted Fenra's power.

"Wake up, Fenra. Don't lose the direction of help, bring it back out with you. Wake up."

She took a deep breath and opened her eyes.

"Do you still feel it?"

Her teeth flashed white. "I do. I wonder what other abstract concepts I can find?"

"Experiment later, rescue first."

"Wonderful." Metenari passed his hands through air that a moment before had contained Fenra Lowens and the Albainil man. Or had it? "Check the outside room, is anyone there?"

"No, Practitioner, it's empty. What happened?"

"A dimensional gate, it has to be." Metenari grinned. "But how did the carpenter know it was there? That's the question."

"He knew about the vault," Noxyn pointed out. "And how to find the workroom."

"Correct." Metenari believed in giving credit where it was due. "It seems more knowledge was passed down in the Albainil family than I'd thought. More, in fact, and less corrupted than I would have believed possible considering the span of time." A careful look around the vault made plain that there were no chairs. Metenari spotted a low table. It would serve as a bench until he could have some chairs brought in. He was more tired than he'd expected. It seemed the gate *forran* used an unusual amount of power. He might set a couple of the junior apprentices to finding it. A gate could be very useful.

"Practitioner?" Noxyn's voice was tentative. "This place looks remarkably like the workroom." The young man gestured at the open doorway, kept open, Metenari felt sure, by his precaution of having Predax, his second apprentice, stand in it. "Though there's not as much open shelving. More drawers and closed cabinets." Noxyn was still talking, giving his observations as though this was a test or exam. Just as Metenari smiled at this thought, Noxyn reached out a finger, as if he meant to check the shelf in front of him for dust.

"Freeze!" For a moment Metenari thought he'd actually frozen him, then saw that Noxyn still breathed. With a hand to the boy's sleeve, he drew him away from the shelves. "Be careful, my boy. If we know one thing about this place, it's that there are likely to be other traps. If I had only been more careful, more suspicious, we would not have lost Fenra Lowens. I knew she wasn't up to the challenge of dealing with the *forran* of a master like Xandra Albainil."

"Was the dimensional gate a trap?" Noxyn looked back over his shoulder at Predax, standing with his shoulders hunched and his lips pressed tight together. The boy's nerves could use some work.

"Very likely. Why position it in the doorway otherwise?" Predax shifted his feet. Metenari gave him a reassuring smile. "Not to worry, my boy, I've neutralized that *forran*, for the moment, and we've nothing more to fear. No, the difficulty was that unsealing the vault as Fenra Lowens did, clumsily, almost by accident, she triggered the guarding mechanism, and the gate. If the unsealing had been left to me, there would have been no harm done. As it was, pushing them out of reach by shoving them through the dimensional gate was the best I could do. It was either that, or let them die right there as they stood."

"But you'll be able to get them back?" Predax said from his post at the doorway.

"Oh, undoubtedly. We'll continue our examinations of the old documents and I'm sure we'll find the *forran* we need. After all, we may need Arlyn Albainil again, who knows what other knowledge he has." Metenari rose to his feet. "Now, let's see what we have here." Standing in front of the nearest bank of drawers, he placed his hands together, fingertip to fingertip, and concentrated. Start with the simple. One of the first precepts of the White Court. He drew his hands apart until only the fingertips were touching. Even more slowly, he drew his fingertips apart, nodding at the lines of orange light stretching between his hands. He pointed his left index finger at the pull of the drawer.

"Hah! The pull resists even the greatest pressure I can bring to bear. I wouldn't have expected anything less." He clapped his hands and smiled. The smile vanished as he sat down again, much faster than he'd intended.

"Practitioner?"

Both apprentices took a step toward him and he held up his hand, this time taking no chances but using power to stop them in their tracks. Suddenly out of breath, he coughed and cleared his throat.

"Stay in the doorway, Predax," he said. "We wouldn't want it to close on us, would we?"

"But Practitioner . . ."

"Not to worry. Use of magic tires even one as experienced as I am. Learn that lesson early. Don't ignore the demands of the body."

Somewhere in here, he thought, in one of the drawers, on a shelf, or behind a cupboard door, he'd find the Godstone. But not today. He'd have to come back rested, and well fed. He'd need all his strength and energy. He pushed himself to his feet.

"Now, Noxyn, Predax." Both boys straightened up. "What is to be done about the opening? How would you leave it?"

Predax chewed on the inside of his lower lip, throwing glances at Noxyn. That was perfectly understandable. He was newer, and less experienced. As Metenari expected, Noxyn answered him.

"For the doorway to the workroom," he began, "we don't need to do anything. Now that we know it's there, we'll always be able to see it. As for the opening to the vault, it looks like we'll have to leave someone here to hold it open, at least until we've found a *forran* to neutralize the seal and the trap both."

"Good, good. Well done, Noxyn. Predax, I'm afraid that duty falls on you for now. I'll send Noxyn back with some food for you, and perhaps a scroll for you to study as you wait."

"Yes, Practitioner."

FENRA

We walked in the direction from which help was coming. Sometimes the practice is like that, it just comes to you, without patterns or *forrans*. Medlyn once said that just as every practitioner uses a particular focus point, every practitioner has a gift peculiar to them. He said the White Court didn't focus on individuality anymore, so it wasn't nurtured the way it once was.

I could not help thinking that if Xandra Albainil was the product of a focus on individuality, they were probably right to abandon it.

We were soon covered with dust. I looked down at my knee boots and grimaced. Not only were they filthy, but they were entirely the wrong footwear for walking on hot uneven ground. My feet and calves felt baked, and the soles were meant for City pavements, not this rough ground. Arlyn's boots were thicker-soled, and for the first time since I was issued them, I regretted my practitioner's colors. Black isn't the best choice when there's a strong sun overhead.

There were birds high above us. Circling.

"Vultures a bit premature, aren't they?"

I shaded my eyes and looked upward. I welcomed a reason to stand still, even for a moment. "Not vultures," I said. "Eagles."

"Eagles? What are they doing up there?"

"Floating."

Just as I began to think the afternoon would never end, Arlyn touched my arm and pointed. Neither of us had spoken for at least two miles. We could not afford to lose the moisture in our mouths. I narrowed my eyes and focused on the distance. A plume of dust. A wagon or cart at the least. Here was the help I had called to us. Without speaking a word, we both stopped.

"Sit down," Arlyn whispered.

"You," I said.

"Take turns."

I did not have the willpower to argue when I knew he was right. One of us would rest while the other remained standing as a target for the help that was on its way.

A wagon, not a cart, drawn by two horses looking bored. I had ample opportunity to examine them, as they did not stop moving until they had put noses on me, snuffling and blowing. Their breath smelled of hay, and was no hotter than the air around us. I ran my hands up their faces, stroked their necks, and gently tweaked their ears. They stopped looking bored.

The man holding the reins had his head tilted to one side as he watched my face and my hands. He was dressed in trousers so faded I could not tell the original color. His thin shirt might once have been white. He wore an old limp leather vest open over the shirt. The vest had embroidery on the front panels where lapels would be. He was so slim I could not be sure of his height. His skin color, though as dark as mine, was at least partly the result of the sun. White braids hung down out of his wide-brimmed black hat. His eyes were a very startling green, examining me as thoroughly as I was examining him.

"You're for the sheriff." It wasn't a question, but I nodded when Arlyn did. "Water here." He indicated the bed of the wagon. "One there, one up front."

"Why the sheriff? Are we trespassing?" Arlyn asked, holding me back. I almost bit him. I needed to sit down on something that wasn't the ground, and I swear I could smell the water.

The carter was shaking his head. "You're new. Sheriff sees every-one new."

"Fenra, you sit up front."

I do not know if I would have been able to climb onto the seat next to the carter if Arlyn had not been there to shove me from behind. Where he found the strength to hop into the back I will never know.

Once he had turned the horses around, the carter pointed to the bed of the cart with one bony thumb. "Water."

Arlyn searched a bit—there were burlap sacks filled with what smelled, of all things, like rock—before he found a water skin and passed it to me. I helped myself to two long swallows before offering it to our host.

"No need." He glanced at me with approval. I had done something right. I wondered what it was.

I passed the skin back to Arlyn. I knew I should not drink any more just now, but I carefully watched Arlyn take his own two swallows, keeping my eye on the water skin until he put it away behind the sack where he had found it.

"I'm Fenra Lowens," I said to the carter. "Thank you for picking us up."

"John," he said. He twitched the reins and the horses stopped looking over their shoulders at me and walked forward. "Coming this way. Felt you."

I knew that couldn't be the strict truth, but perhaps it felt that way to John. "This is Arlyn Albainil." I tilted my head behind me. John nodded, but did not turn around. We had turned into what it took me a while to recognize as a track—it certainly couldn't be called anything grander. I started to rub some of the tension out of my face. I could feel fine dust on my skin under my fingertips.

"Do you have a family name, or is it just John?" Arlyn asked from the back of the cart.

"John for work, Bearclaw for family."

"You mean Bearclaw is what your family calls you."

At this he turned his head to look at me. His smile showed even white teeth. "I like your hat," he said, glancing upward at the cravat rolled around my hair.

I could not help it, I started to laugh. Unfortunately it turned into a cough, and Arlyn had to pass the water forward again.

While I was drinking John looked upward, narrowing his eyes.

"Eagles," I said. "Not vultures."

I thought John would smile again. "I know."

Conversation, such as it was, died away after that. It was just too hot and too dry, even with frequent returns to the water skin. I kept offering it to John, but he always refused it. He held the reins loosely in his hands; clearly the horses knew where they were going. I began to worry how much farther they could go without water when I saw smudges on the horizon that eventually turned into a line of trees, and bushes, and flowers, and houses for them to surround. The track we had been using turned into something more like a road as we drove closer to town. There was gravel, there were gutters lined with rock, which meant the road was shaped to let rainfall run off—though as far as I could tell, there couldn't be enough rain here to warrant it.

"What is this town doing here in the middle of nowhere?" Arlyn's voice drifted from behind me.

"Water underground," John said. "Have wells."

"Who does?"

"Town people," he said, the word "idiot" in his tone, though his eyes still sparkled with warmth.

"You are not from there?" I asked.

He pointed back toward where he had picked us up, back and to the right.

"There's another town?"

"Mine."

A mine, he meant, not that the town was his. "You work there?"

"I own there." No doubt about it, he was laughing at me. I did not mind. He had rescued us, watered us, and brought us to town. He could laugh all he wanted. Luckily Arlyn seemed to feel the same way. He had settled back onto the sacks again, his hat tipped over his face.

We finally left the desert behind entirely. All the trees helped even the edges of town feel cooler. There were people on the streets, including a double-line of what were obviously school children. Several people raised a hand in greeting when they saw John, some even calling him by name, though they called him by his profession, "Carter," not his family name. The town people were dressed similarly to John, if cleaner and fresher. The men wore trousers, shirts and jackets, wide-brimmed hats with a variety of crowns, some trimmed with feathers or with the furry tails of animals. As for the women, some were dressed similarly to the men, though their suits were more formal looking somehow, closer fitting, and in a wider variety of colors. Others wore skirts, cut full to make them easier to walk in—and to ride in, as I soon saw. They wore a similar style of hat clearly meant to shade the face from the fierce sun.

Men and women both wore their hair long, though many pinned it up under their hats. Many of the men had elaborate beards and mustaches, which I had to admit gave them a very rakish air. Beards hadn't been in fashion in my lifetime.

"Does it rain much here?"

"Doesn't rain, until it does. When it does, drown you if you're not careful."

"How many people live in this town?"

"Don't know. Ask the sheriff." We turned off the tree-shaded lane that brought us to town and started down what was evidently the main street. Here raised wooden platforms to either side of the road allowed people to walk above the dirt in front of several stores, a stable, a smithy, and an inn with a taproom. Second stories tended to overhang the first, creating deep shadows cooler than the streets. Halfway down the road John Bearclaw stopped the wagon and climbed down in front of a building standing by itself. He circled in front of the horses, giving each of them a stroke on the flank as he passed. When he got to my side he reached up his hands to help me down from my seat. I was grateful for this unusual attention, stiff and tired as I was.

Arlyn met us on the building side of the cart. For the first time I saw two people standing deep in the shadows, one to each side of the building's door. One wore a pistol of an unusual shape on his hip, the other carried two pistols in holsters across his chest and what could only be a sword hanging from his right hip, as if for his practitioner's hand.

"Newcomers to see the sheriff," John Bearclaw said. "Found them up along the trail toward the salt flat."

"Thanks, John," the man on the left said. "I'll see if he's receiving." With that he opened the door just enough to slide through and shut it again. The remaining guard stood in that seemingly relaxed way that killers have. The first man came back, this time opening the door wide.

"If you will, ma'am, ladies first."

"We would prefer to come in together," I said.

"Nevertheless, ma'am, it's the sheriff's preferences have the weight here."

A curiously formal way for soldiers to speak. I found myself more interested in meeting the sheriff.

The practitioner-handed swordsman led me through an anteroom several degrees cooler than the outside. I could barely make out the shapes of two desks and chairs in the gloom left by shuttered windows. The room smelled of wax and honey-scented candles.

"May I tell the sheriff your name, ma'am?"

"Fenra Lowens," I said.

"Any title?"

"Practitioner."

"You a doctor or a lawyer then, ma'am?"

"Not exactly." Though I supposed I was more a doctor than anything else.

My escort opened a door on the left in the rear wall and held it open for me. "Practitioner Fenra Lowens, sheriff."

The man behind the desk stood up as I entered, the paper he had been reading still in his hand. "A practitioner, what do you know." His smile spread his mustaches wide. "Just exactly what I need." He realized he still held the paper, and tossed it down as he came round the desk, his hand outstretched. His right hand. We were almost exactly the same height. Dark hair hung in loose curls just past his shoulders, though neither as dark nor as curly as mine. His mustache, full over his upper lip, narrowed to stiff points. The tuft of hair on his chin made his whole face appear longer. His leather trousers were beaded on the front, and tucked into boots that came just above the knee. He wore a short jacket, with the ornate cuffs of his shirt showing at the end of his sleeves. Altogether his clothing was more elaborate than that of his men, but I thought I had seen similar styles in old paintings.

"Practitioner, a pleasure." I could see from the sparkle in his eye that it was. "You're from Ibania," he said as he took my hand in both of his. He used the old pronunciation, eeBaynia. "A girl from Ibania broke my heart once. Are all the women from there so beautiful?"

"Yes," I said, too surprised not to answer with a smile of my own. "How did you know where I am from?" How strange it was to meet someone for the first time and feel so immediately at ease, as if I had known him my whole life. I felt my smile fade. I had a feeling I knew the answer.

He laughed then, drawing me to the seat in front of his desk by the hand he still held.

"I'd be pleased to tell you the tale, but sit, sit. What can I have brought? Wine? Tea?" He settled one hip on the edge of his desk. "It's a long while since I've met with a . . . compatriot, shall we say. I couldn't be more delighted."

"Whatever you have ready."

"Of course, of course. Lugg!" An older man I had not seen before stuck his head through the door behind me.

"Sir?"

"A jug of cold tea for my guest, and some biscuits if we have any left."

"And if we don't have any left," the old man said, lifting his eyebrows.

"Then get some."

"Yes sir, certainly sir," he muttered as he closed the door. "Like I've got nothing else to do."

The sheriff laughed again. "It's tea flavored with what we call lemongrass, a nice bite and very thirst quenching."

"Thank you, that sounds wonderful." I hoped there were biscuits. "How is it that—I am sorry, I do not know your name?"

He leapt again to his feet, first bowing deeply with his hand over his heart, and then offering me his hand again as he straightened. "I beg your pardon, madame, I am Elvanyn Karamisk, once Guard Lieutenant of the White Court, and now High Sheriff for the Dundalk Territory." He sat again. "There are people here, in the Territory and elsewhere, who might interest you. Now that etiquette has been properly served, what were you about to ask?"

The White Court? Guard Lieutenant? That answered several questions, and posed several more.

"May my companion join us? I'm sure he would appreciate some tea as well."

Just at that moment the tea entered, carried on a tray with three glasses and a plate of biscuits by the old man who was still muttering under his breath. Elvanyn Karamisk watched him with a fond smile. "Here on the desk please, Lugg."

"Well, I wasn't going to put it on the floor." The old man shook his head as he set the tray down on top of the papers covering the desk. He very pointedly poured me a glass, which he presented with a smile full of crooked teeth that would frighten a child. "Here you go, miss, nice and cool."

"Thank you."

"Which is more than he ever does," Lugg muttered on his way out the door.

"Thank you, Lugg," Elvanyn called after him. He lowered his eyes to me. "We've been together a very long time," he said.

"He helped raise you?"

"Something like that." He poured himself a glass of tea and raised it to me. "To the beautiful women of Ibania." He tossed back half the glass. I took a more careful swallow of my own, and then another. He was right, this was refreshing.

"And my companion?"

"Of course, of course. I'm afraid I don't know his name."

"Arlyn Albainil."

I am not sure I would have seen it if I had not been watching him so closely. The sheriff had gone perfectly still, just for a moment, before his features moved again, and his mouth smiled. But now his eyes were shuttered in a way they hadn't been before. In that moment I had seen a colder, harder man. A man who could kill. *He is a sheriff*, I told myself. *He was a guard. Of course he can kill.* I had a sinking feeling that I knew why he had reacted that way to Arlyn's name.

"I am sorry, what has upset you?" I might as well get it into the open.

"You saw that? Astonishing. The name startled me, that's all. I haven't heard that name in a long time. Knew someone with that family name once, but as I say, it was very long ago." The clouds in his eyes grew darker.

I shut my own eyes. It had to be Arlyn he remembered, or more properly Xandra, that much was clear. How had Elvanyn Karamisk kept himself alive all these years? Did time pass differently here? Whatever had happened between them, I hoped they would be able to set it aside. We had to find a way back, and Karamisk, though not a practitioner himself, might know of one.

"I think you had better have Arlyn come in," I said.

"Wyeth!" he called. The man who had escorted me opened the door and looked in. "Bring in our other guest, would you?"

"Sure thing, Sheriff."

The door opened again and Elvanyn was on his feet so fast I did not see him move. He touched his left hip with his right hand as if reaching for the sword that wasn't there. The shock on Arlyn's face slowly dissolved into a look of deep sorrow and, yes, I could see shame as well.

Elvanyn's hands relaxed. He took in a deep breath and let it out. He looked from Arlyn to me and back again. His face was ever so slightly flushed, though whether it was anger or something else I could not say. He sat down behind his desk, suddenly a tired man at the end of his day.

"Well, well, Xandra Albainil." His smile was empty, his eyes cold. "Better late than never, I suppose." His tone was light, but there was steel under it. My stomach plummeted.

"Elva, I—"

The sheriff held up his right index finger and pointed to the other chair. "What do you want?"

Arlyn sat down. It was somehow a surrender. He stayed silent for a few more heartbeats. "I need a way back, Elva. I need a gate."

"Don't we all." The sheriff nodded. "Will she be going back with you?"

I began to speak, but Arlyn answered first.

"She has to. The gate won't work without her."

Again Karamisk looked at me. Whatever he saw in my face relaxed him. "Lucky for you," he said before shifting his gaze once more to Arlyn. "I know the spot I came in. I know it exactly." For a moment something like resignation colored his voice. "You never used it, so it might work."

Arlyn opened his mouth, shut it, and swallowed. "If you could have someone show us the way?"

Karamisk shook his head. "I'll have to do it." He got to his feet and glanced around, as if looking for something that should have been close at hand. "Wait a bit." He turned to me and said, "Practitioner, would you like to bathe and rest first?"

I was tempted. Very tempted. "I'm afraid if there is a way back, we should use it at once."

He nodded, and smiled at me a little sadly. "Help yourself to more tea and biscuits," he said finally. He then circled around my side of the desk and headed for the door. Arlyn stood and intercepted him, laying his fingers on Karamisk's arm. The sheriff froze, the paleness of his face making his beard and mustache stand out.

"Please don't touch me," he said in an even voice.

Arlyn lifted his hand, and Karamisk went out.

I put my glass of tea down on the tray very slowly, very deliberately. I turned toward Arlyn and took a firm grip on the arms of my chair with both hands.

"You left him here. I can see it in your face." And I could; it was there as plain to read as one of my own *forrans*.

"I told you. I had no power." His voice was low, but firm. "You don't understand. He was my best friend—the only person who ever looked at me and just saw me. Not something dangerous, or valuable, or awful. Just me. Do you think I would have left him here if I'd had any other option?"

I stood up. "And there was no other practitioner with power you could use—as you are now using mine? You could not have given someone else the *forran*? Not even to save your 'best friend'?"

The most horrible thing about the look on his face was that this had obviously never occurred to him.

"Are you questioning my authority?"

"Yes, as a matter of fact, I am. The Red Court is very clear on my requirements on this point." Ginglen spoke as calmly as he could. No point in showing even an apprentice practitioner how angry he was. "You must know that I can't release the personal property of a guest in my hotel without authorization either from the guest, or some civil authority. *Written* authorization," he added when it seemed the young man was about to argue his point.

"Very well. I'll report your lack of cooperation."

"I'd expect nothing less." Easy to see what must have happened. A new apprentice, full of zeal and eager to impress his mentor, thought he could take a few shortcuts.

So Ginglen wasn't surprised when the boy, now with an entirely different air and dots of pink in his embarrassed cheeks, returned almost an hour later with the properly issued authority, from Practitioner Santaron Metenari no less, for the effects of Practitioner Fenra Lowens and her companion, Arlyn Albainil. Ginglen took as long as he could reading over what was really a very short document.

Doesn't mention the horse, he noted. *It's like she knew.* Good thing he'd hidden her saddlebags as well.

"And my payment," he asked, looking up from the document. "There's the room for two extra days."

"That has nothing to do with me." And how happy the silly boy was to be able to say that. "You'll have to make application to the White Court yourself."

Ginglen took his time packing up the few things his guests had left behind in a small, shabby trunk, abandoned by another guest. He wanted the apprentice gone, but not at the expense of doing things properly.

"Well, I suppose that means goodbye to our fees." Itzen came up behind him once the boy had gone, slipping his arms around Ginglen's waist and resting his chin on Ginglen's shoulder. Ginglen laid one hand on his husband's arm, and reached up to pat his cheek with the other.

"Practitioner Lowens paid us four days in advance, so we're actually still a day to the good. Though I wasn't obliged to tell the White Court *that*." He sighed. "We'd be in her debt anyway, you know that."

"What do you think this is all about?" Itzen followed Ginglen into their private rooms, a small suite off to the left of the entrance hall. "Think it might be . . ." he tilted his head in the direction of the stables, "Chispa's niece?"

"I'd guess not, since we weren't asked to come along as well."

Itzen took the authorization from him, and whistled when he read it. "Santaron Metenari, no less. The newest member of the White Court council. Soon to be the next Headmaster, they say."

Ginglen turned to look his partner in the face. Itzen was perfectly serious. "I don't like this, I don't like it at all."

"They have the authority, you couldn't refuse." Itzen rested his hands on Ginglen's shoulders and gave him a little shake.

"I know, I know. The White Court is getting more and more high-handed and it's giving me a bad feeling."

"We could throw ourselves on the protection of the Red Court."

Ginglen felt a little better once they'd finished laughing.

Five

ARLYN

I COULDN'T FEEL my hands and feet; sounds were far away. Had to be shock. Elva should be dead. Long dead. The tone of his voice when he told me not to touch him, like a punch to the gut. The stony look on Fenra's face. How could I have been so stupid? How did it never occur to me —I swallowed over and over. I thought I was going to vomit. I examined her face for understanding, for sympathy, didn't see any. Somehow that strengthened me. Straightened my back, squared my shoulders. I pushed my hair back off my face.

"Who am I kidding?" I said. "If I'd known ahead of time that losing Elva was the price I had to pay for sealing away the Godstone, I'd have paid it gladly. There are more important things than one person. Elva would have said the same."

Fenra coughed again, as if the muscles of her throat were tight. "How long has he been here?" She unwrapped her cravat from her head, smoothed out the material and tied it around her throat, regardless of the heat. She wouldn't meet my eyes. I couldn't blame her.

"I don't know," I admitted. "I'm not even sure how much time has passed in our dimension." I rubbed at my eyes. "There's a lot I don't remember." I watched her process this information.

"Long enough that you did not expect to find him here," she said finally. "Otherwise you would have warned me about him before we got this far." She took in a deep breath. "Well, at least he is going to help us. Even if he will not forgive you."

Another blow. How real that sounded, coming from Fenra's mouth. "I don't know what he's feeling."

"Do not be ridiculous. Of course you do. And if you are in any doubt, I can tell you."

I pressed my lips together, looked away.

The door opened. Elva stepped back into the room. He now had a sword belt slung around his hips, a blade I didn't recognize hanging from it. He used to carry a single pistol on his right hip; now he wore a double holster, with a pistol under each arm, a make I'd never seen before. Over his left arm he carried a dark red cloak. I wondered whether he'd chosen the color so as not to show blood. He glanced once at me before settling his attention on Fenra. From the set of his shoulders, he felt at ease with her. There was something in her he recognized, and trusted.

"If you are ready, Practitioner Lowens."

Without another word Fenra stood, dusted down her trousers, followed Elva out of the room. On the street outside were three horses, saddled, with packs and canteens. Elva immediately mounted the large black beast on the left. Fenra approached a smaller gray animal, making sure to let it see her before she caressed its head, murmuring to it. Elva watched her from the back of his own mount, his expression soft, almost smiling.

This left a chocolate brown animal which paid me no attention whatsoever when I hauled myself into the saddle, though it shifted its feet as I settled myself.

Elva nodded at a thin, hatless man leaning against the outer door-frame of the sheriff's office. The man nodded back, touching an index finger to his forehead.

"Is it very far?" Fenra said as the horses moved forward. "It's just that except for those biscuits I have not eaten in almost a day, and if my power is needed . . ." She fell silent when Elva shook his head.

"About an hour's ride," he said. "You'll find dried meat and fruits in the right-hand saddlebag, hard cheese and road bread in the left. Water in the canteen."

At first we rode single file with Elva in the lead, Fenra next. Once out of the town proper, Elva slowed enough that he rode at her side. When I tried to join them he clicked his tongue, and my horse fell in behind them again, ignoring any further instruction from me.

He should let me explain, I thought, my lips pressed tight. This wasn't fair. I knew he'd understand if he'd only listen. It wouldn't change his situation, but he'd be happier with me. Anyone else would have been demanding an explanation—why wasn't he?

"Elva!" I called. I can't say that he stiffened, but he definitely sat up straighter.

"Arlyn, stop." Fenra spoke quietly, but she flicked her fingers at me. She didn't need to do that. I would have stayed quiet if she'd just asked me to.

⟡

FENRA

Any other time the surprise on the sheriff's face when Arlyn actually stayed quiet would have been funny. Right now, the roiling of my stomach prevented my feeling amused. I wrapped the piece of road bread I had been nibbling back into its waxed paper. An unfamiliar sharp and bitter taste that caught at the back of my throat lurked somewhere among the nuts and dried fruits.

"How did you do that?" he asked.

"I am a practitioner," I reminded him.

"*He's* a practitioner." After the first time, the sheriff had not called Arlyn by name—either of them.

"Not anymore. He has no power, he cannot function as a practitioner."

"Did he tell you that?" Incredulity underlined with bitterness. "And you believed him?"

I shrugged. "He told me he left you here. Was he lying about that?"

"He told you? That's a surprise. He never liked people to know when he was in the wrong. Did he tell you about the Godstone? Ah, I see he did. The lengths his 'curiosity'—huh." Elvanyn swung his head from side to side. "Arrogance is the real word. When he wouldn't listen, I told him I'd go to the council." He waved around us. "This is where I went instead."

"He learned you were right, but sealing away the Godstone took all his power. He could not come back for you."

"And he's told you that as well, I suppose? No, don't." He cut the air between us with his hand. "I know how charming he can be. He could persuade a bee to leave a flower. Look, if what he says about his power is true, why didn't he ask someone else to help him? Didn't want to admit he'd made a mistake? Wait, oh, that's right, he'd be that arrogant."

I did not want to tell Karamisk that I had asked Arlyn that myself. I wondered if I should tell him about the lowness. *Not my secret to tell*, I decided finally.

"Sheriff Karamisk," I began.

"Elvanyn. Call me Elvanyn. Or Elva." His smile only reached his eyes for a moment, and he seemed again the light-hearted man I had first met. Almost.

"Then I am Fenra," I said. "I have had similar thoughts myself, but not speaking to him now may be a bad strategy." They both needed to be shut of this.

This time Elvanyn shook his head while staring at the sky. "My head's spinning like I don't know which way is up. I've been here such a long time . . . At first I was so angry. And hurt. And frightened. If he'd come back in the first few months, even in the first few years, I probably would have done my best to kill him. Only he didn't come back. And . . ." His voice died away.

"A person cannot stay angry indefinitely," I put in.

"Exactly. Though it's hard to remember that right now." He switched his reins from one hand to the other. The horse took no notice. "When I realized I wasn't aging, I was happy, thinking I'd still be here if he came back. Then I was horrified, at the thought that he might not." He quirked up an eyebrow, whistled without making a sound. "Some days I'm still not sure. But I've made a new life here. I've done well." He gave me a twisted grin. "There are others, not like me, but they age very slowly. We have to move along before people notice, but gunslingers travel, so no one's surprised."

We rode a while in silence. "So," he said finally, "you're saying that he came to his senses in time, and used all his power to lock the Godstone away?"

"Yes."

"Is it true? That he has no power? Can you tell?"

"I have spent most of my life working as a healer. That is a mistake I could not make." I shrugged again. "But I only know Arlyn. Xandra Albainil is someone I have never met." *And I do not think I want to*, I said only to myself.

Elvanyn was silent for a long time. "I'm coming back with you."

This left me speechless. "But your life here?" I finally managed to say.

"Even if he has no power, it doesn't mean he isn't dangerous. You think you can trust him? You can't. He's using you right now, isn't he? What if he has a plan to take your power from you?" I opened my mouth to protest, but Arlyn *was* using me, that much was true. "Or maybe he thinks he can get his own power back from the Godstone—again with your help," Elvanyn continued. "You can't be sure—what if he's more Xandra than you think? His arrogance—his carelessness—exiled me, and if there was anyone he loved in the world, I was that person. If he could make the Godstone in the first place—"

"What?"

"Ah, so there's something he *didn't* tell you? No, I see from your face he didn't." Elvanyn's crooked lips said "I thought so" as loudly as if he had spoken.

No, Arlyn had not told me. I thought he had found it. I pressed my lips together. He had said he could not remember everything, but surely this was not something he could have forgotten. He had chosen not to tell me.

Just as he had chosen to seal the Godstone away. I *had* to believe that the Arlyn I knew was not the man who had made the artifact in the first place. Not the one who had deserted his best friend.

"The Godstone is in his vault," I pointed out. "When he found he could not destroy it, he sealed it away." I clung to my conviction. I had to.

"Or when he found he couldn't use it in the way that he expected."

I drew in a breath through my nose. "I must trust my own instincts," I said as firmly as I could. "What can I do to convince you Arlyn is a changed man?"

"Nothing comes to mind."

Movement drew my eyes upward. I saw the uneven 'v' of a flight of geese. Was it autumn here, or spring? "No offense, Elvanyn, but if what you say is true, what can you do this time, that you were not able to do before?" He was quiet long enough for me to think that I *had* offended him.

"Of course I couldn't stop him then. A guard lieutenant against the most powerful practitioner? I was here before I knew it. Now, if what you say is true, he can't do that again. But he's up to something, he always is, and you need an ally, whether you know it or not. I'm coming." That was the tone of a stubborn man who has made up his mind. I felt an unexpected sense of relief.

"You put your trust in *me* so quickly?" I knew that if I smiled I would definitely offend him.

"Am I wrong?" This time his smile was warm and genuine.

We rode a while longer in silence, though Elvanyn remained at my side. When I looked over my shoulder I saw Arlyn slumped in his saddle, his eyes closed. I felt an almost overwhelming wave of exhaustion sweep through me.

Elvanyn tapped me on the thigh and pointed to my right-hand saddlebag when I looked at him. "Eat something," he said. He was right. Not the bread though, my stomach still rebelled against that. But dried plums and dried apples were the same in both dimensions, it turned out.

What would I be able to do, even if we found the right spot? *Could* I open this gate? Where would it take us? I looked back again. I was reluctant to wake Arlyn just to ask him. We had no other option, and I would know soon enough.

⟋

ARLYN

I didn't realize I'd fallen asleep until the horse stopped moving. We'd left the road while I napped, and come cross-country several miles, as far as I could judge. Nothing looked familiar. There were hills around us now, thick with trees and underbrush, though the trail we followed showed regular use. I tried to examine the ground more closely; my horse snorted at me, wriggled until I sat straight again. Not all the hoofmarks I saw belonged to horses.

Finally Elva called a halt, dismounting. Fenra followed him, favoring her bad leg while Elva looked on with concern. He knew better than to offer a practitioner his help. Neither of them noticed me clinging to the saddle when my legs threatened to give out under me. My horse looked around at me, showed me his teeth.

"We'll leave the horses here," Elva said, shifting canteen and pack from the animal's rump to his own shoulder. I did the same with mine before he asked. That would surprise him.

Fenra shot him a look I was familiar with and stopped massaging her knee. "They are worried about their safety."

Elva looked from her to the horses and back again. "They can't come with us, the trail's too rough. And they've always been safe before."

"Perhaps so." Fenra began picking up small stones, a few twigs. "But they have not *felt* safe. I will make a ward for them. It will only take a moment."

"You must have a very strong connection to know what they're feeling."

Fenra set down the last twig and dusted off her hands. "They are real to me in the same way you are."

Elva held out both hands, palms outward, and smiled. "No offense. Anyone who doesn't respect an animal that helps keep him alive is an idiot."

I dusted off my trousers and joined them. "How long a walk have we?"

"Not long." Elva answered me, but looked at Fenra. I told myself he was worried about her leg. We should have asked for a cane while we were still in town. Without discussion Elva and I divvied the load from Fenra's packs between us, leaving her unburdened. She walked with her hand on my shoulder, and I felt a part of me relax. As usual, once we were moving, she seemed to have no trouble with her leg at all.

Elva was right, however. Not more than a mile later we passed through a rocky ravine that showed signs of spring runoff. Soon it leveled out, plants and trees put in an appearance, and we found ourselves in a clearing. Someone had been keeping the area cleared, not obviously, but the signs were there if you knew how to look.

Frowning, Elva turned slowly, until he was facing the sun. "Here," he said, spreading his hands before him. "This was my entry point."

"The sun would be in a different orientation at a different time of day," Fenra pointed out before I could.

"That's why I'm not using only the position of the sun." Elva kicked some growth out of his way and sat down on a rock jutting out from behind a bush. He set his hands on his knees and looked for the first time at both of us. "This is what I first saw." He gestured behind us. "There's been some changes, of course, young trees have grown larger, and others have fallen and rotted away. But I've kept this rock clear of growth, and it's exactly where it's always been. I tripped over it when I arrived." He patted it as though it was a favorite dog.

"Can you clear this growth away, Fenra? I'll need a spot to write my pattern."

"I have paper." Elva patted his pockets, stopped when Fenra shook her head.

"I believe we will need something very much larger." She looked at me; I nodded. "I think there is something I could do. Arlyn, how big a clear spot do you need?"

"From that poplar to the rock, from that shaggy bush to this sapling."

She paced out the area I'd indicated, stepped into the center, crouched down on her heels. She rubbed her hands as if to clean off her palms, pressed her hands together, pulled them apart slowly. A glow formed between her two hands, instead of the lines of light I was expecting. Elva stared at her with narrowed eyes, trying to see what only a practitioner could see. I remembered that whenever I could, I used to explain to him what was happening as I was doing it, the changes as we traveled from Mode to Mode. Not now. Fenra couldn't be distracted. Besides, Elva wasn't listening to me anymore.

Fenra placed her hands palms down on the ground. The glow seeped out of her hands and spread over the area we needed before sinking into the ground. Her lips moved, but she wasn't speaking aloud.

"She's singing."

I glanced at Elva, but he wasn't talking to me, just thinking out loud. I considered Fenra again, took into account her facial expressions and how her breathing moved, realized he was right. She *was* singing, but not for us to hear. She stroked the plants, singing directly to them. Kneeling now, she sat back on her heels, rested her hands on her thighs.

At first nothing happened. Elva took a step toward her, but stopped when she lifted her hand without opening her eyes. When I looked more closely, I saw it. The plants were withdrawing into the ground, their normal growth reversed. Elva almost jumped backward, hissing through his clenched teeth.

In much less time than seemed possible, Fenra was kneeling in the middle of a patch of dirt. "Can you work with this?" she asked.

I shook my head. "It needs to be smoother, or the pattern might twist."

She nodded, got to her feet, stood straight with her hands palms down. The ground beneath her smoothed, following the slow movement of her hands, small rocks and pebbles sinking into the ground, larger stones rolling aside.

"What about the plants? What happens to them?" Elva sounded genuinely concerned.

"They will grow back almost as quickly as they withdrew."

"How did you make them move?"

"I did not 'make' them. I asked them. I promised it would not be forever. I also promised that you would no longer come here to cut away the growth. I undertook that on your behalf."

Elva laughed, and the sound of it, the tilt of his head, the sight of his smile took me back for a moment to the last time we'd been drinking together, and I'd refilled the glasses without touching the jug of wine. He'd laughed in the same way.

While they were talking I'd broken a fairly straight branch from a nearby tree and peeled it until I had a stick roughly the length of Fenra's cane. It occurred to me that Fenra's cane would have made a perfect instrument, considering that it was a tool used by a practitioner, even though not a practitioner's tool. I focused my attention, centered myself, reached out with my drawing stick, and began to make the pattern. By the time I was finished, sweat trickled down my back, and my mouth was terribly dry. I hadn't lost my pattern when I lost my power, but my strength wasn't what it used to be.

Fenra and Elva were sitting close together on the rock. She gnawed on a sausage, and Elva held a canteen for her, watching her face as she ate. Fenra's clothing was now dusty enough to match his, making them stand out against the greenery behind them. When she saw that I'd finished, she pushed herself to her feet.

"Do you need more rest?" Elva stood with her, hand reaching out for her elbow.

"I think eating has made up for it. If not—" She shrugged. "Our time is limited to what the grasses allow. We need to be finished before they grow back. Arlyn, what should the colors be?"

"The sage and the periwinkle will be enough," I said. "Power it, and it should rise to form a gate."

FENRA

Once I had the colors down, sitting back on my heels was easier than standing. I did not want to chance another wave of dizziness. We had more time than I had led the men to believe. The grasses had never met anyone like me, and they were curious, watching. As long as they continued to be, I could take all the time I needed. I know that there are practitioners who do not believe in the sentience of the natural world. They think it's some kind of primitive outer Mode superstition—like the existence of a Maker—but then, they do not live on the outer Modes. Even in the City, the grass and trees and suchlike are only dormant, and will speak to you if you speak first. In this at least, the New Zone wasn't so different from our world.

When I felt I had gathered my strength I got back to my feet and studied the pattern.

"I see similarities to the pattern that sealed your vault," I said to Arlyn.

"If you used yours more often you'd have noticed the same thing. The theory is that these similarities represent an individual practitioner's own pattern." Arlyn's voice came from behind me. It seemed he was staying as far away from Elvanyn as he could.

"You mean each of us has a unique, recognizable pattern? It's not just a family marker?"

"Like a fingerprint."

At this I looked at Elvanyn. He held up his right hand and touched the tips of his fingers with the index finger of his left hand.

"Here they've discovered that every person's fingers have a pattern unique to them. They call them fingerprints."

"How could they know such a thing?"

He shrugged. "Observation. Testing. Experiment."

"Fascinating as this is, perhaps we could discuss it once we're back home?"

Elvanyn looked sideways at Arlyn, one eyebrow raised—and I realized why it bothered him so much when I did it. At least they were finally interacting.

I took a deep breath, centering myself on the earth and concentrating. I lifted my arms to shoulder height and held my hands palm down. For the first time I could feel the power flow up through the ground and move through me. I waited for the pattern to become three-dimensional. Nothing.

I shook out the tense muscles in my arms and shoulders and tried again. Nothing.

After five more tries, I rubbed my eyes with the heels of my hands, feeling the trembling in my wrists and arms.

"It's not working, is it?" Elvanyn's sympathy irked me.

"Stating the obvious isn't helping either," I said, lowering my hands. The two men were standing on the far side of the pattern from me. "What about using my own pattern?"

Arlyn shook his head. "You don't use it enough. Without the written *forran* to guide you, it would be too dangerous." He drew down his brows. "I could try writing it for you . . ."

"I can barely lift my hands," I said, "let alone perform a strange *forran*. Who knew that achieving nothing could be so tiring."

"So we stay here?" Elvanyn said. He glanced up at Arlyn but quickly looked back to the pattern on the ground, as if he expected to see something this time.

"We can't stay here," Arlyn said finally. "We've got to get back. If Metenari finds the Godstone . . ."

Shaking my head, I bent over to place my palms on the ground. The grasses had been patient enough. At least for now, they could have their space back. Everything would look different in the morning. Fortunately, telling them this took almost no power at all.

The locket Medlyn had given me fell out of the top of my shirt, and swung loose. I closed my right hand on it, leaving my left on the ground. The locket was in no danger of sliding off, but somehow touching it made me feel better. As if my old mentor stood by to help me.

I straightened so quickly I could feel my vertebrae crack into alignment, and my vision darkened.

"You've thought of something." Arlyn had hold of my upper arm.

"Now who's stating the obvious." Elvanyn lowered the hands he'd reached out to help me and sat down on his marker rock.

I tapped the locket with my fingernails. "Could I open Medlyn's vault from this world?" I said.

Arlyn nodded slowly as he thought through my question. "Maybe," he said. "The vaults exist in a space of their own . . . but reaching his vault from here won't guarantee we can reach our own dimension from it. We don't know enough about how the locket works."

I thought I understood his expression. "Nor do we have time for study and experiment," I told him.

Elvanyn stood up, straightened his sword belt, and checked the hang of his revolvers. "What are we waiting for?"

I suddenly found myself on the ground, with Elvanyn on one side and Arlyn on the other.

"Fenra? Can you hear me?"

"Yes. Stop patting my hand. You would make poor healers, both of you."

"You fainted."

"Once again, the obvious. Thank you."

"Sarcasm. She must be feeling better." Elvanyn took a step back and again checked the balance of his weapons.

Each of them took a wrist and an elbow and helped me to my feet. I scrubbed my hands over my face, stifled a yawn. "Arlyn. Is there any reason for us to remain here in this spot?"

I felt rather than saw him shake his head. "Before you try anything else, you need rest. Lots of rest. And food. We can't do anything until the morning."

I did not argue with him. After resting for a few moments, I moved off the new grass and began choosing stones with care and setting them to one side of the cleared space. I wanted the grasses and insects to know where I planned to put a fire, in case they wanted to avoid it. After a moment Arlyn began gathering dry twigs and sticks.

"So Elvanyn tells me you made the Godstone?" It was easier to ask when I could not see his face.

"You told her." Arlyn turned to him, but Elvanyn just went on pouring water from his canteen into a pot he'd pulled out of his pack.

"He told me." I set down the largest stone I had found and began aligning the others to it. "The question now is, why did *you* not tell me?"

"To be honest, I can't remember making it." Arlyn squatted on his heels on the far side of the fire ring from me. It was light enough to see his face, though it wouldn't be for long.

"I remember," Elvanyn said under his breath. Arlyn pretended not to hear him.

"I know I must have done it. I remember the planning, the research. I remember the panic and the fear, the horror of trying to stay ahead of it, so I could seal it away. Wondering if I was strong enough to do it, or if it was already too late. I'd caused some damage I was too ignorant to foresee." He stared down on the small pile of twigs in the center of the ring of stones as if there was already a fire there. "I can remember having it, using it."

"What did you intend to do?" My curiosity got the better of me.

"He thought our world was supposed to be different," Elvanyn said when it became clear Arlyn was not going to answer. His voice was surprisingly gentle.

"Different how?"

"He thought it was artificial, the City, the Modes, all the changes that you practitioners see when you travel. He'd been here, and he thought our world was supposed to be like this one. All of a piece. As if there were only one Mode."

At first I did not understand him. Then I tried to imagine what that meant. "Could that be why the gate won't work for me? Because this world is so different? But the animals, the growing things—they know me." The plants had felt odd, just as the bread had tasted bitter, but they had responded.

"No." This time Arlyn spoke. "The practice works the same way— though artifacts don't, now that I remember. Here, take your earrings off."

I put my hands to my ears, felt the familiar facets of the emeralds in my lobes. "These are my final ones, they can't come off, that's the whole point of them."

"Humor me."

I started with the left one, unscrewing the tiny back from the metal post that pierced through my ear. Though I had not believed it was really possible, the earring parted, and in a moment I had the two pieces in my hand. I felt absolutely nothing different. Which was, considering what we had all been told, a very strange thing. A recklessness came over me and I removed the other earring as well, making sure I re-attached the backs before slipping them into the watch pocket of my vest. "Just a symbol," I said. "Nothing more."

"Don't say it like that." Elvanyn pulled out the lapel of his coat to show me a silver star pinned to his vest. "They're your badge, and a badge tells people who and what you are."

I turned back to Arlyn. "Is this how you removed yours?" I asked him. "Here?" His face clouded over and he touched his own ear.

"No," he said, lowering his hand again. "When I woke up, after— afterward, they were gone."

Gone with his power, is what I did not say aloud.

"I did not know the earrings could be circumvented," I said instead.

"They can't," he said. "Many people have tried it, for one reason or another, and failed."

"Because it hasn't been done doesn't mean it can't be done," I said.

Elvanyn chuckled, but not as though he had heard something funny. "He means he *knows* it can't be done," he said. The tone in Elvanyn's voice was that of a person who has been proven correct, when they did not want to be. This told me much about his real feelings.

"You designed them," I said to Arlyn, remembering what Medlyn had told me.

"Specifically so they couldn't be removed."

I thought about the locks on his vault. Even how his furniture was structured. "You designed the jewels by harnessing the power inherent in the stones."

"I'm afraid so. There's no *forran* to circumvent because it wasn't done with one person's *forran*."

"Commoners like myself might think it a fine idea to keep track of practitioners, but I can't imagine why practitioners themselves welcomed the idea," Elvanyn said dryly. "You weren't exactly acting in the best interest of your own kind, were you?" Again, the undertone of distrust and cynicism.

Arlyn paused so long I thought he was not going to respond. "I *wasn't* acting in their best interest. I was acting in the best interest of mundanes like yourself." This last phrase came out in a voice I had never heard Arlyn use before. "Power gets abused if it's not controlled— "

"You'd be the best judge of that." Elva's hand shifted on his sword hilt.

"Stop." We had no time for this. "We wear the earrings so that if one of us goes missing, or goes mad, they can be quickly found, and . . .

helped." I am not exactly sure why I hesitated over the last word. "So they *are* to our benefit."

"And finding a practitioner who's gone mad has obvious benefits to mundanes as well." Arlyn was determined to have the last word. His increasing irritability was worrying me more and more. I'd leveled him only hours ago.

"Why did you ask me to take them off?" I said.

"I wanted you to see what would happen for yourself. Artifacts have no power, but the practice itself actually works a little better here, *forrans* seem to take less energy. Remember you were able to call for help to come? That's why I thought the two worlds were supposed to be more alike. I thought a world where the practice worked more easily had to be the correct form."

"So you tried to change our world . . ." The concept was too big. I could not visualize it. I did not really want to.

He nodded. "I made the Godstone to do it. At first I tried to find one. I spent a long time looking before I realized that it wasn't there to be found. Then I thought, all I had to do was make one . . ." He shivered.

"What happened when you used it?"

He shook so much that at first I could not see him nodding. I reached across the still-cold fire pit and grasped his wrist in my left hand. He tried to pull away, but not as though his heart was in it.

"What happened?" I said again, willing him as much calm as I could considering my own exhaustion.

"I don't know. There was a tower—I don't remember." Elvanyn looked up at this, but Arlyn shook his head. "I remember the panic and the horror. Of wanting it all to stop, of hoping it wasn't already too late."

A cold thought rose to the surface. "Was it?"

Now he shook his head. "No. Not really. Except for me, of course. Except for . . ."

"Do not stop talking now," I warned him.

"I think our world jumped a Mode." He waved at my clothes. "Suddenly people were dressed differently. Dressed the way you are now."

As far as my experience went, people had always dressed the way I was dressed now. "You mean, before the Godstone, people in the city dressed like Elvanyn? Like in the old paintings?"

The muscles twitched at the side of his jaw. "Yes. More like that. The Road wasn't paved as well, and no one had pistols, just muskets."

"And swords." It wasn't that there were no swords anymore, just that people saved them for formal dress. "And since then? Nothing else has changed?"

"Nothing."

"And the world is developing as it should?" Again I remembered what Medlyn had told me. There were fewer Modes now, and no new ones were forming.

Arlyn pressed his lips together and the muscles around his mouth jumped. I supposed that answered me.

"We have to get back," I said. "Metenari may be pompous, but he is stubborn, and not in any way stupid. As you pointed out, your vault is now open. If the Godstone is there to be found he will find it. We cannot count on his not using it, once he has it in his possession."

Using a thin glass rod, Santaron Metenari edged the folded parchment on the top of his worktable closer to the light coming through the open window. It wasn't raining—it never rained in the White Court during the day—but an overcast of cloud obscured the sun. Still, natural light often told you more than artificial, no matter the brightness. Noxyn had very properly brought his find straight here, the room where Metenari attended to administrative details he could neither ignore nor delegate. He shifted his grip on the glass rod, holding it between the index and middle finger of his practitioner's hand, swinging it back and forth with the movement of his fingers. The parchment was dirty and discolored along one edge, as if it had been sticking out from a pile of similar documents. The inner side was a soft, clean, off-white still, protected by whatever it had been pressed between.

"How sure are you?"

"I unfolded it just enough to glimpse the heading." Noxyn shifted his feet, his hands reaching forward slightly, fingers curled, as if he wanted to pick up the parchment again. The boy wore fine silver-mail gloves lined with silk, exactly the precaution Metenari would have expected from his senior apprentice.

"Here," he said now, indicating the parchment packet with the glass rod. "Go ahead and open it for me. Spread it out fully, I'll get the glass." He noted the flush of pleasure that colored Noxyn's expression. He deserved it; the boy worked hard, and was coming along well.

The panes of glass Metenari wanted were in their own cupboard on the far side of the room from the window, so he had to turn his back on the boy. Which would show him how much he was trusted. It was with these gentle touches that apprentices were rewarded, keeping their spirits and morale high. The panes of glass rested in the third cupboard from the left, each on its own shallow shelf, the wood cut with finger space so the edge of the glass could be easily gripped. He stopped with the doors open, his hand hovering inches away from the top piece as an idea occurred to him.

He had his own way of organizing his materials; most practitioners did. He used the same kind of organization with practically everything he owned. Surely Xandra Albainil had been the same? And surely, since his vault was almost a copy of his workroom, a highly skilled practitioner should be able to deduce the most logical hiding place for the Godstone from a close examination and comparison of both rooms.

"Practitioner?" Noxyn's voice brought Metenari back to the present moment.

"Yes, of course, Noxyn, forgive me. I had a thought that may help us later." He pulled out one of the larger panes of glass. They couldn't be sure of the size of the opened parchment without touching it directly and assessing thickness and weight, but experience gave Metenari a fairly good idea. "Ready?"

Even though he was wearing the gloves, Noxyn touched the parchment as tenderly as possible, using only the tips of his fingers, on the very edges of the document. It wasn't so much necessary from the point of view of personal protection, but to interfere as little as possible, even by light touching, with the *forran* they believed the parchment held. Once Noxyn had the document spread out as much as he could on the tabletop, Metenari held the glass over it, letting his apprentice take one side so they could lay it down directly, and with great gentleness.

A common *forran* had been used on the glass itself to keep it clean, and to encourage clarity of vision to anyone looking through it. Metenari rested his hands on the edge of the table, glanced up at Noxyn, and smiled. Whatever could be done to help them read and under-stand this *forran* had been done.

"Tell me what you see," he instructed.

Noxyn cleared his throat, but didn't answer right away. He'd stepped back once the glass was down, showing proper respect for his mentor. It didn't matter that Noxyn had found the item; if his mentor wanted to look at it first, it was his right. Now, encouraged, he took a step closer to the table and did what he'd wanted to do from the first moment he'd been sure what he had. Now was his chance to actually examine a *forran* created by one of the legendary practitioners of old. To decipher it, to participate in its reconstruction.

First he examined it from above, taking a general view of the whole. Handwritten, of course, the writing quaintly old-fashioned and ornate.

"There are more symbols than one would normally see in a more modern specimen," he began.

"Are you sure? Look again."

Ears hot, Noxyn leaned over the document, his hands clasped behind his back, and scrutinized the line of symbols he thought he could see across the bottom of the page and turning to go up the side. "Ah," he said finally. "Some of *these* symbols, at least, are actually letters, some grouped, and others standing alone."

"Exactly," his mentor said. "Well done. We'll consider the significance of that in a moment. Continue."

Leaning forward again, this time letting his fingertips rest on the edge of the glass, Noxyn cleared his throat. "There *are* symbols, however," he said. "Here is the symbol for pattern, here is another for size and . . ." Noxyn's voice dried in his throat.

"What is it, my boy?"

Noxyn hadn't been this reluctant to answer since his first week as Practitioner Metenari's apprentice, when he'd been sure then that one wrong answer would result in his expulsion. Now that he knew how rare practitioners actually were, he knew his talent wouldn't have been wasted. But he also understood that to continue as the apprentice of someone as advanced as Metenari—and therefore keep his own advancement at top speed—he needed to keep his wrong answers to the minimum.

"I believe this parchment may have been re-used." The look on his mentor's face made Noxyn wish he'd been more ambivalent. He would have lost marks for lack of confidence, but that face told him he was in danger of losing more than that.

"Explain."

Noxyn swallowed. He might as well continue, better wrong than a coward. "If you look right here," he pointed at the third row down, on the side of the page to his own right, "there's a rough area on the surface, as if the parchment has been used before, scraped clean, and then used again."

His mentor squinted, bent until he was looking at the parchment from an angle. "I see what you mean," he said. Noxyn felt a glow of pride. "But I would have said such a thing was impossible. For a *forran* of this significance, to use anything other than a fresh piece of parchment would be unheard of."

"Um."

"Go ahead, Noxyn, tell me what's on your mind."

"What if the new *forran* was essentially the same?" Noxyn leaned over the glass in his eagerness to explain. "Say this opens the dimensional gate, but then Xandra Albainil wanted to change something about it, like, maybe, how long it stayed open, or the size. Would he make a whole new *forran*, or would he just alter the one he had?"

"He was remarkably skilled for such primitive times, that's true." Metenari walked slowly around the parchment, scrutinizing it from all angles, at one point brushing a speck of dust Noxyn couldn't see from the surface of the glass. Finally his mentor straightened, still focused on the *forran*. "I believe you're right, Noxyn. Well done. The *forran* was indeed altered. If I'm not mistaken, the alteration imposed a one-time-only factor to each gateway."

"You mean the gate in the vault could only be used once?"

"Exactly, but there could be a wider significance. It might be that while no one could use the same gateway to follow you, you yourself could not use that same gateway to return. There never would be another gateway in that precise spot again."

"But how would you get back, then? How are we going to get them back?"

"Obviously we can't do anything until we determine whether making a new gateway in a different spot would enable us to return. Someone would have to volunteer to try it, taking the *forran* with him, and hope to make it back again. Unfortunately, it would take someone of advanced study and power . . ." Metenari pulled himself upright and studied Noxyn with his head to one side.

Noxyn pulled his shoulders back, and tried to look eager rather than frightened. He would settle for determined, he thought. "I could go, Practitioner. I'll try it." *Or I'll find some way to trick Predax into going.*

Metenari smiled at this, and gave Noxyn a sharp, approving nod. "I expected nothing less of you, my boy. However . . ." He took his chin in his hand and tapped his cheek with his index finger. "No." He waved his hand over the parchment, casting shadows on the glass. "I'll need you with me. Let this be our last resort, if we find we need the carpenter after all. Intriguing as it is, I'll set Predax on it. Perhaps it's more straightforward than it appears."

Noxyn tried hard not to show his relief.

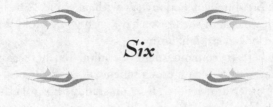

Six

ARLYN

I LAY THERE a long time with my eyes closed before I knew for sure I wasn't going to fall asleep. My mind spun. Anger. Guilt. Worry. Anger again. I told myself we had done everything we could. Fenra's locket would work. I could stop worrying, sleep. Anger again. Trees seemed unnaturally silent. I wondered if Fenra had asked them to let us rest. Fenra slept the sleep of the innocent, the exhausted. I wished I'd been able to do this without her.

Elva rolled over. Was *that* why I couldn't sleep? Yes, I'd left him here, but I hadn't *abandoned* him, it wasn't on purpose. I had been protecting him, keeping him safe. And it hadn't been my fault that I couldn't come back for him. And he wouldn't let me explain. Most of the time he acted as though he and Fenra were alone.

I sighed and sat up. This was ridiculous. I pushed myself to my feet. Needed to clear my head. Looking up, I saw unfamiliar stars. If we had to stay here, what would we do? How would we live? I'd be able to make furniture, but could Fenra still work as a healer? Even if the people here were past the time of belief, like in the City, she'd probably be able to manage. You didn't have to believe in the practice for a cure to work.

"Ow!" I'd walked right into Fenra's ward. I thought it was supposed to keep things out, not close us in. A shadow rose up to my right, and I jumped.

"What are you doing?" I asked.

"Checking what set off the wards, what do you think?"

Maybe the first thing Elva had said directly to me, though he was looking toward where Fenra lay sleeping. Between the stars and the waxing moon there was enough light for me to see his face. I'd seen that look before. I unclenched my jaw with an effort. "She has more in common with me than she'll ever have with you," I told him. For the first time he looked straight at me.

"I'm not here to romance her, you idiot. I'm here to make sure that your latest stupid, dangerous scheme doesn't kill her—or strand her in some other dimension." Elva pressed his lips together, as if he was sorry he'd spoken. "Besides, if you think you've got more in common with her than I do, you really are an idiot. *She's* still a human being."

"Elva, you have to forgive me." I spoke through clenched teeth. This animosity wasn't helping us.

"No, actually, I don't." His ever-so-reasonable tone only made me angrier.

"Listen, I did what was needed, what I had to do—" I began. He talked over me.

"What you *thought* you had to do," he said. "It turns out you were wrong, and I was right."

"I acted to save the world!" I glanced at where Fenra lay asleep and lowered my voice. "As soon as I knew what was really going on, I acted."

"If you had listened to me in the first place—"

"Not helpful. Listen, Elva, I'm sorry, I'm really sorry." I'd make the supreme gesture, one of us had to. "I meant to keep you safe, I made a mistake, afterward I couldn't undo it. I couldn't get you back. I couldn't . . ." My throat clogged, and tears came to my eyes.

"Xandra, stop," he said. "You think I can believe anything you say? You can't leave well enough alone, you never could. You put it in your vault, didn't you? No one can even find your workroom by himself, so no one could have found your vault. All you had to do was stay home, and the world wouldn't be in danger *again*. But you couldn't, could you? You have to solve everything yourself, because no one else is smart enough." He sighed. "Your apology doesn't change anything. You haven't changed. I'll help you now, but you'll have to accept that help without forgiveness."

"Both of you shut up and go to sleep."

FENRA

"Are you ready, gentlemen?" They had been considerably subdued since we woke up, and I wondered if I would ever know what had happened between them in the night. I pulled the locket out from under my shirt by the chain, and closed my left hand on it. Gold, lapis lazuli, never living, but part of the natural world. Would it be enough?

Arlyn tucked his left hand under my right elbow, facing in the same direction I did. Elvanyn tucked his left hand into my left elbow, facing backward. It would help that we were all using our practitioner's hand, even though neither of them were, at least at the moment, practitioners.

"This is a very strategic stance," Elva said. "Where did you learn it?"

"Don't distract her," Arlyn said. "If we live through this, one of us will tell you."

I did not need this reminder that the locket might not work at all. Perhaps nothing would happen. Perhaps we would be moved into an unknown dimension, or set down smack in the middle of the Patio of Horses in the White Court—whatever, we had to be prepared for danger. I could not protect us and open the way at the same time. Elvanyn looked at me, drew the gun from under his left arm, and nodded.

The locket felt cool on my forehead, as if it hadn't been all this time inside my shirt, against my skin. I opened it . . . and we were surrounded by a fog so dense I could not even see my companions, though I could still feel their hands gripping my arms.

"Fenra." Arlyn sounded as though he knew something he feared to tell me.

"Calm. Stay calm." The steadiness of my voice surprised me.

The temperature plunged, and suddenly I could smell the sea again. A dark shadow came toward us in the fog, something alive. I could feel it. No matter how much I blinked and strained, I could see nothing, but the sensation of size and danger hung just inches out of my sight. The hairs on the back of my neck stood up. Suddenly the grip on my arms tightened and I knew I had been about to fall.

"It's the Maker," Arlyn whispered. I knew he was still beside me, but his voice echoed with distance.

"No," I said. Did it have me by the arms, or did those hands still belong to Arlyn and Elvanyn? A dark shadow and a light shadow, moving closer and then farther again. I concentrated, telling myself I was not terrified.

ARLYN

There was a floor under me, not dirt. When I heard Elva gasp I tightened my grip on Fenra's arm. If we hadn't been holding her, she'd have fallen. Her eyelids fluttered, her breathing shallow.

"I do not know whether I am exhausted or relieved," she said, as she braced her feet, stood swaying very slightly. I didn't let go, and neither did Elva. "Are we here?"

I'd known we weren't in the New Zone any longer, but it wasn't until I saw the empty shelves, the wood-paneled walls, dust-free tables and carved chests, that I knew for certain we'd made it to Medlyn Tierell's vault. I felt a thickness in my throat that I hadn't felt in a very long time. I had known she was more powerful than others suspected. I hadn't known she was more powerful than *I* suspected. This was the work of a first-class practitioner. For the first time I felt real hope.

"You need to sit down," Elva was saying. "Can you make it to that bench? I can carry you."

Fenra shifted her hand so she had hold of my arm instead of the other way around. "I will sit, yes. Thank you." She walked on her own, but kept her grip on us until she reached the bench. She released Elva's arm, but still didn't let go of me. "Give me your other hand," she said.

When I realized what she had in mind, I pulled out of her grip before she could stop me, put both hands behind my back like a child caught with contraband cookies. "No," I said. "We need you fresh. You can level me later, if you think I need it."

Elva looked from her to me and back, his left eyebrow raised in that annoying way I'd almost forgotten. Fenra lifted her hand to him, fingers slightly spread. He relaxed, nodding. I clenched my teeth. He trusted *her*.

"You will have to be more careful, then," she said to me. "Relax as much as you can. Resist your anger. Above all, remember you will not always feel the way you feel in the moment."

I knew she was right, tried to follow her advice, took a couple of deep breaths, looked at a particularly pretty burl in the grain of the wood panel closest to me. Walnut. Walnut? No, zebrawood. I still felt resentful, but I also knew it wasn't real. At least in a manner of speaking.

Elva took a step back from her. "And now?"

Fenra turned to me. "Now we find out whether this space has a strong enough connection to our world for the locket to move us there. Or whether it will only return us to our starting point in the New Zone." She shut her eyes and rubbed at her forehead, as though she had a headache.

Elvanyn checked his sword and the lay of his pistols in a sequence I'd seen him do a thousand times. "Whenever you're ready, Fenra." She smiled without opening her eyes.

"In theory," I said, "the connection with Medlyn's work or home space should be stronger than any other connection. The path between here and there should be the most used, therefore the easiest to access. Since we've been in his office most recently, that one should be the easiest of the two."

"There's a few too many 'shoulds' and 'in theories' in all you're saying," Elva said. He turned slowly, looking around the room. His brow furrowed, his hand on the hilt of his sword—as if he could sense a danger he couldn't see.

"Take your time," I said to Fenra, meaning every word. "Be sure you're ready. We only have to save the world."

"Thank you, Arlyn." More sarcasm than patience. I smiled. She held out her left hand. At first I held back. Then I realized what she meant, took her hand in mine. Once again the room came to life around me. The books, the tapestries, the manuscripts. Even the tarnished silver lamps hanging from the ceiling, filled with bright gas.

"What would the old guy store here?" I wondered aloud. "Is there anything we might use?" Fenra picked up a skull from a nearby shelf. She shook her head, a wistful smile on her face. "Medlyn used to keep this on his desk, until one day it was gone. He never told me what it was. I can tell it's not an animal, but it's too squat to be human. It looks as though it's melted somehow. Pulled out of shape."

"It's human," I said. "But very old. Older almost than the world. It came from the outermost Mode, when the City and even the White Court was just a collection of mud walls and reed roofs between the rivers."

"Before your time?" Fenra asked. Elva looked away, but not before I saw the corner of his mouth curling into a smile. Elva looked from Fenra to me and back again. He couldn't see the skull. I was glad.

"Just," I said. "I wouldn't have thought the old man had so much power in him."

"He was a modest man," Fenra said.

I squeezed my eyes closed. I hadn't been modest, is that what she was saying?

"You have to be fairly powerful to have apprentices," Fenra was pointing out to Elva. "You must have had quite a few of your own," she added, setting the skull back in place before turning to me.

"Yes, 'Arlyn,' tell us about your apprentices."

Before I could answer, Fenra held up her right hand between us. "Stop this. Focus on what we are doing, both of you. I do not want to hear about *before* unless it helps *now*. Is that clear?" She stood up. "I have rested enough, let us try the next stage."

"Fenra . . ."

"Yes?" She turned her face to Elva's.

"Nothing," he said finally. "I'm ready." We resumed our grip on her arms. She closed her eyes, lifted the locket to her forehead, and the empty room around us became even more empty, the wood grain in the paneling growing indistinct, the colors fading, the glow of the lamps darkening into clouds. Fading away to nothing at all but a hot mist, thick as smoke and just as dirty. I thought something brushed up against my leg, but I must have been imagining it.

Then I saw walls around me, empty shelving, furniture, and I thought we were back in the vault again, until I realized that I could see more—

"Ow!" Fenra let go of my arm and rubbed at her knee. She must have bumped against the metal-reinforced corner of a large trunk. Its lid was laid back, and packing straw stuck out of the open top. When I looked around the room I saw other, smaller crates, padded wooden boxes covered with leather. A familiar hat and cane were propped in the corner by the door, exactly where someone who had found them would see them and remember to take them away when they left the office.

The shelves weren't completely bare. There were books, a long silver dagger-shaped tool with a braided blade. I recognized a glass safe lying sideways in the corner. Whoever owned it would need to turn it right side up before the glass decided to break.

Someone had obviously begun moving into the old man's office.

"Who are you? How did you get in here?"

"We've mastered sending objects through," Predax was saying as they walked through the gardens where people sat on stone benches under the trees. Moving, shaded water made the air feel cooler. In the week since the discovery of the vault, spring had turned into summer. "But you were quite right, Practitioner, when you theorized that the gate would close immediately, and that we wouldn't be able to open another gate in the same place."

Metenari knew that by "we," Predax meant himself, and two of the senior students. "And what is your next step?"

"Well, Jordy wanted to send animals through, but I pointed out that we wouldn't have any way to know whether they'd made it or not. Then I said—" The boy broke off and his ears turned red.

"What did you say?" Metenari pressed his lips tight to hide his smile.

"I said we should send Jordy through, and then Konne pointed out that we still wouldn't know for sure whether it had worked."

"And then you said, 'No, but at least we'll have gotten rid of Jordy.' Oh come now, Predax, don't look so alarmed. That's exactly what I would have said when I was an apprentice. Might have done it, too, if I'd had the chance." He laughed at the look on the boy's face. Predax relaxed, but only a little. "So, your conclusions thus far?"

"We'll have to send a practitioner through, sir, someone who can work the *forran*. There's no way around it."

"I agree. But it's not yet time for that. We have proof of theory—the *forran* does what we thought it did, but not easily, not smoothly. Let's exhaust every other option, before we send anyone after the carpenter. That remains a last resort."

"But Practitioner, isn't the experiment worthwhile in itself? I mean, don't we want to be able use this *forran* when we want to? Not just for this occasion?"

"Of course, and we will, when we have more time. Just now, we must return our attention to the original project. I'll need your help for that, and unless you think Jordy and Konne can be left to experiment alone—"

"Absolutely not, sir, no."

"So I surmised. No, we'll have them go back to relieving the others at the doorway, while you and Noxyn assist me in the vault." Predax only nodded at this, but his smile told Metenari how pleased the boy was.

He and Predax left the garden, and strolled down one of the wide yew-lined paths leading toward the older buildings. Several people nodded as he passed, and Metenari was careful to acknowledge every greeting, sometimes with a lifted hand, sometimes a nod. Only once did he stop to exchange a few words. It paid to be gracious to everyone. You never knew when you were going to need a favor.

When he recognized a particular set of shoulders coming around a hedge toward them, he deliberately turned his head away. He hadn't spoken directly to Ronan Sedges since he'd beaten the man out for the last council seat, and he wasn't going to start now.

The stairwell that hid Xandra Albainil's workroom was now blocked off. In fact, that entire end of the building had been put out of bounds. Several of Metenari's students who hadn't yet reached the apprentice stage were posted at the hallways connecting to the old tower, to keep people out, or divert them away, citing Metenari's authority when they had to.

They reached the tower and turned into the stairwell. The workshop door was no secret to them, now that they knew where it was. Inside, Noxyn was speaking with a student whose name Metenari couldn't remember, who sat on a stool placed in the open doorway of the vault next to a small table where he could work and eat. As he and Predax came in, the student got to his feet, and Noxyn turned to greet them.

"Now, I haven't been wasting my time while you boys have been working. I've made a careful analysis of Xandra Albainil's workroom and I believe I have enough understanding of how the man thought for us to find the Godstone more quickly than we'd hoped."

Metenari took a long look at the room and nodded, satisfied. Parchments, scrolls, artifacts, and tools were laid out neatly on shelves and tabletops as if they expected their master to come back. Oddly, there were very few books. "Now, all practitioners keep certain things nearest to hand, but Albainil was unusual in his organization. Can either of you spot it?"

Predax cleared his throat. "Everything in here is in a circle, around a central seat. Not laid out in rows like most people do."

"Very good, Predax. In fact, excellent."

"Well, I've been stationed here quite a bit and . . ."

"And you've been watching me, of course, and so you should. Never be shy of watching a master at work, particularly your own."

The two apprentices nodded respectfully.

"All right, then." Metenari returned to the student at the door. "Just sit here quietly, will you? Remember, don't step out of the threshold." Metenari laughed and patted the youngster on the shoulder. "Noxyn, follow me. Predax? Bring in that stool, would you?" The boys followed him as Metenari passed through the door and then stood to one side in the old practitioner's vault. The first thing Metenari did was point out where he wanted Predax to place the stool. As far as possible, he wanted it in the same spot it had occupied in the workroom. Metenari oriented himself with his back to the door.

"Slightly to the left. A little more. There, I think that's it, don't you?"

"Relative to everything else, yes. But Practitioner, the large cabinet to the left, there's no equivalent for it in here."

"I noticed that myself, Predax, but thank you. You may not have noticed, but the cabinet you speak of holds glassware—drinking glasses, bottles, jars and pitchers and the like. I imagine there'd be no need for that kind of thing in here. A vault is essentially a large cupboard, and while you might do actual work in a cupboard, you don't spend any time eating or drinking in one." Again, both apprentices nodded.

Metenari sat down on the stool, folded his hands in his lap, and looked slowly around. Would Xandra have put the Godstone into a place near to the central seat, or far? In the workshop, items near the stool were commonplace ones in constant use, or everyday items like the glassware cabinet that had no duplicate here. The chances were equally good, therefore, that something as unusual and precious as the

Godstone would be in one of the further cupboards, or on one of the shelves that were covered with doors in here, though they were open in the outside room. But the Godstone was more than unusual and precious.

"Now, something dangerous," he said to Noxyn. "Close to hand, or farther away?"

"Would it be likely to act on its own?" Noxyn said.

"An excellent question, my boy. Excellent." Metenari nodded. He should have thought of that himself. "If it was, he'd keep it nearby, don't you think? So he could react quickly. If it had to be activated somehow, then a corner cupboard would be just as likely. Here, you take my seat, tell me what you think."

When Noxyn was seated, Metenari looked around the room. "If it were nearby, what could you reach if you had to spring at it in a single bound?"

His lower lip between his teeth, Noxyn looked around him carefully. "I'd say nothing more than two paces away. Circular orientation, so we can't go by what direction he'd be facing . . ." Metenari nodded his encouragement, as Noxyn stood and paced slowly around the stool. "So it could be anywhere in this circle." Noxyn gestured. "I'd start by trying that cabinet there, the one that looks like a clothes press."

Metenari approached the cabinet with caution. The boy was quite right, any practitioner could cross this distance in one step, as it were, even if startled and otherwise unprepared. And it *did* look like a clothes press. About as tall as Metenari himself, it stood on four nicely turned legs, complete with matching stretchers. Above the legs were two drawers, about three hands deep, with brass pulls, and above them the cabinet proper, two doors that met in the middle with matching brass knobs. Not one to overlook the obvious, Metenari first tried the doors. They were closely fitted, and there was no sign on the outside of the astragal that they closed to. Nor was there any physical lock visible. But nevertheless, the doors did not open. Nor did the drawers when he tried them.

Metenari smiled. He would have been disappointed if they had.

"Watch my surroundings carefully, Predax. Now, Noxyn, let's consider this a quiz. Which *forran* would you use to open this cabinet?"

The boy took a moment to consider, which pleased Metenari very much. Impulsiveness was not a trait he considered useful in a practitioner. That, in his mind, had always been Fenra Lowens' problem.

"I'd say Llono's *forran* for locks would be the place to start."

"And why is that?"

"It's simple," Noxyn said, spreading his hands. "And it's a good principle to eliminate the simple before advancing to the complex."

"Exactly. Well done. Besides, Albainil would know that people would expect him to use a complicated *forran*, so he might have bluffed them by using a simple one."

Metenari pressed his palms together. A breath sufficed to fix the details of the *forran* in his mind. He slowed his breathing down, listening as his heart slowed to match. At the right moment he spread his fingers and drew his hands apart. He couldn't see any light with his eyes closed, but the familiar feeling of the power coursing from fingertip to fingertip told him he was ready. Still without opening his eyes, he turned his palms toward the cabinet. His hands heated as the power poured through them. In his mind he drew the pattern, in the manner indicated by Llono's *forran*, and waited.

And waited.

"Nothing, Practitioner."

Metenari sighed, letting his hands fall and the power dissipate harmlessly into the room. "Well, not unexpected, so don't let's allow ourselves to grow discouraged. What's the next *forran* you would try?"

But the next one didn't work either, nor the one after that. Metenari paused for reflection. Usually he would eliminate all possibilities before moving on to the other pieces of furniture. "Finish what you are doing before moving on" had always been his motto and advice. But in this case, there was good reason to eliminate the simple *forrans* for every piece before trying other, more complicated ones. He explained his reasoning to Noxyn and Predax, lest they think they were now allowed to apply this approach willy-nilly.

"Each item of furniture is likely to have a unique locking *forran*, therefore let us eliminate the *forrans* themselves by trying them on each lock," he concluded. "That way we can be sure not to miss anything obvious." He coughed and rubbed his forehead. He shouldn't be this tired, he'd only performed three *forrans*, and simple ones at that. He glanced around again at the seemingly innocent room. He wouldn't be surprised if someone so clever and devious as Xandra Albainil should leave a *forran* running in his vault that would subtly drain an intruding practitioner's power.

"Noxyn, why don't you try the first three locking *forrans* on this cabinet here; Predax, on that chest of drawers." No need to tell the boys anything yet, especially not that he was using them as guinea pigs. *Let's see if they tire as quickly as I did.*

Two hours later Metenari had his answer. With less experience, it took longer for apprentices to perform three *forrans* on the second cabinet and the chest of drawers, but they got just exactly as far as Metenari had with the first cabinet. With the droop of their shoulders, and the pallor of their faces, the boys looked the way Metenari felt.

"I think that's all for today, boys. We may have to rethink our approach a little."

"Perhaps one of the juniors should join us," Noxyn suggested, glancing at the student watching them round-eyed from the doorway. "Jordy, or even Konne. It's unlikely that any of these simple *forrans* will open anything," he continued as Metenari waited, "but we still need to check. That said, it needn't be any of us doing the checking."

"You've hit on something there, Noxyn. A very good idea. But let me improve upon it. Using the junior apprentices, we could even work in shifts. You, Predax, and I could be trying the next set of more complicated *forrans* on these initial cabinets while the others test the remaining furniture and shelving. We would follow each other in sequence until the correct one was found." He still thought the nearer ones the most likely.

"Like a round song," Predax said.

"Yes, I suppose it is, in a way. Though we must remember that just because a particular *forran* opens a particular artifact, it needn't follow that it will open all of them. As I said, everything we've seen so far tells us that Xandra Albainil had a complicated and even paranoid mind."

Metenari was just pushing himself to his feet when Jordy appeared in the doorway, clearing his throat.

"Excuse me, Practitioner? Konne says Practitioner Otwyn is at the foot of the tower. She needs to speak with you urgently."

Metenari sighed. Of course she did. Every new practitioner believed that every little thing they thought or did was the most important thing in the world. She was a nice girl, careful and usually considerate. He hoped she'd get past this stage quickly.

"Tell her I'm unavailable, but that I would be happy to see her in my office first thing in the morning." A headache began to throb behind Metenari's eyes. Maybe there was someone else he could palm her off to. Too bad Medlyn was faded, he'd been so good at this kind of thing.

"I'm sorry, Practitioner, but I told her that already. She says it's an emergency. Three people just appeared in her office out of nowhere, pushed her aside, and escaped before she could do anything. She's very worried." The boy rushed through the last few words and looked thankful to be silent.

The headache disappeared. "Out of nowhere? *Three* people? Is she sure? Never mind, I'll come down and question her myself."

⬦

FENRA

I wished we had not frightened the new practitioner who was taking over Medlyn's office. I could tell from the way she had treated the furniture that she had valued and respected him; she might have been willing to help us. As it was, she had undoubtedly gone to spread the alarm. Someone would be looking for us soon, and the stares of a few of the people we passed in the courtyards made me well aware of my stained and dusty clothing. One or two seemed about to speak, but something in either Arlyn's face or Elvanyn's kept them from approaching us.

"We're attracting too much attention," Elvanyn said. I had my hand in the crook of his elbow. Without my cane, left behind in our escape from Medlyn's office, I found I was tired enough to need his help. "Anyone looking for us will find plenty of witnesses to recognize our descriptions, and tell them which way we went."

"Your suggestions?"

"Get out of the White Court?" Elvanyn said. "Though it won't be easy. Perhaps we'll have to fight our way out." There was a smile in his voice, if not on his face.

"And go where?" I said. "We are here to deal with the Godstone, not to escape and wait in hiding for Metenari to find it and use it, because firstly, he can, and secondly, he will."

As we were talking, I had steered my companions into the Patio of Horses, one of the largest courtyards in the palace, large enough to have mature trees in it, and deep arcades on both long sides. We found a bench in a private but not too secluded alcove deep under one of the arcades and sat down out of the sun. Elvanyn put his hand down on top of mine before I could withdraw it from his elbow.

"Without me, the two of you could pass as visitors from an outer Mode," I said into the silence. "The danger lies in our being together."

"I hope you're not suggesting that I leave you," Elvanyn said. "Because I won't."

"No," Arlyn said. "Of course not. But the two of you can leave *me*."

"What?" Elvanyn and I spoke at the same time, though in entirely different tones.

"I'll go to the vault. I imagine that Metenari has kept it open, since he doesn't have the key." Arlyn licked his lips. "I know where the God-stone is, and the lock should respond to me, whether I have power or not. Like the earrings, the locks aren't dependent on outside sources of power."

"No." Elvanyn's voice was so cold I felt a shiver pass through me. "I don't trust you alone with it, no matter what you say about not having any power."

"Then *you* come with me," Arlyn said. "We'll leave Fenra on the outside—out of it completely. If we're caught, we'll claim we tricked her, and she can say the same."

"It's one thing for visitors to be on the grounds, especially during the day, but you won't get into the building without me," I argued. "Metenari is sure to have posted guards. I am the only one who can get past them. Tell me where the stone is, give me the key . . ." I trailed off as both men shook their heads.

"We don't even know for sure that there are guards," Arlyn began.

"That's easy enough, I'll go and see," Elvanyn said. "No, it *should* be me. No one knows me here." I thought I detected a flatness in his voice, as if he was thinking about all the people who would have known him. "I'll check if there are guards, how many, their schedule. Once I've reported back, we can make more informed decisions."

He had been a guard himself, I remembered. "You're not a practitioner," I pointed out.

"Nor do I need to be, to see the number and quality of any guards who may have been left to prevent unsavory strangers from entering.

My guess is that if anyone sees me, they'll simply tell me to move along, whereas either of you two are bound to be detained."

"Very well. Then I suggest you go now. Fenra and I will wait for you here."

"And if you have to move?"

"Then in the Sun Dial Patio. If things go badly, we'll try to get out that way."

ELVANYN

As soon as he was out of their sight, Elva's footsteps slowed, though he tried to look as much as possible like a commoner here on an errand. If things hadn't changed too much, the guard passed commoners in over the East Bridge, and wouldn't bother those already inside the grounds unless they actually caused trouble.

If things hadn't changed too much.

He didn't know how he felt—or even what he should be feeling. The first few years of his life in the New Zone, he'd dreamed of being right here—home, the White Court, the stone of the buildings, the patios and fountains, his own small room and the dining hall of the guards. His duties, his friends. The smell of wet earth and stone when he had night watch on a rainy evening. He would have traded anything to be back where he belonged.

But over time that homesickness faded, leaving only the occasional pang, fewer and fewer as the years went by, when something that looked or tasted or smelled like his old life suddenly struck him. He'd stopped thinking of his life here, and started living his life there. Now that he was back, everything looked and smelled just different enough to throw him off, to unsettle him.

Or did everything feel different because of Fenra Lowens?

As soon as he'd heard the word "practitioner" his heart had started beating faster, and it didn't slow down when Fenra walked into his office. She'd been so solemn, so obviously exhausted—so dusty—that he'd only thought of what he could do for her, not what she could do for him. Xandra walking in swept all that away, but later, when she'd worried about how the horses felt, and how the grass felt, he'd known that the feeling he'd had —that she was somehow important in a way he'd never thought of—he knew that feeling was real.

He shook away these thoughts and focused on the here and now. Evidently his sheriff's clothing was not considered odd enough, when he was walking by himself, for anyone to look at him twice. Or maybe it was the way he walked: with just the right amount of swagger, one arm swinging and the other hand resting on the hilt of his sword, as though he had every right to be here, and what's more, had an important errand. Still, it was odd to see all these unarmed people. In his day even practitioners routinely wore swords, though of course on the right hip, not on the left like commoners. He suddenly remembered a summer afternoon teaching Xandra to fence right-handed. No one would expect that from a practitioner, any more than Elva's own opponents had expected him to fight with his left. Another memory he didn't have time for right now.

There were more people here than there had been in the courtyard— more than he'd ever seen in this particular area before. People here were on their way somewhere, supper perhaps; it was the time of day for it. Very few were lingering to speak to one another, and small shops along the way were closing, or at least bringing in their wares. There didn't used to be so many commoners living in the fringes of the White Court, away from the better patios and gardens. Of course, someone had to bake the bread, provide fish, fruits, and vegetables. The meat and the spices. Practitioners were too busy to do their own marketing, or to cook their own food, though travel on the Road taught them all how. Still, he thought this section, here in the older part of the palace, was more crowded than he remembered.

He'd taken the long way round to reach the tower, and now that he was closing in, Elva shifted direction subtly, making sure no one watching him could tell it was his destination. He also remembered the arched entrance of the tower as having both an outer and an inner gate, but the openings were empty, and the whole area appeared deserted. Nevertheless, Elva knew that a guard alcove was tucked into the right-hand section between the two now-nonexistent barriers, where men could stand without being seen from the street outside. That couldn't have been changed, not and leave the entrance intact.

As he had expected, a young man in apprentice gray but without the black cap stepped out of the alcove as Elva drew even with it.

"I'm sorry, Dom, this area's off limits today."

"Really? Whyever for? I was hoping to see the view from the battlements." He spoke at his most charming, trying to appear totally fascinated.

"There's been a wall collapse, sir, and we're keeping people out until repairs are completed. Should only be a day or two, if you'd like a tour later on."

Elva drew down his eyebrows. Turned out his clothing *was* odd enough for the boy to take him for a tourist from an outer Mode—someone too naïve to know about tensions between the Courts. "Two days? I wouldn't have thought it would take practitioners two days to fix a broken wall." As he expected, the boy took the bait, crossing his arms and leaning his shoulder against the stone, ready to gossip with a friendly and interested stranger.

"Well, we don't want to fix it with practice, you see. Then someone's theories would always be in use, keeping it up, no pun intended."

Elva grinned, rolled his eyes, and gave the boy a light punch on the upper arm. This child was an insult to real guards everywhere. "Whatever do you do, then?"

"Well, the council sent to the Red Court for the Court Engineers to come look at the problem, and find some mundane solution. That's what's taking so long, they're in no hurry to get here, and then each and every one of them has to have their say."

"Makes sense," Elva said. He remembered Xandra using that word, "mundane." It meant people who weren't practitioners. Commoners. "Practitioners are far too valuable to waste on buildings, that's for certain. Or for guarding, for that matter." He smiled his warmest and most winning smile, knowing that there would be a twinkle in his eye. "I'd love to hear more about what's going on in the White Court. What time do you get off?"

The young man grinned back, and ducked his head. "Not until sunset, Dom." When visitors had to be gone from the Court.

"What a shame. Well, thanks for the information. I'll try to come back once the tower's open again, if I'm still in the City. Fair day to you." The salutation was out of his mouth before he thought of it, but the apprentice didn't blink. Either it was still in use, or the boy put it down to his being from outside the City.

Except for two people who smiled at him, one female and one male,

Elva's walk back to the courtyard where Fenra and—he had to start calling him Arlyn—were waiting for him was uneventful. The crowds thinned out as he left the older areas of the Court, to where red brick replaced white stone.

Fenra and Arlyn still sat on the bench where he'd left them, though he thought they'd been sitting closer together. Within the framing of the arch they sat under, they didn't look at all like fugitives, but like two people taking advantage of a cool place to sit. The girl looked better now, he thought, her natural color restored, less ashy gray than it had been when she'd brought them through from her old mentor's vault. *That was some trick*, he thought, wondering whether even Xandra in his day would have been able to do it.

Which endangered Fenra all that much more, since Xandra— Arlyn—needed and coveted that power. Elva made sure he sat down between them.

"They're taking no chances," he said. "The whole tower is closed, and there's a guard at the lower entrance to prevent anyone going in. Mind you," he continued, cutting Arlyn off as he began to speak, "he wasn't a real guard, but an apprentice—or maybe a student, since I don't imagine there's so very many apprentices that they can be used as guards. In my day commoner guards were used against outsiders, since we weren't much use against even the youngest apprentice."

"Metenari started mentoring apprentices almost on the day he became a practitioner himself," Fenra said. "So he might have a few to spare."

"Did you ever have one?" Elva gave her a more genuine smile than he'd given to the young apprentice.

"No." She smiled back, but shook her head. Unlike the apprentice, she saw right through him. "Very few practitioners want to live in the outer Modes, once their apprentice tours are complete, so I have always been alone."

"I've never had one either."

"That's because you're a prick," Elva said without turning around. Though he grinned when Fenra laughed.

"Perhaps we should get back to the matter in hand," she said, still smiling.

"I see two approaches," Elva said. "First, the soldier's answer: we kill him."

"No." Fenra and Arlyn spoke together.

"Fine. Second, he's silly enough for us to trick him by pretending to be with these engineers he spoke of—"

"No engineers have been sent for, really," Arlyn said.

Elva spread his hands, palms up. "I know that, and you know that, but I'd judge the boy on the gate doesn't, not from the way he was talking."

"I have a suggestion." Fenra looked from him to Arlyn and back again. "I can go myself and ask to see Metenari. I will say you tricked me, and that I have only now managed to get away from you. While I distract the guard with my story, the two of you will sneak in."

"I don't like it," Arlyn said. "I don't like it at all."

"Neither do I," Elva said. "Unfortunately, it may have the most chance of success."

Seven

FENRA

THE ONLY THING we were able to agree on was that we preferred not to kill the guard.

"Think." I knew I was right. "You cannot risk capture. Elvanyn has no status here at all. Whereas *I* can charm the guard."

"With what?" Arlyn said while Elvanyn looked at him sideways.

"This." I broke off a small twig bearing three new leaves from a nearby orange tree, freeing the scent of the blossoms. I smoothed it between my hands. "The boy at the gate will take me to where Metenari can be found," I told the twig, breathing on it twice. "Once we have cleared the gate, the two of you can take advantage of the now unguarded doorway to enter the tower. Once your business there is concluded, I will rejoin you."

"We?" "How, exactly?" Elvanyn and Arlyn not only spoke at the same time, they wore identical expressions.

"You'll be in custody," Arlyn added.

I tapped the spot where Medlyn's locket lay under my shirt. "Will I?"

Elvanyn's eyes narrowed. "Metenari will take it from you."

"To him it's a piece of jewelry," I said in my most certain tone. I myself had not known the locket was a key when I first saw it. "And no amount of examination will show him anything different. He would be ashamed to take a piece of jewelry from me, especially if I have come to him humbly, asking for his help."

It took some time, and further arguing, before they finally agreed.

"And if Metenari is in the vault? If the guard takes you there?"

I shook my head even before Elvanyn finished his question. "Let me in to see what he is doing? Absolutely not. Metenari would never run the risk of having to share credit for some discovery by having another practitioner present. Even one he believes is only a third class."

"His apprentices don't count," Arlyn added when it appeared Elvanyn would ask. "He takes credit for their work, not the other way around."

Elvanyn finally nodded, but I could tell he was not entirely convinced. "So he might leave an apprentice in the vault? What do we do then, if Arlyn hasn't any power?"

I ignored the "if." "You are in no more danger from apprentices than you would be from anyone else—less, in fact. Metenari disapproves of training in arms, and that is the only way they can be a danger to you, since apprentices are bound from using practice on others without permission until they become practitioners."

"Since when?" Arlyn said, his voice sharp.

"Since the Red Court insisted on it."

Arlyn pressed his lips together and shook his head. I refrained from pointing out that too much autonomy and too little oversight was what had brought us to this place.

"In any case," I continued, "I do not imagine that apprentices who cannot use any practice on you would be a problem for the High Sheriff of the Dundalk Territory."

Elvanyn touched the hilt of his sword, and the grips of his two pistols, and bowed.

Even though it was my own idea, I trembled as I approached the tower entrance alone. I did not try to hide it, as I thought it would fit with my story. I jumped when the youngster at the gate stepped out to meet me, and like a fool he let me grab his arm. I hung on as if I were about to fall over from exhaustion and fright, and tucked my three leaves into the cuff of his shirt, where they touched his skin.

"Practitioner Metenari?" I asked, breathless. "I was told I might find him here."

"He is, yes, but—"

"I must see him immediately, a matter of great importance that pertains to his present work."

Annoyance, curiosity, worry, all flashed over his face in the few seconds it took him to make up his mind. "If you follow this wall, Practitioner—"

I moved my hand to his cuff and squeezed. "Please, apprentice, I am in no condition to walk any further distance without assistance. I need your help."

This time only worry showed on his face. Worry over what Metenari would say to him if he made the wrong choice. Finally, just as I was thinking the leaves would not work, his face cleared completely and he smiled. "If you would come with me, Practitioner . . . ?"

"Lowens, Fenra Lowens. Your mentor and I were apprenticed in the same year," I added, and he became even more relaxed, more certain he was doing the right thing in abandoning his post.

"Then if you would, Practitioner Lowens, please do use me as your staff."

I did not have to fake my exhaustion very much, and my experience with my leg was a great help in using the young man as a cane. I leaned on him more than I needed to, to give him the idea that I was even more tired and weak than I appeared.

He led me into the building, past the main staircase, and down the long hallway to a bench at the foot of the rear staircase, lowering me with great care and courtesy to the seat.

"If you will wait here, Practitioner Lowens, I'll see whether Practitioner Metenari is upstairs."

I had no option but to watch him trot up the stairs without me. It would have been out of character for me to insist on coming with him, considering the emphasis I had put on my exhaustion and pain. At least I had cleared the door and the main staircase for Elva and Arlyn. Still, I knew how long the apprentice should take, and how long I would wait before going up after him, regardless of how it might look. Just as I pushed myself to my feet, the youngster came skipping down the stairs, with Metenari descending more sedately behind him.

"Santaron! Thank you for seeing me. Please help me—you must—I have made such a foolish mistake."

"Of course, Fenra, of course. Come this way—Konne, please help Practitioner Lowens, can't you see she's limping? Come, Fenra, let's get you sitting down and you can tell me all about it."

His concern was genuine. That made me feel a little shabby.

I had hoped that we would go to Metenari's office in the other wing, but my old classmate had taken over a suite of rooms in the old tower itself. They had been cleaned, and supplied with a worktable and comfortable chairs, but the paint on the walls was cracked and crumbled with age, and new carpets and bowls of peonies floating in water couldn't quite cover a lingering musty smell.

The young man, Konne, helped me as far as an overstuffed chair near the window and held my elbow while I lowered myself into it. I gave the leaves under his cuff a final squeeze. A casement window to my right let in cool air, and when he noticed me shivering, he closed it. Though the panes of glass were small, and many were rippled, they didn't block the light.

"Tea, please, Konne." Metenari took the seat across from me. "My dear Fenra, please, tell me how I can help you."

"I am so embarrassed," I said. "I thought I was helping a poor villager safeguard his rights, and his inheritance, and I was tricked . . ." I rested my forehead on my right hand and shook my head.

"You thought his rights were going to be abused? That the Courts were going to ride roughshod over him?"

I sneaked a peak at him from behind my hand. "Something like that, yes."

"Why don't you begin at the beginning?"

Exactly what I wanted to hear.

Sometime later another youngster came into the room with a fresh pot of tea, clean cups, and new napkins and tidied the table before carrying away the used things. I had managed to answer all of Metenari's questions, and I hoped I had answers to the ones I was waiting for.

"One thing, Fenra. How were you able to return from this 'New Zone'?" His smile didn't fool me. His eyes were hard as diamonds.

I hoped my own smile looked more real. "That was where Dom Karamisk saved us. He knew of a place where legend spoke of a stranger appearing from nowhere, and took us there. As it happened, the gate had been left prepared, ready to open, not unlike the one in the vault. It was tricky, but all I had to do was activate it."

Metenari nodded and smiled in satisfaction. "I surmised it had to be something of the kind," he said. "I knew you would never have been able to create a gate yourself, without the *forran* in hand."

"And perhaps not even then." I hoped my own expression was as humble as his had been satisfied.

"And so this Karamisk fellow was only helping you? He had no other interest?"

I shrugged and wrinkled my nose a little. "He's an adventurer, and you know what they are like." I selected a chocolate biscuit from the plate. "Always looking for a new way to die." I shuddered and bit into the biscuit, savoring the crumbly texture. "He has only stayed with Arlyn now so that I could get away and warn you."

"You couldn't control a mundane?"

This was the tricky part. Metenari still thought Arlyn was an ordinary mundane. I had to be careful not to give him away. "I made a vow," I said, shrugging as if a little embarrassed. "When I made the village my base, they made me swear I would never practice on them without their free will and consent. You know how superstitious the outer Modes can be." This appeal to his superior knowledge sat well with him, as I had known it would.

"Oh, Fenra, you should have known better. This is precisely why we've instituted review by the Court Council, to prevent this kind of carelessness."

I lowered my eyes. I knew he would see it as submission, but I was really hiding my reaction to his words. *When did the council start "reviewing" practitioners?*

He leaned back into his chair, saucer in right hand, cup in left, and stopped with the cup lifted halfway to his lips.

"Getting back to the Godstone for a moment," he said. "Did he ever mention . . . that is, does this Arlyn know where in the vault it is? I mean," he added when I looked at him with my most stupid face, "was the precise location part of the knowledge passed down to him?"

<hr>

ARLYN

"It's time," I said, getting to my feet. Twice I'd tried to get Elva to talk to me while we waited, but he refused. Stubborn. "Lend me one of your pistols." I lowered my hand only when Elva didn't move.

He looked at me with one eyebrow raised. "I'm not going to arm you," he said finally. "I'm not here to *protect* you. I'm here to make sure that you do as you've promised. That this isn't some scheme to use the Godstone after all, or to get your own power back from it. One finger flick out of line and I shoot you. And these," Elva pulled a pistol out of his left-hand holster, "will kill you for certain. These are revolvers. Six bullets. Six chances to make sure you're dead."

Now I could see it up close, his "revolver" only superficially resembled the pistols I'd seen being used in the City. Where the shot would normally be placed, the powder charged, was a cylinder. I reached out to touch the gun with the tips of my left fingers, but Elva moved it away. Not that touching it would have told me anything.

He studied me for a moment longer, his dark eyes unreadable, until he finally reversed the pistol again, and slid it back into its holster. "Let's go."

I wanted to say something, though I didn't know what. Elva believed what he believed. Bringing it up again wouldn't help. Nothing would change.

I expected him to slow as we approached the gate to the tower block, but he strode right up to the archway as if he had important business and didn't expect to be denied. Sure enough, the alcove to the right of the entrance was empty. Fenra had done her part. Now it was our turn.

"This is what the man gets for using apprentices instead of guards," Elva muttered under his breath.

"What do you mean?"

"There should be two guards here, and another inside," he said, as we started up the main staircase. "Post guards only at the perimeters, and anyone who manages to get past them is home free."

"I remember you explaining that to me once before," I said.

When he didn't bother to respond, a hot spot of anger in my chest flared up and I swallowed it down. Fenra was right, I didn't have time for this. The corners of my mouth trembled as I clenched my teeth.

The door to my workroom was closed. I had barely brushed it with the tips of my fingers when it was yanked open from within. The young man staring at us wore an apprentice's gray jacket but no black skullcap. His light brown hair was brushed back off his face, tied at the nape with a blue ribbon. He was a little taller than me, blue eyes made brighter by tanned skin.

"We are to wait here for Practitioner Metenari."

For the first time, as if I was listening with the apprentice's ears, I noticed a change in Elva's accent. I'd heard that tone before, though, the confidence of somcone in authority, who was too polite to prove it.

So I wasn't surprised when the boy unconsciously moved back half a step, giving us room to enter. Elva walked immediately into the space.

"Dom, I don't think—"

As he passed the boy, Elva's right elbow shot out, striking him on the side of the head. As the boy fell, eyes rolling up, Elva caught him and cradled him in his left arm, the fingers of his right hand at the boy's throat.

A movement, and a blond boy watched us from the opening of my vault. I could see the whites of his eyes.

"Help me. Quickly," Elva said to him. "He's fainted."

"I can't. If I leave the doorway, it will close."

"You." Elva turned to me. "Don't just stand there, take over so the apprentice can help me."

I dashed forward and took the blond boy's place. I wasn't sure what Elva was going to do, but then I hadn't known he was going to knock the first boy out. The blond helped Elva lower the first boy to the floor and knelt beside him. Elva drew the pistol out of his left-hand holster. I opened my mouth to stop him. The sound of the shot would be bound to bring others. But Elva only reached over and knocked the boy behind the ear with the barrel of the gun.

"Bring one over here," I said.

Without speaking, Elva dragged the smaller of two boys by his collar, careful not to knock him against anything as he pulled him around the furniture. I resisted telling him to hurry. Once he had the boy lying on his side in the opening, Elva straightened, looked around the room for the first time.

"Not much changed," was all he said.

"Come on," I said.

Elva drew his second gun. He tapped the unconscious boy with the pointed toe of his boot. "He makes a fine doorstop," he said. "But from here I can watch both rooms, and act if necessary."

I knew what he meant by act. I drew in a lungful of air, but finally I said nothing. Just turned back to my vault.

Everything looked the same. Everything *was* the same. Except me. I wasn't Xandra Albainil anymore. None of the items stored in this room were mine, not in the way old Medlyn Tierell's things were his. And nothing would change that for me now. I'd never be that person again. I took in another deep breath and let it out slowly.

"These boys won't sleep forever," came Elva's voice from the door.

Flash of irritation. Even more so because he was right. What if this didn't work? What should I do? I took another step, stopped. My body felt heavy. I wished I could just turn around, go home. I wished Fenra were with me. I hoped she was all right.

I gave myself a mental shake. Four more slow steps took me to the cabinet that held the Godstone. It looked like an ordinary piece of pine furniture, tall, shallow, made beautiful by its simplicity and its perfect proportions. My cousin made it for me. The real Arlyn.

I ran my hands over the surface. Cool. Smooth. The lock looked like a simple brass thumb latch. In a way, that's all it was. No *forran* simple or complex could open it. Like Medlyn Tierell's locket, it took something special to trigger it. I licked the tips of my fingers and thumb, reached out. I pulled my hand back without touching the latch.

Would the Godstone know me again? I'd left it dormant, so I should be able to carry it out. Should. But I thought I'd have Fenra with me, able to act, if anything got out of hand. I hadn't really thought past this point. All I'd thought about was making sure no one else got the stone.

"Xandra."

Something in Elva's voice made me peer around the edge of the cabinet, get a better view of the door. He stood braced against the wall, as if keeping himself upright by will alone. That would have been alarming enough in itself, but his arms hung down at his sides, as if the guns were suddenly too heavy to hold. He licked his lips and swallowed. His lips said "Xandra," but he made no sound.

I took a step toward him and caught hold of the side of the cabinet with one hand. My legs felt thick and heavy. I took two more steps. This time I couldn't lift a hand to steady myself and I fell against a chest of drawers and slid to the floor.

I hoped no one heard what Elva had called me.

FENRA

Metenari believed my story—at least to the extent of bringing Elvanyn straight to me when he and Arlyn were caught in the vault. We were taken to another building, and shown into a pretty sitting room, obviously part of Santaron's actual suite of rooms. Arched windows looked out into a balcony screened with thin, intricately carved panels of wood that kept the place shaded against the worst of the sun—and obscured any view of the outside.

"All he had to do was stay away," Elvanyn said, and not for the first time. He stood at one of the windows, but he was looking at me.

"I am not so sure," I said. I sat in a green-velvet-covered wingback chair, a little low for me, but otherwise comfortable. "Once a practitioner like Metenari learned the Godstone existed, he would never stop trying to find it. Could Arlyn take that chance?"

"But . . . Arlyn," he said the name with difficulty, "hasn't changed. He still thinks he knows better than anyone else."

"He and Metenari have that in common, then."

Elva leaned his shoulders against the window frame, crossing his arms and looking at the floor with a stiff face. "If he'd stayed away, you'd have been safe in your village—"

"And you would still be High Sheriff of the Dundalk Territory."

At this he looked up, pushing his fingers through his curly hair. His dark eyes stood out against the pallor of his skin. He looked a little strange until I realized his guns and his sword had been taken away. Until now I had never seen him unarmed. Not even in his sleep.

"I'd have stayed in Dundalk forever, if it kept this world—if it kept *you* safe from the Godstone."

Irony, I thought, *that's exactly what Arlyn said he was willing to do.* What, in fact, he thought he had done, exiled Elva forever. I pulled my knee up and rested my chin on it. I had heard all of this already. "Tell me, how is it you understand the Modes?"

Elva sat down at the end of the sofa nearest me, a small half smile on his lips. "Xandra told me about them. How no one but practitioners can see them."

"Why would he do such a thing?" I had always felt it would be cruel to tell people about things they could not see for themselves.

"He kept nothing from me, in those days. At least, that's what I used to think. Since then, I've had plenty of time to reconsider. Eventually I realized he'd also been conducting an experiment, seeing if he could make me aware of the Modes."

"Whatever for? Did he think he could make you into a practitioner?"

"Could be. He never actually said. It isn't something any normal person would want, no offense."

I was not sure whether I was offended or not. For me, being discovered as a practitioner had been a salvation. I had never wondered whether other people might see it differently. Of course, I *had* wondered if everyone who could see the Modes revealed themselves. "But the Godstone," I began.

"When he was sure I wouldn't ever see the Modes . . . well, he got interested in other things. First he discovered what he called the New Zone—and a few others, he said, though I never saw them—and there he found something unexpected."

"No Modes."

"Exactly. No matter how far you travel, in whatever direction you take, you are always in the same Mode."

I rubbed at my eyes and lowered both feet to the floor. This was too much. When was the last time I slept? "Perhaps there *are* Modes and not even we can see them."

"Stop. Take a deep breath. Here." He fetched a pitcher of water from the sideboard and poured me a glass. Slices of lemon made the water tart and somehow more refreshing. I handed him back the empty glass.

"Thank you."

ARLYN

I didn't know where we were. Could have been a private sitting room in any gentleman's club. Had that look of impersonal comfort. Everything matched everything else, but nothing, not wing-backed chairs, not small tables with beautifully carved legs, not richly colored carpets, not even ornaments, seemed to be anyone's personal choice. Eyes shut, I leaned my head against the high back of my chair. Couldn't hear anything, windows and doors closed. Could smell flowers in the vase on the table next to me. Fenra could have told me what they were.

"You do understand, Arlyn, that you wouldn't be able to do anything with the artifact even if you had been able to find it."

Stopped being Dom Albainil while I was unconscious, I guess. Could think of half a dozen responses, but none worth making. Too much trouble to lift my head, open my eyes. How had I become so low so quickly . . .

"Arlyn."

Idiot tapped me on the knee. I slapped his hand away, energized by sudden anger. "You don't know what you're talking about," I said. "Why do you think it's locked away? This isn't some experiment in a lab, you clown, it's dangerous. It could destroy the world."

Now my eyes were open I could see his face change color. Red didn't suit him.

"I assure you, my knowledge of the Godstone is far more extensive than yours, no matter what lore has been passed down in your family. Come, tell me which cabinet the artifact is in, you'll save me a good deal of time. Then you can go home to your village."

Knew I should tell him the truth. But the thought of having to convince him—there was just too much. Didn't know where to begin. Wouldn't believe me anyway. And really, if he destroyed the world, was it such a bad thing? Wouldn't it be better if everything just stopped? Elva could stop hating me, Fenra could stop leveling me, following me around, helping me. She could lay all of that down, she could rest forever, if this clown destroyed the world.

"Where's Fenra?" Couldn't ask about Elva. He'd want to know why I cared, wasn't going to tell him.

"Practitioner Lowens is resting. You led her on quite a chase, you know. If you had only told her the truth about the artifact in the first place, all this ado could have been avoided."

Ado? What kind of practitioner said "ado"?

Now he looked at me with his head to one side. "Arlyn, are you well? Was Practitioner Lowens helping you in her capacity as a healer? Here." He held out his hands as if it never occurred to him I would refuse. "Let me check you," he continued. "I don't specialize in healing, but Fenra's only a third-class practitioner. I may be able to do more for you."

Easier to go along. Put my hands in his, gripped them. Warmer than Fenra's, smoother. Soft where hers had callus and muscle. Felt nothing, no leveling, not that I expected any. Didn't know if I was disappointed. He grunted, flexed his hands to free mine.

"Strange," he said. Now his head was tilted to the other side. "There's an emptiness in you. Is that why you wanted the artifact?"

"Why do *you* want it? Why did you want it badly enough to forge a testament?" Again, anger gave me energy.

"I'm going to finish what Xandra Albainil didn't have power or courage enough to finish." Metenari sat up straight. I swear he pushed out his chest. "I'm going to repair the world."

Would have been funny, any other time. Let my forehead fall into my palm, hand propped on the padded arm of the chair. How comfortable this chair would be, if this was another day, another place. Saw everything that was coming, every step. Couldn't see a way to stop him. Wasn't sure I wanted to. Fenra would understand. Elva—well no, he wouldn't. Didn't the first time, wouldn't this time either. Wish I was a soldier. Then we wouldn't be here now, would we?

"What is it you find funny, Arlyn?" He was rubbing the palms of his hands together, as if they were suddenly cold.

⟍⟋

Fenra

One of Metenari's older apprentices, a young man called Predax, came in with Elva's weapons, the sword carried by the hilt, and the guns by their holsters.

"They've been checked for *forrans*," he said, before I could ask him. I suppose the practitioners felt themselves in no danger from mundane tools. "Please, finish your lemonade and cookies. I'll come back in a while to show you to your rooms." His nod was a little formal, but his smile was warm enough as he left the sitting room.

"We should go." Elvanyn finished buckling on his holsters, hung his sword on his belt, shrugging until everything felt right. "Before they realize they haven't told us to stay."

I stayed in my seat on the sofa. "Where?" I said. "No, I'm not disagreeing with you, but if Metenari gets the Godstone—"

"He won't." Elva went through his weapons ritual. "First, he can't make Arlyn tell him where it is, and second, even if Arlyn tells him, Metenari won't be able to open the cabinet. No one can open anything the great Xandra Albainil has locked. Never have, and never will."

An odd note of pride lifted his tone.

I shook my head. "I keep telling you, Metenari is not stupid. My mentor told me there's a way to use a family member to open a sealed vault. Metenari admitted that was his intention. So, we cannot leave without Arlyn."

Elva got down on one knee before me and looked up into my face. He took my hand. Such dramatic gestures. So many of his gestures were, I realized, as though he acted always for some unseen audience. "Is there another reason?" His eyes searched my face.

"Arlyn is my friend," I said. Elva's hands were warm, and I could feel sword calluses. So different from Arlyn's. "And besides, the village children like him."

Elva looked at me for a long time, and finally smiled. "In that case, we'll rescue him." He raised my hand to his lips and kissed it.

ARLYN

"Xandra didn't lock the thing up because he was afraid," I said. "It doesn't do what he designed it to do." Metenari looked thoughtful. I continued, "You know he was a great designer. Look at all his *forrans* that are still in use."

"If it doesn't function properly, how can it be so dangerous?"

Clenched my teeth. Screaming at him wouldn't achieve anything. Too smug. "It's *because* it doesn't function properly that it *is* dangerous. It doesn't just sit there. It changes the world all right, but not in the way Xandra planned."

"Come now. It's obvious what happened. No one as acclaimed as your ancestor would want a story of failure passed down in his family. No, he'd tell a tale of his brave and fatal action to save the world from an artifact too dangerous to use."

Rubbed my face with my hands, massaging around my eyes with stiff fingers. This fool would never listen. Not to a mundane. Never. Closed my eyes, felt the world fall away. Would be peaceful, wouldn't it? My worries, everyone's worries, gone.

"I'm not Arlyn Albainil," I said aloud, opening my eyes.

His face sharpened, eyes narrowed, lips pressed together. "Who are you, then? How did you know how to find the workroom?"

"I'm *Xandra* Albainil."

A flicker of emotion crossed his face too fast to identify. Fear? Doubt? Uncertainty? Then he closed his eyes and sighed. "I suppose I should have expected that. If you are the great and powerful Xandra Albainil, how is it you're sitting here, unable to leave until I allow it? Why haven't you wiped me and my apprentices from the face of the earth?"

My turn to sigh. "Even in my day that kind of thing would have been frowned on. Maybe I was more powerful than anyone else, but I wasn't more powerful than *every*one else."

"But Xandra would be stronger than I am, right now, wouldn't he? Why not get up and walk away? Take back your workroom and your vault? Rejoin the White Court? Why don't you stop me?"

"Fenra said you had no imagination. Think! If I could have, I would have destroyed the Godstone and that would have been the end of it. As it was, it took everything I had just to lock it safely away." Almost told him what I'd felt at the time. That horrible fear that it wasn't completely dormant, that I had only tricked it into sleep, that without trickery I would have failed. "Do you understand me? I was the most powerful practitioner in the White Court, and it took *all* my power. And you think you'll just free it and use it?"

"Arlyn, you must see how utterly ridiculous your assertion is. Nothing removes a practitioner's power—gods know enough people would have tried it. Practicing isn't an easy life." He slapped his hands lightly on his knees, looked over his shoulder at the apprentice standing at the other end of the room.

"Noxyn, has there been any further progress?"

"I'm afraid not, Practitioner. We think we've narrowed it down to three cabinets, but we can't open any of them. Perhaps if you . . ." The boy arched his eyebrows and glanced my way.

Metenari nodded, looking at a spot on the carpet before he got to his feet. "Bring him," he said as he turned away from me.

The light looked different. No windows, so not that. Was it so many others in the room? Somehow I expected to see Fenra and Elva, don't know why. Three, maybe four, of Metenari's boys scattered about. All boys, huh. Some angling closer out of curiosity, some inching away for other reasons. One stayed in the doorway, chewing on a fingernail.

"We'll begin here." Metenari stopped in front of the wrong cabinet. Lucky. One in three. "I'll trouble you to put your hand on the latch."

Don't know what, left to myself, I would have done. As it was, two held me while Noxyn took hold of my practitioner's forearm and hand. Metenari brought his hands, lightly pressed palm to palm, up to his lips and began blowing air into them. When he pulled his palms apart, a faint glow, like breath on a cold day, hung between them. Seeing his mentor ready, with his own hand holding mine, Noxyn placed my fingers on the latch, just as if I were intending to open it. Glow settled over my hand. Curled my fingers through the handle, put my thumb on the lever.

Pressed down.

I laughed at the look on their faces. Just to myself, quietly. Tears in the back of my eyes.

"It's a low day," I said out loud.

FENRA

Elva eased the door closed again. "Not locked," he said, "but there's a guard in the corridor, a real one this time. Not in uniform, but no one stands looking for that long at a picture he must have seen a hundred times already." He swept off his hat and dusted it with the back of his hand. It had been clean to begin with.

"If we could get him to come in, I could knock him out."

"Can't you do it from here? That's what Xandra would have done," he added when I frowned.

"What Xandra Albainil might have done isn't the guide I would use myself," I said. "I give people sleep all the time, if they are in pain. But usually I touch them. Casting such a *forran* from a distance . . ." I shook my head. "It doesn't feel right."

Elva settled his hat on his head, tugging down one side of the brim, and pulled out one of his strange pistols. "I can easily shoot him from here," he said. "But I'll bet the noise will bring someone running."

I lifted one eyebrow. "You have convinced me to try. And stop smiling."

I placed my hands palm to palm and thought about the *forran* for sleep. Restful sleep. It wasn't the first time I had seen my pattern when practicing, but working with Arlyn had me looking at it in a different way. I mentally traced it, watching it strengthen. I pulled my hands apart. The pattern moved, forming a cat's cradle of light between my fingers. I blew on it, Elvanyn opened the door, and with my practitioner's hand I flicked the light in the guard's direction.

The man shook his head, as if a fly bothered him. Points of light bobbed around him. I repeated the *forran*. He leaned against the wall, and slid down until he was sitting on the floor.

Asleep.

"That was well done." Elva led the way out of the room. "Will he wake up?"

"Of course."

"Ah. That wasn't always true when Xandra did it."

"Again, not all that interested in doing things his way."

ARLYN

I could have told him the lock wouldn't open. Wasn't for the reason he thought, however. I remembered Elva's telling us about fingerprints. My pattern in my hands. Yes, that might have worked. Maybe on the right cabinet, if I wanted it to. Which I didn't.

"There, you see how easily I can prove your story is nonsense? If you were in fact Xandra Albainil, the combination of the *forran* and your skin touching the latch would have opened the door. Fortunately for me, there is another way. Though it's unfortunate for you, I'm afraid."

Sounded sincere. Must have been telling himself he was working for the greater good. That the sacrifice of one mundane was worth it. An argument I recognized.

They held me down, cut my right wrist, lengthwise, not across the veins. Clots slower that way. Watched Metenari weave his hands in the air. Knew what he was doing, though couldn't see it. Extract my pattern from my blood. There, for anyone who knew how to look.

Should have banned this *forran*, along with some others I designed.

When he lowered his hands wet surfaces glinted in the light. Wet red surfaces, as though he wore gloves made of blood. He wouldn't frown so much if he knew it made him look like an angry pig. I shivered. Cold floor. Never noticed it before. Voices over my head as they lifted my practitioner's arm. Too much work to open my eyes. Shifted me over to another spot. Warmer. Closer now.

Felt the light and the heat when the door opened. Heard voices calling, feet thumping.

Think I smiled.

FENRA

The gas lamps were hung in such a way that they highlighted the beauties of the courtyard, though their light didn't penetrate very far into the shadows under the arcade. Elva heard someone coming long before I did and drew me with him into an even deeper shadow.

"If it looks like they can see us, I'm going to hold you against the wall and kiss you."

I almost smiled. "I've read that book," I told him. "Actually, I think everyone has read that book. Trust me, if they are guards they will check." I was about to tell him there was another way when his lips brushed against my forehead and I felt his breath warm against my skin. He smelled like chocolate.

The entrance at the far end of the court was flanked by two enormous doors, perhaps eight or ten feet tall, deeply carved, with huge iron hinges, and bolts as thick as my forearms. They'd never been closed in my memory, and for some reason this was the time I picked to wonder if they could be. We were off to one side, and had only an oblique angle on the corridor beyond the doors. Which meant whoever was in the corridor would have the same angle on us. And we stood in shadow, while from the look of things, they carried lights.

The glow came nearer, and we saw the silhouettes of three men passing the open doorway. I thought the shorter one might be Predax, but he went by before I could be sure. Normally people carrying lamps would be at the front and at the back of any group moving at night, but their hands were empty, and the light came from behind them.

"What light is so bright?" Elva's breath tickled my ear. I just shook my head and tightened my arms around his neck. Something practical, for certain. No natural substance could give off that bright a light. Not gas lamps, and certainly not oil.

As the light neared the doorway I found it hard to look directly at it, and my watering eyes made it even more difficult to identify. I blinked. That couldn't be right. It looked like a person, walking slowly. It had Metenari's stocky shape and size, but it walked taller, with shoulders straighter and pulled back. Elva's left arm held me closer while his right hand loosened and drew out a pistol, holding it close to my side, where the skirts of my coat would hide it.

As the glowing body reached the midpoint of the opening, it turned its head, and seemed to look directly at us. For a moment I thought it was about to raise its hand in greeting. I did not need the tightening of Elva's arm to keep me from moving. But the body didn't stop; it kept its steady pace across the open doorway, heading to Metenari's suite of rooms.

"The Godstone," Elva whispered against my ear. As if it could be anything else.

"Did you see Arlyn?" I whispered back as the last of the attendants passed by and the opening turned dark again.

I felt Elva shake his head. "If they left him in the vault, or the workroom, we may already be too late."

"We still have to look."

This time I felt him nod.

I do not know how a man carrying a sword, two pistols, and belts full of bullets crossed over his chest can move so quietly. Even his footsteps were silent.

We found the entrance to the old building unguarded, and Elva gave me a pointed look. "Not a good sign," he said. "It's telling the world there's nothing of importance left inside."

"From your reaction, you've seen the Godstone before?" I had to ask.

"Once. I wasn't sure at first, but I recognize the light. It's a . . . I've seen it look like a crystal." He held his hands about a foot apart to show me the size.

"Was he carrying it? Metenari?"

"Or it was carrying him." We passed a window, and I saw Elva's eyes flash in the moonlight as he glanced at me. "That's one of its dangers."

"Did it always glow like that?" I tried to imagine how it could have been kept a secret all these years.

"Not at first, and then not unless Xandra activated it."

"So it's activated now." I tried to keep the rush of panic from overwhelming me. Metenari never actually said what he planned to do. He might have some benign idea, but so had Arlyn, and look how that had ended.

"Here we are." Elva put his hand to the latch of Arlyn's workroom door and hesitated. "Can you tell whether they left him in the vault?"

I concentrated for a few moments, and shook my head. "I should be able to. I have leveled him so many times I have a feel for his presence. Once, when he was lost, I found him in the woods where the villagers hunt for boar." I shook my head again. "But I sense nothing now."

"As you said, we still have to look."

"And the vault?"

"If he's inside, then it should open for me—unless he changed it, you know, afterward." He laid his palm flat on the door. "It was set up so that I could come in, but only if he was inside." He shrugged.

"He must have trusted you very much," I said.

"He did. Unfortunately, he didn't also listen to me."

Elva pushed the workroom door open with the barrel of the gun in his right hand and started to step in. I caught him by the sleeve, and jerked my head back. He grinned and made a flourish with his hand, inviting me to enter.

I smelled blood as soon as I cleared the doorway. Smelled blood and heard the faintest of breaths.

"Sweet holy god," Elvanyn said. I could tell he was cursing, though the idiom was unfamiliar to me.

I ran to Arlyn's side and knelt in the blood. His blood. I took a wrist in each hand, and instantly I was in a thick, briny fog. *Of course*, I thought. *Low* and *bleeding to death*. I ignored the approaching dark form. I did not have time to waste. I swept the fog away. No time to be gentle. I found Arlyn sitting up against the rock he normally sat on. He grinned when he saw me. His lips and gums were pale.

"Fenra," he said.

"Shut up." I reached into him and found his pattern. Faded like an old tapestry. Some spots seemed completely worn away, and my heart sank. I had seen the thing recently; would that be enough? Practitioners' memories were carefully trained.

Somehow I felt Elva turn as he straightened to his feet, pulled out his guns, and headed for the door.

"You'd better hurry," he said.

Eight

ELVANYN

WITHOUT LOOKING UP, Fenra said, "It takes as long as it takes." She was speaking to him, but she sounded far away. "People are coming," he told her. "We don't want to be here if it's Metenari. If they could leave *him* for dead," he gestured at Arlyn, "imagine what they'll do with us." Elva flexed his hands, tightening and releasing each muscle in his arms and shoulders, shaking loose each foot and checking the lay of his pistols. "And what happens to our cover story? They've already caught me here once."

Fenra looked up at him. She still had Arlyn's wrists in her hands. "I can't move him yet."

All the blood on the floor was gone, as if the body—as if *Arlyn* had re-absorbed it. When he had time, he'd ask her how she'd managed that. He had the feeling that there were many questions he might never have time to ask.

"You'd best figure out a way. If they reach this door they block our escape. We have to move *now*. We can carry him up the tower stairs. These aren't soldiers or hunters, they won't think to look up. That may buy us some time."

She let one of Arlyn's wrists go, and pulled the locket she wore out of her shirt with her left hand. It seemed impossible, but like the floor, and her hand for that matter, it was clean, without any trace of blood. She looked at him, back at Arlyn on the floor, and then back up at him, her lips very slightly parted. He understood. With Arlyn unconscious she couldn't hold on to both of them and the locket at the same time.

"Take him," he said. "Go."

"I'll come back for you."

"I'll be here." He turned away so as not to see her go. He drew both guns and took up position in the doorway. The enemy would be coming up from his right. He'd see them before they saw him.

He knew this tower well—he'd been a guard in the White Court most of his life before the exile—and the echoes told him they were at least two levels away, approaching slowly. He could hear the sound of boots, and voices, and then, the soft click of a pistol being cocked. With his revolvers he had the range on them, so he'd aim high. *Avoid killing when you can*, is what he'd always told his deputies. He'd have to hope that the range of *forrans* hadn't improved.

Shadows appeared against the curve of the staircase wall. One more turn and they would be in sight.

FENRA

I could not hurry. Healing takes its own time, always. But I knew that every minute I spent on my knees in Medlyn's vault, holding Arlyn's wrists in my hands, was a minute of danger for Elva. If he was right, and the Godstone rode Metenari—a possibility Arlyn had never mentioned—then anything could happen. I began to hum, and to sing under my breath, in part to support the *forrans*, and in part to help myself relax.

After what seemed like an eternity, the cuts on Arlyn's wrists closed and disappeared, leaving only the old marks I had never been able to erase. I had returned most of his blood to him while we were still in his workroom, but until the cuts closed I could not let go of him lest the blood pour out again. Now I was able to set him down on the well-padded sofa, and cover him with a throw woven from silk and linen. I wondered why my old mentor would have had such an item here, but I chased that thought away. I had to concentrate.

Most of his blood wasn't all of it, and I had to find a more mundane way to replace what was still missing. Using real liquids was fastest and least tiring. Most people do not realize it, but practice is, more often than not, only a part of healing—albeit a large part. The rest is common sense and everyday medicines. I thought I had seen . . . yes, a pitcher and a small tray holding three glasses sat on a table in the corner. If it had ever held liquid, I should be able to call it back again. It's the nature of jugs to hold liquids, and the nature of liquids to be held by jugs. I took a deep breath and forced myself to let go of Arlyn entirely.

The jug was not only already full, it was cool to the touch. And if I needed any further evidence of the breadth of Medlyn's practice, the contents smelled like juice, not water or wine. I dipped the tip of a finger into it and tasted. Vegetables, root vegetables to be precise. Carrots, parsnips, and yes, beets. No parsley, salt, celery, nothing to help the flavor. I made a face. Well, I was not the one who had to drink it. I grabbed the jug and a glass and took them to Arlyn.

I propped Arlyn up on the arm of the sofa, using two soft pillows and the folded lap rug. His eyes opened, just slits, but I thought I could see some recognition there. "Drink this," I said, hoping that he was conscious enough to help me. Pouring liquids into the mouth of an unconscious person is trickier than most people think. I set the edge of the cup against Arlyn's bottom lip and tilted it slightly. I thought about the many hundreds of times I had done this, with children, with adults old and young, even on occasion with animals, though I rarely used a glass for them.

Arlyn's lips moved and he managed to swallow most of what was in the glass without spilling too much down the front of his shirt. I would need to find him clean clothes, first thing.

No. First I had to go back for Elva. I fed Arlyn another two glasses of juice, as much as I thought his stomach could take at one time. I pulled up his left eyelid and examined the whites of his eyes and the color of the lid's underside. I pulled down his lower lip and examined his gums, and I pressed the palm of his hand with my thumb and watched how long it took for the color to return. The results of my practice were better than I had hoped. I could do no more without draining the strength I would need to retrieve Elva.

"Fenra."

I took Arlyn's groping hand in mine. "You do not need to tell me Metenari has the Godstone. We saw."

"Sorry."

"So I should hope. I hope you have learned your lesson and will give your word never to do such a thing again."

The shadow of a smile passed over his face. "I promise."

"Good. Now if you think you can rest here quietly, without trying to get up, or anything else as foolish, I have to go—" Arlyn's grip tightened, but before he could say anything, I continued. "I must go back for Elva," I told him. "He held the door for us, when I could not bring you both at once."

"Help me sit up, and get me a pistol. Just in case."

"No one can get here without me," I said, though I helped him anyway. He would do better sitting up.

"It's not impossible for someone to overpower you, force you to bring them here," he told me when I had him sitting up.

I opened my mouth, but closed it without speaking, shaking my head. I did not have time to argue. I pulled two of the pistols out of the display on the wall above the worktable. There wasn't much chance of coming across weapons like these in the outer Modes, but I remembered how they worked and brought Arlyn shot and powder as well as the guns. As soon as he had them in his hands, he could see them, as he had previously seen the book he was holding. While he loaded and charged the first pistol, I looked into a walnut clothes press standing against the opposite wall from the gun display, just in case. My clothes were filthy.

There were clothes in the press, but oddly there appeared to be only one set. I picked up the shirt and held it against myself. Just my size. I pulled off my boots, tossed them to one side, and pushed down my trousers.

"That's very interesting," Arlyn said.

"You've seen me change my clothes before," I said, pulling on the new trousers and buttoning up the flies. It was a pleasure to be wearing something clean.

"But I've never seen you without practitioners' colors."

I stopped and looked down, holding my hands up at shoulder height as I twisted to see myself from all angles. The trousers were a deep rich blue, cut quite narrow, with a small strap across the instep to hold them in place over socks of the same color. I inspected the other clothing. The brocaded waistcoat was white, as was the shirt, the cravat brilliant rose, the jacket a green so dark as to be almost black. And the jacket itself hung swallow-tailed to the knees, held shut at the waist with a single button, with lapels and tails cut so the waistcoat and trousers were revealed. I pulled on the pair of black, medium-heeled shoes that came with my clothing and put on the wide-brimmed hat with a tilt to the left. Elva would have approved.

"Would have been better if the shoes matched the socks," I said, standing up and stamping my feet to set the heels. I tied my cravat and straightened the set of my waistcoat. I shrugged into the coat and examined myself again. "The colors still haven't changed."

"Might not mean anything. We've already established things work differently in this vault. Who knows what colors you'll be wearing once you're back in the real world."

"Wait," I said, with the locket in my hand, "Why didn't Elva's 'revolver' change?"

Arlyn glanced up at me as he tamped down the ball in the second gun. "It's not from this dimension," he said. "Artifacts don't change, remember?"

That wasn't exactly what he had said, but I decided there would be plenty of time to ask questions once I retrieved Elva. I opened the locket.

I was a little disappointed when I found Medlyn's old office empty. I would have liked a chance to explain to the new occupant—and maybe apologize. On the other hand, would she believe anything I told her? As it was, the sun was already well up, and shopkeepers, artisans, and school children would already be out, beginning their daily work. I had to dodge several other practitioners—luckily no one who knew me—before I reached the Singing Tower. Guards and watchers had not been replaced. My stomach sinking, I ran up the stairs as quietly as I could, slowing as I got closer to the door to Arlyn's workroom.

There was blood on the stairs.

I flung open the door, and found the room empty. No Elva. No one left here to capture me. So Elva had not given me away.

Either that, or he was dead.

ELVANYN

"Tell me again why I shouldn't hang you."

The guard captain's uniform resembled Elva's old one in color only, a royal blue cropped jacket instead of tabard, gray trousers instead of breeches. Elva sat as much at ease as he could, pretending that he wasn't being held in his chair by some *forran*. Metenari watched from behind his desk. Elva wondered why he wasn't conducting the interrogation himself.

"What were you doing in that room?"

"Look, I'm just here for the lady. She sends me to go look at a room, I go. I'm a hired gun, as we say where I'm from. I've got no dog in this fight. Hire me, send me on my way, it's all the same to me." He grinned. "Besides, did I kill any of your boys? No, I did not. And why? Because I've got—"

"No dog in the fight, yes, you've said that already. If that's so, why did you attack my men?"

"Hey, now listen, that was self defense. If they hadn't shot at me first, I'd have come along quietly. But a man has to defend himself."

"So the lady, as you call her, hired you?"

Elva leaned further back, testing the limits of the *forrans* that bound him. "Where I'm from a gentleman helps a lady when she's in trouble. I didn't like the look of that guy with her."

"Where you're from . . . ?" Fenra's old classmate straightened in his chair.

"The Dundalk Territories." Elva looked Metenari up and down. The practitioner's shoulders were further back, his posture straighter, and Elva could have sworn the man was slimmer. His head cocked to one side, he turned one of Elva's revolvers over in his hands, probing the mechanism with the tips of his fingers, in a way that tugged at Elva's memory. He'd never seen Metenari handle a gun before, so why should it seem familiar?

"What kind of pistol is this?" The captain had the other revolver in his hand. There was nothing but plain curiosity in his voice.

Elva shrugged, leaned forward until the *forrans* stopped him. "Just a regular gun. A Pope 45, long barrel, if you want to get picky. A top quality gun."

"Why didn't it change, now that it's in a different Mode?" Metenari sounded as if he was asking himself.

"I don't know what you mean. Why should it change?" Elva decided to answer as if the question had been meant for him.

Metenari stood up and came around the table, approaching Elva from the right, still holding the revolver in his hands. He spun the cylinder, watching it move with his eyebrows raised. When the cylinder stopped, he pulled back on the hammer and pointed the gun at Elva. From this distance even a blind man would hit him. Elva smiled. In the heat of the moment, he hadn't counted his shots. The question was, had he fired six shots from this gun, or only five?

Shot with my own gun. Lucky thing his deputies would never know about this. He could hear Lugg laughing now.

"I could shoot you," Metenari said. There was nothing pompous about the practitioner now. His face was cold, calculating.

"You could." Elva was pleased at how steady his voice sounded. "But if you wanted me dead, you'd have killed me already."

Without any change to his expression, Metenari pulled the trigger. Elva blinked. He heard a gasp behind him, but he couldn't turn to see which of the apprentices it was. The gun was empty after all. He let his breath out slowly, smothering another smile. Lucky they didn't know how to reload, even though they had the gun belts.

Metenari rubbed at his left temple with the heel of his hand, face scrunched up as if in pain—or as if there was an itch inside that he couldn't reach. Finally, still holding the gun, he lowered his hands and held them out in front of him, brows drawn together, perplexed. Still holding his hands out, he looked at Elva, head tilted to one side. A glint sparkled in his eyes.

"Do you know me?"

"I met you the other day. You went to magician's school with Fenra." He deliberately used the Dundalk word for practitioner.

"But you knew me before that?"

"How could I?"

"Yes, of course, how could you." Metenari set the gun down and Elva relaxed. The practitioner clasped his hands together and tapped his lips with his index fingers. Elva suddenly felt cold, his skin crawling under his clothes. He knew that gesture, knew it very well. He'd seen it a thousand times. But it hadn't been Santaron Metenari he'd seen doing it. Could this be an effect of the Godstone?

"Look," he said, doing his best to keep both fear and suspicion out of his voice. "It's like Fenra told you. She thought that Arlyn fella was her friend. I don't know what he was trying to get out of his great-grandad's hidey hole, but he took advantage of her to get at it."

"You became her champion very quickly."

"She reminds me of my sister. Look, all I want is to take the lady and go home. You can do that, can't you? Send us back?"

"It seems we have a somewhat common goal. You would like to rescue Fenra Lowens, I would like to find Arlyn Albainil. You believe where we find one, we will find the other. So, we will work together."

"Does this mean I'm on the payroll?"

"You may have a position in the guard—Captain?"

The other man shrugged. "It's true he didn't kill anyone when he could have. There's no question of his skills. If the rest of the men have a different opinion," he turned to look at Elva, his eyes steely cold, "you'll just have to deal with that yourself."

"I'm a sheriff where I come from," he said. "I know how things are."

"One more thing." The smile on Metenari's face was difficult to see. "Do you know where Arlyn Albainil is?"

"What, we're back to that? I told you, I was up in the tower looking for them when your boys showed up and started shooting."

Fenra

All the blood was on the stairs outside of the workroom. Which meant that Elva might still be alive. I began to relax. If he was free, where would he be? We had agreed to meet in the market by the East Bridge if we were separated. There wasn't much hope he would be there, but I had nowhere else to look. I had managed to pick up some dust on the sleeve of my coat, and I froze with my hand in the air before brushing it off. My clothes had not changed. They were still what they had been in Medlyn's vault—which raised still more questions about my old mentor that I had no time to explore.

I brushed off the dust and, feeling a little more confident, went back down the stairs and out into the street. On my way to the bridge, two practitioners I did not recognize walked right past me, only an arm's length away, without even looking at me. At first I was warmly pleased at how well my disguise was working. But slowly, as still more people passed me by without seeing me—not even those who pushed past—I began to feel frightened, and yes, a little resentful. Since I had become a practitioner, I had always been deferred to. What a difference clothing made.

Of course, my clothes would change the moment I crossed into another Mode. At least I thought so. I discovered I was not eager to find out.

I was careful to maintain my pace as I neared the bridge. The gates there had been open for several hours, and tradespeople had already set up their booths of food and other goods that weren't grown or made within the White Court. The market had started years before, when an enterprising spice merchant had set up tables against the wall of the old lecture hall. Soon others came, and tables and stalls spread into the space between the hall and the bridge.

I strolled casually across the small square, watching for Elva, staying out of the way of people carrying things on their shoulders, and occasionally stopping to examine the merchandise on display. I made a mental note of the location of the booth selling meat pies and dried fruits. With luck I would be able to come back and buy some. As I looked ahead, a group of people in the unmistakable blue uniforms of the White Court Guard came out of the alley opposite.

My path would take me right past them, but I counted on the anonymity of my clothing to keep me unnoticed, though I was acutely aware that, unlike everyone else around me, I carried no marketing basket. A familiar movement caught my eye, and I saw Elva among the guard. My heart leaped to my throat, but I kept walking. He appeared to be armed, with his own pistols and what looked like his own sword. He was listening with a smile to something one of the other guards was telling him. I turned my face away, examining a selection of preserved fruits, and walked into the alley the guards had just exited, heart pounding.

He was not a prisoner. Not armed, in uniform, and walking about so freely—though how free he truly was surrounded by guards I could not say. Were they using him as bait?

I also had to consider another, less palatable possibility. Elvanyn might have switched sides. I could not—I did not want to believe it. I had only known him for a short time, but I had come to depend on him, on his humor and on what I had taken to be his sense of honor. But . . . did he hate Arlyn so much? I pulled the locket out of my shirt, and in the first dark corner, when I was sure no one was looking, I opened it.

ARLYN

My arms and legs felt like heavy weights that took all my strength to move. Tried not to worry about Fenra and Elva, about how long it was taking her to come back, but my mind kept circling and circling, as if this was a problem I could solve. Not low, I told myself, just overtired.

Had to rest. My mind continued to spin. My skin was tight, tingly. A fizz of energy swept through my body, as if my blood bubbled. As quickly as it came, the sizzle of energy left me, my body sagged. Let my right hand slip off my chest.

I don't know how long I'd been asleep when Fenra's touch on my forehead woke me. My head felt clearer. I tried to sit up, got black spots in my vision.

"Lie still," she said. "You have not recovered enough for so much movement."

"Where's Elva?" I swallowed. It had taken me years to stop thinking about him, wondering if he was still alive, what he was doing, if he ever forgave me. Now I was worrying about him again. "Dead?" I forced the word out through clenched teeth.

"No." She didn't seem very happy about it, though. "I saw him walking with the guards, armed and free, laughing." She spoke without looking at me, and I answered the question she couldn't bring herself to ask.

"Absolutely not." I spoke as firmly as I could manage. "Elva would never betray us—not even to punish me." I hesitated, suddenly unsure. "Even if he could do it to me, he would never betray *you*." There. On safe ground again.

"I am not a child, Arlyn. I know that 'I want it to be so' is not proof."

"But this is. Elva has the same goal he's always had, stopping the Godstone. If it looks like he's changed sides, he's only pretending."

Still without looking at me, she shoved a warm meat pastry into my hands. My mouth began to water. "Medlyn always told me, 'Plan for the possible.'"

"Then you'll just have to take my word for it that this is *im*possible. I know him, no matter how much time has passed. His honor means too much to him. That won't have changed."

I bit into the pastry before she could press me. A small part of me whispered that I had changed, changed drastically. The same thing could have happened to Elva. I'd seen no signs of it, but it was certainly a possibility. "In any case," I said aloud after swallowing, "we need a plan. Do we rescue Elva? Or do *we* keep *our* eyes on the Godstone?"

Her head came up, and she turned toward me, an apple in her hand. "If Metenari has the stone . . ."

I reached for a piece of dried apricot. "He doesn't know how to use it. No matter where he looks, I never wrote that down. He still needs me for that."

"Could Elvanyn be of any help to Metenari? Did you never give him any . . ."? She was paring an apple and waggled the hand that held the knife. "Any instruction?"

"What would be the point?"

She leaned back against the edge of the table, the knife forgotten. "It's so unusual for a practitioner to have close friends outside of the practice, I thought it was possible . . ."

"You and I are close friends. When you only knew me as Arlyn, did you ever discuss the nature of practicing with me? Show me any *forrans*, recite any for me? No. You didn't. Neither did I."

"So we stay focused on the Godstone? We forget about Elvanyn?"

"Yes, we do, and no, we won't. We'll watch for him, just as he'll be watching for us. He may even have been out in public in the hope that you would be somewhere around, that you would see him. We won't count on him, though. He might not be in a position to help us without endangering himself." She looked at me, a skeptical twist to her left eyebrow. "Trust me, he'll be looking for us just as hard as we'll be looking for him."

"Can he manage that kind of role?"

"All he has to do is act like a soldier."

This one's so ignorant I don't know what to do. This one thinks he discovered me through his hard work and his research and his cleverness, but it was more accident than intention, and now he's posturing around as the great man, the powerful practitioner—he probably wants that inscribed on all the statues he'd like made for himself. But what now? Now that this one has me, what can he do? *He* left power with me, all the power needed to do whatever I like, but *he* took the knowledge with him. *He* has the *forrans*. Power without them will fail, just as *forrans* fail without power.

This one's only just starting to realize that he doesn't know what he needs to know. That's how stupid this one is. I can't work with this. I need *him*.

And *he* needs me.

Someone knows where *he* is. When I find him, I'll know what to do.

Now that he's alone, this one looks down at his hands. Or are they my hands? I'm not sure. "This can't be good."

⟨⟩

ELVANYN

It took Elva a couple of days to realize that his discomfort didn't entirely come from playing a dangerous role. The barrack room he'd been assigned was so familiar, and yet so different. The beds weren't where they had been in his day, the walls were a different color. The kitchen was in the same place, but the stoves were self-contained in metal boxes, and the forest mural in the dining room had been painted over. He'd spent his first few years in the New Zone imagining what it would be like to be home. Somehow this didn't feel as wonderful as he'd expected.

That wasn't what was bothering him at the moment, however. All his experience told him he shouldn't have been accepted so easily, even if this Metenari was nothing more than a gullible gasbag. He didn't know what or how much the bulk of the Guard had been told, but he could guess. For now, the other soldiers recognized him as a soldier like themselves, even though he was a stranger to them. They treated him as they would have any recently recruited veteran, with reserve, testing his abilities and seeing how far he'd let them go before he pushed back.

Not that he trusted any of this. He'd been given back his guns and his sword, but what better way to keep an eye on him than to surround him with forty other people and give him chores to do, errands to run, and watches to stand? Today, for example, he'd been sent into the City as bodyguard and muscle for Predax, one of Metenari's senior apprentices, on a routine errand to the Red Court. Elva grinned when he knew no one was looking. The hard part would be pretending he didn't know the way.

He found it odd, however, that—bodyguard or no bodyguard—Predax seemed happy, even relieved, to be going out.

After fifteen minutes or so, Elvanyn started to think differently about the seriousness of his assignment. He'd expected them to leave the palace on foot; in his day practitioners had always made themselves as accessible as possible. Today, however, people on the street ignored the young apprentice, their eyes lowering and their shoulders stiffening. Elva had walked these same streets countless times with Xandra Albainil. The way he remembered it, people passing in the old days would nod pleasantly to Xandra, as if they were acquainted. Today they pretended not to notice Predax at all.

Were they angry? Or afraid?

And Elva found the City itself was just enough changed to tickle at his nerves, keeping him checking their perimeter, watching the rooftops, and searching the shadows for enemies. He checked the lay of his pistols, and rested his wrist on the hilt of his sword.

"Is it my imagination or are these people avoiding us?" he finally asked when the fourth person he nodded to nodded back without meeting his eye.

Predax looked around without slackening his pace. "They seem normal to me."

"So it's normal for people to step out of your way without even acknowledging your presence? To be always looking in another direction when you look at them?"

"I don't know what you're getting at." Maybe so, but the young man was clearly becoming uncomfortable.

Elva decided to try a different approach. "My friend, Fenra Lowens, said that practitioners help people." He gestured around him with his right hand. "You doctor them, and invent new things for them. These people don't act like you're helping them."

The boy's face stiffened. "Well, Practitioner Lowens seems a fine person, but she's a third-class practitioner, and that limits her, if you see what I mean. She normally lives in one of the outer Modes, where the mundanes still believe in the practice and respect it. Here in the City it's not just a practitioner and a village elder, deciding between them how things will be done. Here," he gestured around him, "they don't believe in the practice any more. They call us scientists and think we invent things—like gaslights—and then keep them to ourselves until the Red Court forces us to share them."

"And you're not doing that." Elva made it as much a statement as he could.

"No. Well, not exactly." Predax looked up at a nearby rooftop, suddenly interested in the complicated tile work. "There's always been tension between the two Courts," he said finally. "The Red Court thinks it should be in charge—and, well, the White Court thinks the opposite."

"That's rather fair-minded of you," Elva said. He'd expected an apprentice to be firmly on the side of the White Court.

"My mother was an advocate—well, still is, I suppose—and she taught us always to look at things from all sides."

Elva nodded. "So what's this tension like?"

"It's hard to explain to someone from away," the young apprentice began. "The Red Court puts a tax on everything that goes into the White Court—food, materials, that kind of thing. They decided that anyone practicing in the city now needs a permit."

"Let me guess, practitioners retaliate by raising their fees and requiring their clients to get a different kind of permit."

"And it's easy for the Red Court to convince people that this is all the fault of greedy practitioners." Predax shrugged. "I know it sounds petty, but it's a little worse, every day."

"And why don't practitioners put a stop to it? Surely they're powerful enough in themselves?"

"Well, sure, but the only way to really stop it is by a show of force, which will only turn the people against us even more." Predax swallowed and looked up at the clouds as if he were checking for rain. "Though there's a faction of the council that's pressing for that kind of solution." He shrugged and looked at Elva again. "So there's this delicate balance of power between the two Courts."

"I'm not sure I understand," Elva said. "Where I come from, people aren't fond of too much government, as a general rule. They're more likely to side with those opposing, and from what you say, that would be practitioners."

"Mundanes are too easily led." That sounded like the boy was repeating something someone else said, and Elva thought he could guess who. Predax confirmed it with his next words. "Practitioner Metenari says they always want innovation they're not ready for, and then they resist every innovation we introduce, claiming they're being forced against their wills. They claim they'd like to decide for themselves what kind of lights they want in their streets. They don't listen, and no matter what we do, the Red Court puts us in the wrong."

For a person whose mother was an advocate, Predax seemed a bit naïve. Elva wondered whether it was all an act, whether he was being tested. It would be easy to escape from this child. But it wouldn't take long for Metenari to find him, if he wanted to, though Elvanyn wasn't supposed to know that, not if he really was someone from another dimension. That was the nature of the test, if there was one. Given the opportunity, would Elvanyn run? Or would he keep to his new role? Was he trustworthy?

And he couldn't overlook that there could be more to this. He'd made a point to Metenari that he was only concerned about Fenra, that everything he'd done had been for her sake. Did the practitioner think that Elva would be a lure to draw her out into the open? To capture her? Personally, he didn't think that Fenra or Arlyn Albainil, providing she'd been able to revive him, were stupid enough to fall for that kind of trap, but that didn't mean they wouldn't be on the lookout for him, just as he was for them.

Another twenty minutes brought them to the outer guard gate of the Red Court, where Elva experienced firsthand the tensions that Predax had described to him. Like the White Court, the Red was another city within the City, built onto, torn down, and rebuilt as things changed, as nobles came and went. In theory, the Red was made up of noble families, of which each head was a member of the council—or "court," as they called it. Every few years a vote would take place, and the leadership of the court would change, and someone new would become "first Courtier."

He and Predax were waved through the outer gates, guarded for the first time in Elva's experience, with no difficulty, but were stopped at the inner gate.

"Name and business?"

Predax sucked in his breath, posture rigid, but answered calmly enough. "Apprentice Practitioner Predax, White Court cour—"

"We don't have a messenger on the list for today." The man didn't look at any list that Elva could see.

"I'm not a messenger, I'm the regular courier. I was here the day before yesterday."

Elva could have told Predax that tone wouldn't help. The boy should have known it himself. The only way to handle this type was to act as though you were enjoying yourself, pleased to receive the interest and attention, thrilled to answer all the questions they could ask.

If they saw they weren't annoying you, they would stop trying.

Elva might actually have had some fun, if he had been the one answering the questions. As it was, this treatment made his skin prickle. This was malicious, and intended to intimidate. *The White Court should have sent a practitioner*, he thought, *not an apprentice.* These guards might still have been surly, but they would also have been civil.

"Ah, yes, here you are. Building seven. Here's your badge."

"You mean building three," Predax said.

The guard stayed stone-faced. "I can give you a badge for seven today, or you can come back another day."

"Oh, very well." Predax accepted his badge with a twisted mouth. Elva reached out for his own badge.

"Not you, sir, unless you leave your weapons." The tone at least was respectful, guard to guard.

And never see them again. "Why don't one of you gentlemen accompany us? Then you could—"

Predax tugged at his sleeve. "Never mind," the boy said, "you can wait here. I'll be perfectly safe."

Elva prepared himself for some rough handling from the guards. The hard look in their eyes clearly showed animosity. He was surprised, therefore, when after showing him the corner they wanted him to sit in, they turned their attention to the people still waiting to enter, and ignored him.

Obviously they thought this the most annoying thing they could do to him. Elva smiled to himself as he clasped his hands behind his head and stretched out his feet. *Amateurs*, he thought, as he began whistling a tune popular in the Dundalk Territory. The guards looked over at him from time to time, but were otherwise too busy to bother him any further. He had a fine view of the inner courtyard from here, and watched as people, some in uniforms, came and went about their business. Two young people stopped to talk to each other, eventually sitting down on the lip of a large fountain. No water was flowing, but one young woman scooped up a water lily and presented it to the other.

Finally Predax came back, carrying what looked like a map case. The guards examined its seals with care, made the young apprentice remove his jacket and roll up his sleeves. Finally satisfied that Predax wasn't hiding anything stolen under his clothes, the guards allowed them to leave.

"Are you always treated like this?" he asked Predax once they passed through the outer gate into the public streets.

"Oh, yes." Predax's grin was a good effort, but Elva could see the stiffness of forced muscles. "This way they can feel important. It's harmless."

The young apprentice did his best to sound certain, but in Elva's experience there was very little distance between insult and injury. Still, with his errand finished, the boy was more relaxed. Elva decided to take advantage.

"I'm new here. May I ask a question? About Practitioner Metenari? Does he seem himself to you?" he said. "He's a little different than when I met him before, with Practitioner Lowens."

"What do you mean?" Predax hesitated, suddenly tense. "He's tired, of course. He's just finished the first part of a very complicated project."

"Ah, well then. There's always that bit of a letdown when you've been working on something for a long time and you finally succeed, isn't there?" Elva reminded himself he wasn't supposed to know what was going on. *Just a gunslinger,* he thought with an inner smile. *Not from these parts.*

"You're right." The boy relaxed. "It's like passing an exam after you've been studying hard for weeks. You feel a little empty, like you don't know what to do now. Yes, that might account for it."

"Account for what, if you don't mind my asking?" People unfamiliar with practitioners might be surprised by Predax's willingness to talk, but in fact this behavior had been the foundation of Elva's friendship with Xandra Albainil. Practicing was a lonely profession, for all that in the White Court everyone knew everyone else. There had always been a few who lived apart by choice, like Fenra. In his day many practitioners still traveled often, visiting other Modes, going about their business—obscure and esoteric as it was. Nowadays practitioners like Metenari—and even Fenra's beloved Medlyn Tierell—seemed to stay in the City and surround themselves with apprentices.

"No, I don't mind. You're easy to talk to, you know?" They waited until an elderly man, leading a well-groomed horse by the bridle, walked past them near the curb. "He's not really himself."

"In what way?"

"Well, he's always been a bit self-important, you know? I mean, he really is important, don't get me wrong, but now he acts as if he feels it, not like he has to keep reminding himself."

This boy was brighter than he looked. "He's less self-conscious, you mean?"

"Yes, that's it exactly." The boy was pleased at being so quickly understood. "It's like he's stopped checking to see that everyone is watching him."

"He's just achieved something astonishing, something no one else has been able to do—or so I gathered from what Practitioner Lowens told me. That alone should give a man all the confidence he needs."

"You know, you're very perceptive for a—a newcomer. But now Practitioner Metenari is suddenly insisting that he wants the Albainil man. Though he doesn't say why."

For a mundane. That was what Predax had meant to say. "Who can guess what's in the mind of a practitioner like Santaron Metenari?" Elva said. The apprentice nodded. Elva allowed the talk to turn away into less personal areas. As they passed down an elegant street that Elva thought he recognized, he noticed that they were drawing the attention of a man standing out front of one of the town houses. This was especially noteworthy, given how everyone else was still ignoring them.

"There's a man staring at us—don't look!" Elvanyn rolled his eyes up. Practitioners would never learn to take precautions.

"Oh, that's Ginglen Locast." Predax frowned. "That's his hotel. He must remember me from when I came to pick up Fenra Lowens' things. She'd been staying here with that Albainil man."

Elva glanced over his shoulder at the place as they passed by. The man turned his head and kept watching them as they turned the corner.

Fenra and Arlyn stayed there, he thought. *This is the last place the White Court would look for them.*

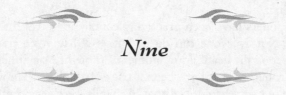

Nine

FENRA

MOST OF THE wooden chests, boxes, and caskets in Medlyn's vault lined up against what would have been an outside wall if this were a real room. At least, it felt like an outside wall to me. I told Arlyn to rest and began going through every piece of furniture I could open. I had already found clothing that fit me, so who knew what other useful thing I might find? The first three chests, each solidly built of several different woods, were empty of everything but the scent of their cedar linings. Next came the clothes press I had already looked into . . . I softly whistled two surprised notes.

"What?" I had seen Arlyn much worse when the lowness was on him, but now, though weak, he looked alert, and interested.

"Another set of clothing," I said.

"Clothing?" He turned his head slightly so he could fix me with one eye. "You've already found clothes, what's the excitement?"

"These clothes weren't here before. This press should be empty, and instead there is another complete set of clothing. Inner wear, outer wear, shirts, boots—look, even a hat. The clothes I found fit me, and I will give you odds these will fit you."

I gathered up the top few items and brought them to Arlyn. He stroked the white silk shirt with his fingertips. Then his mouth spread in the biggest smile I had ever seen. For a moment I saw what he must have looked like as a child. Before the lowness. Maybe even before the practice.

"Do you know what we have here?" His eyes almost glowed.

Obviously he did not refer to the clothes. "I take it you do?"

"The jug." He glanced around and pointed at where the jug still stood on the table. "It had juice in it when we needed juice, and exactly the type of juice we needed most." He turned back to me. "And now it has wine in it."

"Practitioners have been practicing containers to refill themselves for years. Not everyone can do it, but . . ." My voice trailed away as Arlyn shook his head. For a moment I thought I saw lines of silver in his hair, but it must have been a trick of the light. He was so excited I am not sure he noticed that his hands were very slightly trembling.

"The *same* jug, Fenra. The same *jug*. Juice when we needed it, wine when we wanted it. Very likely water when it comes time for that. All in the same jug. Don't you see what this means?"

I started to shake my head, and then I caught a glimmering of his idea. There had been, I remembered, two different liquids in the pitcher in Medlyn's office. "The jug had to have held all these liquids, but obviously not all at once. The *forran* on this jug duplicates not one specific space and time—when the jug held wine, for example—but all the different spaces and times in which the jug held liquids."

"Exactly. And not only that, the jug provides the type of liquid needed at the moment of that need. Without the recipient specifically asking for it." He gestured at the clothes in the chest. "Maybe even before we know we need it."

"The way our clothing and our tools and wagons and weapons and all change from Mode to Mode, without anyone asking for it." I shivered, the world suddenly a strange place.

"As if there is a small modality in every box, every container, every—"

"Every one? Most of these chests and boxes appear to be empty."

"*Appear* to be empty—maybe they just hold something we haven't needed yet. But even if it's only the two we've found, the jug and this press, it's still the most remarkable achievement since, since . . ."

"Since the Godstone."

Arlyn shut his eyes, as if the excitement was draining his energy. "I don't know whether Medlyn Tierell found this *forran* or created it."

I considered, my own intellectual interest sparked by Arlyn's enthusiasm. "I knew Medlyn had been working on something like this, a *forran* to connect various spots in space simultaneously, but I thought it was for transportation."

"Either way, your old mentor had depths I'll bet no one knew about. This is a spectacular achievement. What we couldn't do with something like this."

The healer in me rejoiced to see Arlyn so animated, so much a part of the present moment. The rest of me feared I was seeing the attitude that had created the Godstone in the first place. Medlyn must have had some reason for not sharing this *forran* with the rest of the White Court.

To distract Arlyn, I pulled out the rest of his clothing, tossing the boots to the floor, laying underclothes over the arm of the sofa. Arlyn began to pull his shirt over his head.

"Do you need help?" I asked.

He grinned. "Not since I was three."

"Fine."

Arlyn's suit consisted of fawn trousers, a pale blue jacket cropped at the waist, the white silk shirt, a scarlet cravat, and a dark blue waistcoat. He dressed quickly and was tugging his cuffs into place when I handed him his hat. It looked very much like the flat topped practitioner's hat, except for the flat brim and the thin, pale blue band around the crown.

I closed the door of the press, wondering what I would find in it if I opened it again. "Just think, I can come back here at any time, find whatever we might need, and fetch it out to where we need it." I looked up, frowning. "If only I didn't exit every single time into that woman's office."

"I've got an idea about that."

ARLYN

At first Fenra was reluctant to try tampering in any way with the locket.

"Right now this is our only connection with the world," she pointed out. She walked back and forth in front of the couch, finally sitting down beside me, her elbow brushing mine. A little further exploration had yielded a pie safe containing one ham and one peach pie; what might have been a meat safe held dried apples and some kind of spicy sausage Fenra said we would take with us when we left.

"Look, so far you've been opening it face up, as if you were looking at a book. All I'm asking is that you open it facing away from you."

"Is that all? You half scare me to death and that's all you wanted me to do?"

"Well, that's the first thing I want you to do. I have a couple of other ideas in case that one doesn't work." I bit at the inside of my cheek. "There is something I'd like you to do first, however." She waited without responding for me to continue. I'd seen her use this trick before, when patients couldn't quite bring themselves to tell her what was wrong. I knew she could outwait me. "I want you to go back now in the ordinary way, back to Tierell's old office."

"Why?"

I flexed my hands on my knees. "Metenari's had the Godstone all this time we've been dressing and eating. I'd like to know if our world is still there." I saw the meaning of what I'd said dawn as her mouth fell open.

Fenra pulled the locket free, turned it over and over in her hand. "Would we still be *here*, if our world wasn't?"

"I don't know," I told her. "When I used it—as soon as I saw what was happening, I stopped." I could see her wondering if she should ask me for more details, deciding that now was not the time. "Afterward, my vault was the same, but whether that was because the change I made was quite small, or whether it's something in the nature of the vaults themselves . . ." I shrugged.

She nodded, still staring at the locket in her hand. "Metenari was always very methodical, very careful and precise. I do not see him rushing into using the Godstone right away. If he is still the man he was, he will be taking time for more study, more research, before he tries anything with it."

"I still want you to do a quick round trip, just to be sure there's somewhere to come back from."

"You should come with me."

"No." I started to shake my head and stopped as a wave of dizziness swept over me. "An unnecessary risk for just a test. You're safer without me."

"If I cannot come back for you, how will I deal with the Godstone?"

"The only reason you wouldn't be able to come back is if Metenari has used it. In which case, nothing else would matter."

I could tell she agreed, but from the hard set of her mouth, she wasn't happy about it.

She looked from me to the cupboards and back again. "You cannot see the food. You would not know what was there."

"Exactly," I said. "Which is why you have to come back. One way or another, I'm going to starve to death if you don't."

She passed me the remaining piece of ham pie. "Keep this in your hands. Perhaps if you do not put it down you will not lose it." She got to her feet and without looking at me was gone.

ELVANYN

Predax reported directly to his mentor on his return to the White Court. He handed over the map case and Metenari checked the seal just as carefully as the Red Court guard had. Elva would bet those seals were more than just wax.

Almost immediately, however, the practitioner's expression changed, and he looked at the map case as if he'd never seen such a thing in his life. He stood for several minutes turning it over in his hands. Finally, he took the bridge of his nose between his thumb and forefinger, eyes shut, just like Xandra used to do. Elva swallowed, though his mouth was suddenly dry. *Lots of people do that*, he told himself.

"Do you know me?"

Elva resisted the impulse to look at Predax and check his reaction.

"You're Santaron Metenari, Practitioner." Elva kept his voice calm, uninterested. Servants, he thought, were asked silly questions every day.

Metenari reached out with his right hand, as if he wanted to touch Elva's cheek. Elva stepped back, the way anyone would. The practitioner seemed about to say something else, but at that moment Noxyn, the senior apprentice, walked into the room carrying a shallow willow basket full of scrolls, loose papers, and books, some so old their covers were broken and about to fall off.

"I think I've found something, Practitioner," Noxyn said, ignoring Predax and Elva completely. Elva was happy to be overlooked. He'd be off duty soon, and he wanted to get back to that hotel. It was just possible that Fenra or Arlyn, or both, might return there. He could at least leave a message for them, just in case. He'd have to word it pretty carefully, so that no one but Arlyn or Fenra would understand.

"You may go."

Elva had kept an eye on the practitioner as he spoke with Noxyn, and so he wasn't startled by the sudden dismissal. Predax, however, kept looking back over his shoulder as they headed for the door. As he glanced at the apprentice, Elva saw another familiar gesture out of the corner of his eye, as the practitioner clasped his hands and tapped his lips with extended index fingers. *Xandra?* Elva shook himself. Impossible. And there was nothing he could do about it anyway.

Once out of the room he headed straight for the guard barrack, where he changed into his own clothes. They'd draw attention, but his experience with Predax made Elva think he'd rather be noticed as a stranger in town than as a White Court guard. Besides, his gun belts hung better over his deerskin jacket than they did over the guard uniform. Each piece of his clothing had been designed over years to let him carry his weapons comfortably. His jacket had extra layers of leather on the shoulders, to pad the weight of the gun belts, as well as an extra layer on his left side, where his sword hilt rested.

Elva tugged down on the right-hand brim of his hat until the feathers brushed his shoulder, and walked out the door, across the open space on the far side of the barracks and over the bridge. He thought he could retrace the route he'd taken with Predax without much trouble. As he'd thought, without the uniform—and the apprentice at his side—the few people who bothered to notice him looked him over frankly as he passed them, interested by his clothing, or his weapons, but nothing else.

Elva had once been told that it was impossible to tail someone through a city and not be noticed—if the target in question was alert. So he wasn't surprised when he caught the middle-aged woman who turned away every time he stopped at a crossroads to check the street names. She was good, but he'd been at this game longer.

Now that he'd caught her, how to lose her? He stopped and took his bearings, pretending to check the sky for clouds. He oriented himself right away. He was a street or two over from the clock tower of the market. He took the next right—a textile shop had replaced the barber who had once been on this corner—and slowed down as he reached the old market building. Refaced in red brick, and with what looked like extra skylights, it was recognizably the same year-round market, with stalls, arcades, stands, and booths. And dark corners, and obscured exits. He hoped.

He walked through the main entry doors, looking around as if seeing the place for the first time. From the front entrance, a central series of stalls was flanked by two wide aisles running parallel down the length of the building, with another row of stalls set up against each of the outer walls. Every row of stalls was broken into smaller sections by narrower aisles running crosswise. The cobbled floor hadn't changed, nor the high, vaulted ceiling, but the place was so much brighter that he began to worry. Then he realized the effect was created mostly by freshly whitewashed walls and clean skylights. Someone had also gone to the trouble of removing decades of cobwebs from the rafters. It took a minute to see that, from his point of view, nothing important had changed.

Most people who frequented the market thought the outer stalls were built right up against the plastered walls, but Elva knew a narrow corridor ran behind all those walls, connected to loading bays that allowed stall holders to move their wares in and out without having to force their way through crowds of customers. The place wasn't crowded at this time of day, but there were enough people picking up fish or bread or cheese or meat or carrots or potatoes to distract even as competent a follower as the woman watching him.

At least, temporarily.

As Elva moved along with the flow of foot traffic, slowing down and stopping from time to time to examine the wares more closely, he noticed that the stalls themselves had also been painted a clean white, and each now had a matching sign indicating the name and type of business. Elva found the sameness dull, but he supposed it did cut down on misspellings.

He bought himself a cup of hot cider and strolled along aimlessly, watching as a butcher sliced cutlets for an elderly woman hugging a small dog tucked under her arm. She noticed him watching, smiled at him, and used the small dog's left paw to wave at him. The dog gave him a look that clearly said this wasn't his idea of fun, and Elva gave him a sympathetic look in return.

Elva had been keeping his tail in the corner of his eye. The thin crowd worked against her as well, giving her fewer people to hide behind. She had to keep a greater distance, and let more and more people get between them.

He'd reached the section where the fishmongers had their stalls. Not much had changed here either. There were still the smaller stalls with only one or two people serving, and a couple of others where one senior person was flanked by a handful of junior women and men who acted as assistants, fetching, wrapping, and bagging. There were even a few runners who would be following the customers home with their purchases.

Elva looked for a chance to slip behind an outer stall and disappear down the hidden corridor, but nothing presented itself. He was almost out of the fish section and beginning to worry when two young men a little older than the runners started tossing fish across the aisle. Obviously a well-rehearsed game, one of them would pick up a fairly hefty fish, a salmon or perhaps a large bass, and sling it over the heads of the shoppers, as if the fish were swimming across the aisle. The young man on the other side caught it in his arms and cradled it like a baby.

"Flying fish!" the blond on the right called out. "Pennies a pound for a fish that will fly right into your skillet. Don't let such a bargain fly away from you, Doms! The best prices on flying fish!"

Passersby gathered, laughing, some calling the two young men by name. What had probably started out as a game between assistants bored by the lack of business in the closing hour had turned into a source of entertainment and a close-of-the-day attention-getter.

And gave Elva the opportunity he'd been looking for.

He maneuvered himself between a tall woman still in her City guard uniform and three men obviously shopping together. He watched for the moment when the gathering audience hid him from his tail and stepped backward between two stalls, turning into the back corridor and watching from the shadows as she walked past. The corridor, never very wide, had been narrowed by piles of boxes, crates, and pallets. He waited just long enough to make sure the woman wasn't turning around to look for him before he ran toward the loading docks in the rear of the building and out through the bay doors.

He'd spent more time in the market than he liked. Lamplighters were already adjusting the flames on the oil lamps in the slightly shabby streets behind the market building. The newer gaslights were obviously reserved for the more important areas of the City. He was sorry he'd missed seeing those lit. He'd heard about them in the other world, but they still hadn't reached the Dundalk Territory when Fenra walked into his office and changed his life.

Street lighting did make him easier to see, he thought, but it also meant the streets were more crowded. After zigzagging through several lanes and alleys, Elva turned into a secondary street and followed it to the quiet square he'd walked through with Predax. On the far side, a square-built woman with a stepladder stood lighting one of the lamps. A second-class square, then, where the lamps were lit only after those on the main streets. But still respectable.

And on its way up, he thought as he noted the gas lamp hanging over the door of the inn, already lit. *Business must be good.*

As he expected, the front door was still open. Even if these hotels provided meals, and in Elva's day most did, there were always patrons out for the evening, and it was easier to leave the door open than to have a night attendant. However, here, there was a tall narrow man standing behind a tall narrow desk. For a moment Elva thought it was an Albainil, but then he shook his head. It was only a good copy. A very good copy. No hotel on a street like this one—or very many others for that matter—would have an Albainil in a public lobby.

The attendant behind the desk took one look at Elva's clothing and smiled, opening his book. "Welcome to the Hotel Ginglen."

"I'm afraid I'm looking for information, not a room."

The man's smile didn't fade. "If *I* know, *you'll* know."

Elva found himself smiling back. "I'm looking for two friends of mine who might be staying here. I'd prefer not to mention names." He touched the side of his nose. The attendant almost smiled again. Evidently that gesture hadn't changed its meaning. "One is a man about my height, with darker hair and blue eyes. He may have identified himself as a carpenter." It seemed that the man's face stiffened ever so slightly. "The other is a practitioner, a woman, very dark of skin, and curly of hair, with gray eyes."

"They're not clients of ours at this moment," the man said.

Which meant they had been, and might be again. "I see. Do you know, does the Autumn Rose Tavern still exist?"

The man frowned. "Well, it's still in business," he said. "But I'm sure it's listed in the merchants' association as a chop-house."

"Ah, then things have changed a bit since my—my father's day."

"Did you need directions, Dom?"

"No. No, thank you, I'm sure I remember the way. Well, if you should see my friends, tell them I asked about them, and tell them I also asked about the Autumn Rose."

"I will, of course." The man hesitated. "I would also tell them about the horse."

"The horse?"

"The young, ah, practitioner's horse. She might be worried about it. If I should see her again, I would of course let her know we've been keeping the animal safe."

"I'm sure she would appreciate that." Then Fenra meant to come back. She wouldn't leave her horse otherwise. Elva felt better than he had in days.

Elva took the long way around getting back to his barrack, stopping at two coffee houses on his way. He looked, but this time no one was following him. He crossed the East Bridge and entered the gates of the White Court with a nod to the guards.

"Hey, Karamisk," one of them called to him. "Better hope you're sober, Captain's handing out extra duties."

"I'll just change into uniform—"

"He's already asked about you. Change later if there's time—it's all pistols primed, as they say. Hey, if you get a chance, come back and tell us what's going on."

Elva waved him off and trotted toward the square outside his barrack. When the guard captain saw him, he beckoned Elva forward.

"All right, then, you all know the descriptions of the two we're looking for. Remember, they're not to be harmed, either of them. Is that clear? Just detain them. Once you have your assignments, get going."

As the captain pointed to pairs of guards and gave them their instructions, Elva had an idea. Fenra would need to use the locket, so there'd be one place in particular worth looking.

"Captain? Maybe I should go to Practitioner Otwyn's office? They'll most likely show up there."

"I've already assigned that post, Karamisk."

"Of course, Captain, but they'll be more cooperative if they see me waiting, especially if I'm not in uniform."

"You may have something there. Very well. Are you armed? Fine then, go. Tell Rontin you're there to assist her."

Elva lost no time in saluting and tearing off down the alley. Of course the Captain wouldn't let him wait alone. The man wasn't stupid.

FENRA

I polished the locket on my sleeve. What was the worst that could happen? If Metenari had used the Godstone, and the world as we knew it had completely changed, there was every possibility that Medlyn's office was no longer there. In which case the locket simply wouldn't work, and Arlyn and I would be stranded in the vault forever. I looked at the jug and the pie cupboard. Well, we would not starve to death. And who knew? Arlyn Albainil was still the greatest mind in the history of the practice, I told myself, and power or no power he would probably be able to design a *forran* for a dimensional gate that would work from a vault. We might go to the world where Elva had lived for so long. At least we knew the practice would work there. *Without Elva*, I thought.

"I'll be off, then," I said again.

"Good luck."

I forced a smile, and nodded, and hoped I would not need it.

Even before the locket was completely open, I could feel the change in the air around me. Cooler, damper, the scents open and fresh. Medlyn's vault didn't feel stuffy until you were out of it. I wondered where the air in the vaults came from. I wondered if Arlyn could tell me. I hoped for the chance to ask him.

It wasn't moonlight coming in though the uncurtained window, but lamplight. The new gaslights were evidently bright enough to cast shadows even on the second floor. I shut my eyes for a moment, relieved that the office was still here, the world was still here. Elva was still here. I thought I heard a noise and snapped the locket shut. Was the new practitioner still setting up her office? What could she be doing in the middle of the night? Dusting?

I tiptoed to where I could stand to one side of the door. Halfway there, a blow to the back of my head knocked me to my knees. Strong fingers wrapped around my left wrist and pulled my arm up behind my back. There were *forrans* that didn't need the practitioner's hand free, but in that moment I could not think of any. Then a cloth bag smelling of licorice was forced over my head and held tight around my throat by a fist at the back of my neck. I tried to breathe and started coughing as I sucked in a mouthful of licorice-flavored dust.

Suddenly my arm was free and the grip around my neck loosened. I tore the covering off and took a deep, dust-free breath.

Elva stood over a guard, holding a gun in his right hand. The guard—a woman in the uniform of the White Court—lay on her side on the floor. Luckily she had fallen on a thick carpet. I crawled over and felt the side of her neck. Her pulse beat firm and strong.

Elva lifted me to my feet. "Are you all right?"

Just in time I remembered not to nod. I felt for the tender spot on the back of my head. "You?"

"Better now." White teeth gleamed against dark mustaches. "Where's Arlyn?"

"Still in Medlyn's vault." There was something in Elva's face that made me add, "He is weak, but he will live."

"Metenari—or whoever that is—has the guard out looking for him—for both of you."

"What do you mean 'whoever it is'?" I sat down on the edge of the desk, still rubbing at the back of my head, feeling the pain dissipate.

"Metenari isn't acting like himself—"

"How do you know? You've only met the man once . . ." Elva was shaking his head at me.

"Predax, the younger apprentice, he's noticed it too. Most of the time Metenari seems just like the man we met before, your self-important schoolmate. Some of the time he's a much sharper fellow, walks differently, talks differently. Different in a familiar way."

"Familiar how?" For some reason I felt cold, and I wrapped my arms around myself.

"I didn't know at first, or maybe I didn't want to know, but familiar like he's Xandra—no, wait, I know what you're going to say. It doesn't feel like Arlyn, like the man with us, it feels like *Xandra*. Do you think—is there any way that somehow, along with his power, Arlyn might have left some part of himself behind?"

I swallowed. "Come, you can ask him yourself."

To my surprise Elva did not take my hand. "I should stay," he said. "The White Court is too secure, too full of guards. We need to get Metenari, and the Godstone, out somewhere we can deal with him better. Away from the City entirely." His brow furrowed. "Xandra used to have a tower, in the Third Mode, I think. Anyhow, I know how to get there and Arlyn will as well. It's private and it's secure. If I stay here I can make sure Metenari comes." He looked down at the guard he had knocked out. "Besides, I can't leave her like this. I don't know what they'll do to her."

"Elva." He looked back at me, his mouth forming just the slightest curve. "There are larger things at stake."

Now he did smile. "There are no larger things." He gripped me by the shoulder. "Tell Arlyn once you're back here to go to the Autumn Rose—he knows where it is, and he'll know what to do. I'll make sure your horse is there—"

"Terith?"

"Yes, Terith. Now go, before Rontin wakes up."

Arlyn

"I don't know." I rubbed my face with my hands, the only response I had to what Elva had said about Metenari. "I suppose it's possible I left more than my power behind. All I know for certain is that when I finished sealing the stone away, all my power was gone."

"Was that the beginning of the lowness?"

I looked up at her, elbows braced on my knees. "Probably. I don't remember much of what happened immediately after."

"So it's possible that the Godstone contains some aspect of you, some part of your essence?"

"Not me, Xandra."

Fenra lowered herself into a chair. "Xandra's knowledge and Metenari's power?"

Ideas flashed through my brain. "Not my knowledge, not my *forrans*."

"How can you be so sure?"

I straightened up, taking a deep breath. "Because I still have them. My power's gone, but my knowledge is still with me."

"You hope." Fenra sat quiet for a few minutes, staring at a spot on the floor. Finally she nodded and stood up. "It doesn't change what we have to do. Let us try it before my nerve wears off."

"I don't think that's possible." I'd hoped to make her smile. It didn't work.

We gripped each other's wrists, my left with her right. The locket was already in her practitioner's hand and she placed it for a moment on her forehead. Then, as she'd done before, she breathed on it. As I'd suggested, she turned it upside down and, using the nail on her index finger, she popped it open, face down, as if she was about to shake something out of it.

Fog drifted out until we were completely surrounded. At first I could see nothing, and then shapes began to form, slowly, as though they were coming closer. I thought I could see furniture, a table, but everything was suddenly blown away by a powerful wind that flapped our clothes around us and sucked the breath from our mouths. And we were on a beach, clean white sand, deep turquoise ocean, clear pewter sky.

"This is where I find you when you are low," Fenra said. "Why would Medlyn's locket bring us here? I never thought it was an actual place."

Even though the sun was shining, I felt cold. Regardless of what Fenra said, I didn't remember ever being here before, and yet I knew the place. And Fenra knew the place, but Medlyn most likely hadn't. "It can't be the locket," I said. "It's something in this place, something that diverted us once we were moving—or more likely diverted you." My heart pounded and I licked sweat off my upper lip. How could I be sweating when I felt so cold? "It can't mean anything good. Get us out of here. Fast."

I could see Fenra had questions, but she knew when to do as she was told. I didn't see exactly what she did with the locket this time, but the next thing I knew, we were standing in a sitting room, darkened by drawn shades. All the furniture was covered with dust sheets. Enough light entered around the edges of the shades to tell me that it must be midafternoon on a sunny day.

"It appears Practitioner Otwyn did not take over Medlyn's living space when she moved into his office." Fenra spoke quietly, and her words seemed to hang like dust motes in the stale air.

"Even if she had . . ." I went over to the right-hand window and pulled back the edge of the shade. Medlyn's rooms looked out over the sunny end of the Watchmaker's Gardens. "She'd likely be out at this time of day anyway."

"But her servants would not."

I let my hand drop. I'd forgotten about servants.

"The good thing about the midafternoon," Fenra said as I followed her to the door and into the hallway, "is that it's the busiest time in the more public areas of the Court . . ."

Her voice died away, as we walked down the stairs across the arcade and into the garden. She was right, there were other people in here, but only about a dozen, and all of them practitioners or apprentices. In case any of them glanced our way, I smiled and nodded as if Fenra had said something witty. About halfway down the long central path to the south exit, I noticed she wasn't limping.

"My limp is sure to be part of my description," she said when I asked her. "When I do not limp, no one sees *me*. They will not think of me at all. It is a small, subtle change, but all the more persuasive because of that."

"Also, you're not dressed like a practitioner."

She squeezed my arm. "Also that." She slowed down, tugging me ever so slightly to one side. I was about to ask her what was wrong, when I saw for myself. Two men in guard tunics entered the garden by the exit we were heading for. It was too late for us to think of turning into another path. Fenra stopped and picked a yellow flower from a nearby bush, turning it over in her hands, softly breathing in its scent.

"Your business here?" The taller guard stopped in front of us while the other, shorter one kept a step back to get a better angle on us if he had to use the pistol he now had in his hand.

"Oh, we had to come!" I'd never heard that voice from Fenra, and any other time I would have been vastly entertained by her spot-on imitation of Jonsel Weaver. The accent alone was atrocious, and said better than anything else could have that we were from one of the far outer Modes. So we couldn't possibly be the people they were looking for. "Everyone said, 'If you're going to the City, you've got to visit the White Court—why, they have the best gardens in the world!' And they were right! I mean, look around you—" Fenra waved the flower through the space between us. "But you're probably so used to it you hardly notice anymore."

"The White Court isn't open to the public today," the tall one said. His frown had faded.

"Oh." Fenra did a marvelous job of looking dumbfounded. "But no one stopped us coming in," she added. "No one said anything to us."

"You'll have to come with us." The tall one gestured toward the exit they'd come out of.

"But why? We didn't know . . ." I swear Fenra looked as though she was about to start crying.

"Not to worry, Dom, we're just escorting you out of the Court."

"Oh, thank you so much. That's very kind."

FENRA

"The Autumn Rose is a tavern on the upper west side, across the river." Arlyn took me by the elbow and steered me across the bridge.

"And what is so special about it?" I asked him.

"It's built over the old cisterns of the City, and there's a way to get outside without using the Road." He glanced at me. "Stop playing with your locket. It's not going to talk to you, and we don't want to end up somewhere else by accident."

I tucked the locket away. It felt warm. "Cisterns, you were saying."

"If they haven't been found and filled in. If Elva hasn't arranged for guards to be waiting for us there."

I stopped in my tracks. "If that was the intention, why did he not simply keep me? Why did he allow me to return to you?"

"Because, with no disrespect intended, it's me they want, not you. There's something I have that Metenari needs. More of my blood, or something I could tell him—*something*."

"Perhaps the *forrans* you are so certain the Godstone does not know." I hooked my arm through his and we set off again, taking care to stroll along casually, in case the White Court guards were watching. "Suppose he could not have you? Suppose you were in the other dimension? Or dead? Then whatever he needs from you would be unavailable."

"Setting aside the fact that I'm the only one who knows anything about the Godstone—and that includes Metenari . . ." Arlyn's voice faded away and his steps slowed. "If what Elva suggests about it is true . . ." This time it was Arlyn who stopped in his tracks. "Metenari wants me for what I know, just as he always has. He has the Godstone, but he doesn't know how it works." He looked at me from the corner of his eye. "And the Godstone wants much the same thing, if for a different reason."

"The *Godstone* wants? Are you saying it's alive? Sentient?"

"I don't know. Maybe. I don't remember."

From the look on his face, he remembered perfectly well. Hoping I was wrong, I tugged him into motion. "Again, if you were not here—"

"From what you've said, Metenari's as stubborn as a bulldog. He won't stop looking, and that's supposing the Godstone would let him. We're talking about a man who was able to learn about the Godstone in the first place and find it."

"Research has always been Metenari's strong point," I admitted. "His power level is nothing unusual for a first-class practitioner, but he worked hard, and his knowledge of *forrans* and historical facts is extensive. Even Medlyn praised him for that once." I touched the locket through the silk of my shirt and cravat.

"So even without me, with the Godstone pushing him, he'll eventually find out what he needs to do on his own—if he doesn't destroy the world first. But as long as he's looking for me, we have a hold on him."

"And Xandra's—*your* tower . . ."

Arlyn looked away, brows drawn down. "It's risky, but if we get him there, we may be able to neutralize him. Them." Arlyn pressed his lips together in a parody of a smile. "I've got an idea."

The noise level at the Autumn Rose didn't change at all when we came through the doors. From the description Arlyn had given me as we strolled through the streets, it was clear the place had come up in the world. For the most part the clientele were smaller merchants, traders, and clerks. A few seated close to the door glanced up from their meals, but did not give us a second look.

"How can I help you, Doms?" The host looked to be a man old enough to be my father, stout, with a thick shock of graying hair just starting to recede.

"My brother left my horse here," I said, hoping that Elva had managed it.

Without any change of expression whatsoever, the host waved for us to follow him through the dining room and out a pair of doors at the far end, which turned out to lead into the kitchens. Rather than taking us past the ovens and cooks, however, he indicated a narrow corridor that ran off to the right.

"If you'd excuse me," he said. "I got work to see to. You'll find a courtyard at the end of this passage, you can't miss it. Good day." With a nod, he turned and went back the way we had come.

"Huh," Arlyn said as soon as we were alone. "Elva must have paid him to mind his own business."

The courtyard was one of the largest I had ever seen inside the City, fully as wide as the building itself, and deep enough for several outbuildings. A snort from behind a wheelbarrow upended over some half-filled sacks of flour led me to Terith. He knocked my hat off nuzzling my ears, and rolled his eyes at Arlyn.

"He doesn't like me."

"He has always been a sensible beast. Now, where is this escape route?"

Arlyn led me around a large lemon tree, branches heavy with fruit, a chicken run, and a small raised bed full of herbs. I helped him move two crates and an empty birdcage, revealing a section of wall neatly bricked in.

I ran my hands over the rough surface. From the feel of the mortar, this wasn't recent work. "No way out after all."

"Maybe, maybe not." He laid his hands on the bricks, almost as though he were looking for a secret trigger that would open the wall if he pressed on it. "Remember how you asked the grass and plants to move out of our way in the New Zone?"

I thought I could see where he was going, but—"Those were living things," I pointed out.

"And this," he tapped the wall, "is bricks and mortar, all natural ingredients. How do you know you can't ask them to move?"

I had never done such a thing before, but really, what choice did I have? I had to at least try. I rubbed my hands together, and when they felt warm enough I pressed my palms to the bricks. At first I felt nothing but residual heat, and then I found I could distinguish brick from mortar without moving my hands. Crushed stone, clay, sand . . . all living, though not in the manner of trees, and all seemed to be asking for something. I could smell the place the sand had come from, a beach in the Fourth Mode, and feel the sunshine on the mountainside where the stone had been quarried and crushed.

"Go home, then," I whispered, and Arlyn caught me as the wall I'd been leaning on disappeared.

I can't tell how much time has passed. The buildings all look the same—so, not long enough to affect brick and stone—but lots of other things have changed. The clothing is the least of it. Gaslight. When I was here oil lamps were still experimental.

So why does Elvanyn Karamisk look so familiar? Maybe I knew his grandfather? Or great-grandfather? So he comes from another dimension. I've been there, so it's still not impossible.

Once I convince Xandra to give me the *forrans*, there won't be a repeat of what happened last time. Xandra was a coward when all is said and done. "Unacceptable casualties," is what he kept saying. He just didn't want to play the odds. And the lives of any left would be so much improved, surely it was worth it.

Xandra was trying to make everything change at once. I've been thinking, maybe it would be better to change one Mode at a time.

Now's my chance to find out. Once I find Xandra.

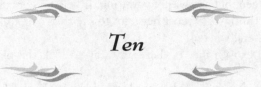

Ten

ELVA COUNTED SLOWLY to sixty before reviving Rontin. As soon as she showed signs of coming around, he went out into the corridor and came back in, as if he had just arrived. With luck, she wouldn't associate him with the blow to the head. Her eyes fluttered and she twisted away from him, pulling a dagger before he could stop her.

"Rontin! Watch it! It's me, it's Elva." He backed off, holding his hands open at shoulder height so she could see they were empty. "Captain sent me to watch with you."

"How long was I out?"

"How can I know? I just got here, and you—"

"Never mind that now. Go after them—maybe they haven't gotten far."

"Which way?" Elva stood in the doorway, looking first to the left, then the right. If he were really who he said he was, he wouldn't know the best way to go.

Rontin beckoned him back and used his arm to haul herself to her feet. "Probably doesn't matter. Probably long gone by now. Sssss." She touched the top of her head with careful fingers. "What did they hit me with?"

"Let me see." He parted the hair and revealed a red swelling, already forming, roughly the size of his gun barrel. Or a walking stick, or just about any other blunt object. "Didn't break the skin," he said. "Did you see anything? Was it them?"

"Who else would it be?" she said. "Thing is, we don't know whether they're still in the Court, or they went back where they came from."

"The captain will want a report. It might be worth my waiting to see if they come back. Are you well enough to go alone?" She was looking a bit green, but she wasn't vomiting. Likely no concussion, then.

"I don't think so." It cost her a lot to say that. She touched the top of her head and winced.

"Here. Lean on me. Is there any way to lock the place from the outside?"

Rontin started to shake her head and hissed again, shoulders hunched against the pain. "No. Can't be locked by mundanes—certainly not against practitioners."

She was leaning heavily on him by the time they got back to the guard barracks, to be met with exclamations and cursing. Another guard led Rontin away to the practitioner on duty for emergencies, but at his captain's signal, Elva stayed behind.

"Markin, Tova, double time to Otwyn's office, stay alert, don't get separated." The captain turned back to Elva.

"Report."

Elva drew himself up and squared his shoulders. "I went directly from here to Practitioner Otwyn's office. I saw no one of interest in the grounds on the way. The halls and corridors near the office were empty. I found Rontin face down on the floor just inside the door. It's pretty clear she was struck from behind. Her hand was on her pistol grip, so she must have tried to defend herself when she realized she was going down. I have no way to know how long Rontin was unconscious before I arrived. She might have been attacked just as she got there."

"Which means they might have arrived as much as an hour ago, which puts us more than an hour behind them." Captain sat down on the edge of the table. "Yet no one else has reported a sighting."

"We wondered if they might have gone back when they found someone waiting for them. If so, it's possible they'll return to that room. Rontin wasn't fit to report back on her own, so I'm afraid we left the place unguarded."

"Not optimum, but understandable."

"Should I go back, Captain? Are two guards enough?" Elva said.

The captain shook his head, frowning over his crossed arms. "It's a long shot at best. It's far more likely that they escaped, and I've rearranged the patrols to take that into account."

"Where do you want me, sir?" With luck he'd be able to lead a few more people astray.

"It's not where *I* want you, I'm afraid. Practitioner Metenari wants to see you. Now."

Against the tight feeling in his stomach, Elva reminded himself that this was what he wanted. He couldn't convince Metenari to look for Fenra and Arlyn in the Third Mode unless he had contact with the man.

The practitioner was alone when Elva reached his sitting room, and he entered with caution. If he was right, and Metenari's prissy, know-it-all persona was being overlaid with someone harder and more arrogant, someone like Xandra . . .

"I know you," the practitioner said, eerily echoing Elva's thoughts.

A cold worm crawled through Elva's guts. "Of course you do, sir. We've met several times."

"No. You're Elvanyn Karamisk, or his great-grandson. Why did you say you were from another dimension?"

Elva went cold all over, his ears buzzed, and the world receded. Only two living people knew who he really was, where he really came from, and neither Fenra nor Arlyn would have told Metenari. He steadied himself, shifting his feet slightly farther apart. This *had* to be the Godstone. Having a few of Xandra's mannerisms wasn't so incredible—he was the first person it knew—but it seemed mannerisms weren't all it had.

The practitioner folded his hands together, index fingers extended, and tapped his lips.

The shock somehow cleared Elva's head. Xandra *had* left more than his power behind with the Godstone. Enough of himself to eventually recognize Elva. What should he say now? Whoever or whatever this was, Elva still needed its trust, needed it to be willing to follow his advice. Decision made, he took a deep breath.

"I'm not Elvanyn Karamisk's great-grandson," he said. "I'm Elvanyn Karamisk."

A puzzled frown replaced the practitioner's smile. "However did you manage that? The real Elvanyn Karamisk must be dust five times over."

"I've been living in the other dimension—the New Zone," he added in case the Godstone knew about it. "Either time moves differently there, or people from here don't age the same."

"Interesting. So all that nonsense about only being here for the sake of Fenra Lowens was just . . . nonsense?"

Elva opened his mouth to agree, but closed it again. A reasonable question for either Metenari or Xandra. What should he say? Whom should he answer? When he'd told Metenari he was only here to look after Fenra, he'd been playing a role, trying to mislead the man. Now Elva wasn't so sure himself. Of course he wanted to stop the Godstone, to save the world. But primarily he wanted to save the damsel in distress. Somehow, Fenra had become as important to him as anyone else. Or more.

What should he tell this . . . *person* who might be Xandra?

"It isn't all nonsense," he said finally. "Of course I wanted to come home. I've been waiting for a chance for . . . I don't even know how long. When Fenra and Arlyn told me Xandra Albainil was dead, I didn't believe them; I thought they might be acting for him. I needed to come back and see for myself." Elva ran his hands through his hair. Even if what he feared was true—maybe even especially—his best bet now was to pretend he thought the being he spoke to was still Metenari.

"And I do care about Fenra." Again he thought the truth, or most of it, would be his best bet. "I don't want her to come to harm. She's innocent in this. And she *is* in danger from him, just as I said." Elva hesitated. "He intends to use her power, that much I'm sure of."

"But Fenra Lowens is no more than a third-class practitioner, second at the most. She wouldn't have the strength to perform the *forrans* needed to manipulate the Godstone." A shadow of Metenari's smile floated over its mouth. Smug was the closest Elva could come to identifying it before the expression disappeared completely.

"With respect, Practitioner, this Albainil character may not know that." Elva wasn't sure how he knew, but something told him that Fenra was no more a third-class practitioner than the Godstone was a rosebush.

"Surely she would have told him herself as soon as she realized his intent?"

"I'm sure she did, but why would he believe her? And remember, she's already succeeded in some of the *forrans* he's given her. Maybe the ones this man knows are designed not to require much power."

"That is a very good point." Its voice softened and its eyes unfocused, as if it looked inward, thinking of something else entirely.

"Practitioner?" Elva waited until he was sure he had the thing's attention again. "I believe I know where they've gone. Xandra Albainil, when he was my friend, had a retreat, a tower, in the mountains of the Third Mode. So much other information was passed down in the family, maybe he left the knowledge of this place too?"

"And you know where it is?" A slight smile touched its lips.

"Yes, sir."

"And he won't know that you know?" The smile broadened. "This Arlyn?"

"Yes, sir."

"So if we don't find them here in the White Court, we could look there?"

"That's what I'm thinking, yes, sir."

"We will give the guard one day, and if the fugitives don't appear, we'll look into your suggestion."

This was the best he could hope for. "Yes, Practitioner."

"Dismissed."

⌇

ARLYN

"I'd forgotten about the steps," I said. "The walkways are big enough for horses, but with these steps . . ." I shook my head.

"Horses can manage steps," Fenra told me. "It's cows that can't."

"Good thing he's not a cow, then."

Her horse managed the fourteen steps just as Fenra said he would. Swear he looked at me sideways when he reached the bottom, his eyes partly shut as if he was laughing at me for doubting him. At the bottom the narrow steps widened into a limestone tunnel, worn smooth in spots from untold numbers of hands and feet. You could hear the sound of water, and smell it, so some of the cisterns remained intact. After we'd walked for maybe an hour, the walls became blocks of limestone interspersed with patches of rock, shored up here and there with beams. Reached a wider space, a vaulted ceiling, made from tiny bricks. Floor ended abruptly on the right, falling away into the darkness that had once held water for the entire City.

"These would have been full of water, when the system was in use," I said.

"I know what a cistern is, thank you."

Swear the horse snickered.

"There's plenty of room here," Fenra said. "You'd better ride."

"I can walk."

"No, you cannot. Remember, I am not just any practitioner, I am the practitioner who has been leveling you all these years."

Knew that tone. In no shape to argue with her anyway. Truth was I probably couldn't walk very far on my own.

"It doesn't like me."

"Perhaps because you keep calling him 'it.'" The horse snuffled at the nape of her neck. "Terith, do not let him fall off."

I had to be content with that.

Tunnel started to slope downward and veer to the right. Cisterns had been built where it was easiest, partly dug out, partly natural fissure, so they weren't the shortest route out of the City. The new cisterns, cleaner, easier to maintain, had been built by City Engineers from practitioners' designs when sewage systems came into wider use. These old cisterns had long been forgotten.

"The ceiling gets lower again up ahead," I told her. "We'll both have to walk, and maybe even the horse will have to duck its head."

"Terith will manage."

Once down off the horse, I felt both better and worse. Less danger of falling off or banging my head. More danger of my legs collapsing under me. Thought of beet juice, gritted my teeth. Wasn't going to ask Fenra for strength, at least not yet. Once I would have, without thought. Would have just taken it, would have thought myself justified.

I always thought myself justified.

"I'd forgotten how boring this is," I said. Too much time to think.

"Why don't you tell me what happened in your vault?"

That would teach me to complain about boredom.

"Metenari cut me." I found I was rubbing at the inside of my practitioner's arm and let my right hand fall to my side. "He knew how to extract my pattern from my blood. Unfortunately, he wasn't sure which cabinet held the Godstone. He'd narrowed it down to three—don't ask me how—but he had to try each one. Every time he tried a new cabinet or a different unlocking *forran*, he needed more blood."

"He did not find the correct one right away, then?"

"No."

Fenra brushed her finger along a black growth level with her head, examined it carefully, and wiped her hand off on a handkerchief she'd pulled out of her sleeve. "Did you see much of what happened afterward?"

I closed my eyes, tried to call up the scene from memory. As practitioners our memories were our greatest assets. That's what made the blank spots in mine so frightening.

"I knew when Metenari found the Godstone, because his boy—"

"Noxyn? The redhead?" She stepped aside to the edge of the walkway, opened the soiled handkerchief, and dropped it. I imagined I heard a distant splash of water.

"Yes, that one," I said. "When Noxyn stopped coming back to me for more blood, I knew that they'd found it." That's when I tried to see. I should be able to remember whatever my eyes saw. "The stupid apprentice was standing in my way at first. For some reason I wasn't worried."

"Perhaps you were lightheaded from blood loss."

"Could be. I knew—or thought I knew—that the seals were composed of layer upon layer, complex, some interdependent, no single one good enough."

"A sum greater than the whole of its parts."

"Yes. Who said that?"

"Lorist Medlyn. So you did not use up your power making the seals, your power *was* the seals." Fenra wasn't asking a question. "Metenari could not have known what he was dealing with. He should have realized that the highest level of caution was necessary. Well, he was always too sure of himself."

"By the time the boy moved out of my line of sight, Metenari already had the Godstone in his hands. In a way. In a manner of speaking."

"What do you mean?"

"We talk about it as if it were a physical object, but it isn't. At least, that's not all it is. It can look like a crystal, true, but I think that's just our minds trying to make sense of it. It's made of light, and power. More of an artificial entity, alive, and unexpectedly self-aware."

"Did he awaken it somehow? Could he use your pattern for that as well?"

Rubbed at my eyes, fingers stiff. "You don't understand. It was awake when I sealed it away." I was afraid to look at her. "It would have been expecting *me*. I'd have been the only one it *could* expect."

"But you were there."

"It didn't sense me. I must have been too empty."

Her eyes were full of concern, calculation. "How did you make it in the first place?"

"I don't remember." Wondered if she believed me. Wasn't sure myself. Never tried to remember. Except the fear. "I remember the fear."

"What about Elva? Have we left him alone with that thing?"

"Elva's in no danger. The Godstone doesn't know him."

She nodded, didn't look happy.

<hr />

Fenra

We did not speak again for a very long time. Eventually there were no more water reservoirs, and the tunnel narrowed again. It appeared to have been cut from granite at this end. Finally it began to rise, but abruptly came to an end as if the people cutting it had just packed up their tools and walked away.

"What powers the *forran* that keeps this door closed?"

"This isn't practice. This is engineering."

"That might have been true, once." I laid my palms flat on the rock and leaned in. "Disuse and time have fused the mechanism almost solid."

"Can you do anything?"

I thought for a moment. I could not release the rock as I had the mortar and bricks. It was already in its natural place. But there was something else I could try. I had a *forran* I frequently used on people, minute vibrations, at a deep level, that encouraged stiff muscles and joints to loosen without damaging them, and stimulated damaged tissue to heal. Here, it might do the same for the old mechanism, shiver it open. Without saying anything to Arlyn, in case it didn't work, I repositioned my hands, hummed to myself, and gently pushed again at the door.

The stars were out, and the third-quarter moon up. I was glad to see them.

"I am starving," I said.

"If you can wait a bit, we can sit down to eat in a civilized way."

Another hour of walking along a well-maintained road brought us to what the locals undoubtedly called "the village" but I called a "town," having lived so many years in actual villages in the outer Modes. A small livery was happy to rent us a cabriolet—in exchange for what he thought was a brooch of three gold leaves—and not so happy that we did not need a horse as well. We went along the street to an inn for a meal while everything was being organized. The hostler called his boy—in this case a girl—to come help him clean the cabriolet and harness Terith. I gave him a look as Arlyn and I walked away, and he blinked at me in answer.

Dinner consisted of a thick soup of pureed vegetables, thin slices of rare lamb off the leg, roasted potatoes, and a treacle tart for dessert. I forced myself to eat slowly, as if this wasn't the first meal I'd had in more than a day. We refused wine or beer, but the coffee was most welcome.

"By carriage we're no more than another day or so away," Arlyn was saying. "Less if we don't stop."

"A night in a cabriolet is not my idea of comfort," I said. "But I have slept in worse places. And Elva said he would try to make sure Metenari follows us, so we will want to be there first."

ELVANYN

It takes time to organize an expedition, even a simple one. The captain wanted to send ten men, and tried to insist, but Metenari coldly stated that he would take only Elva and his two senior apprentices.

"Any idea why the practitioner suddenly wants to go on a journey?" The captain looked at him with narrowed, speculating eyes.

"Not a clue." Elva smiled and shrugged.

"And why take only you? Why not take more guards?"

Elva shrugged again. *Maybe it's the Godstone.* Elva took care to keep the thought off his face. "Want me to ask him?"

Pressing his lips together, the man finally turned his attention away. The set of his shoulders said "damned practitioners" as clearly as if he'd spoken aloud.

Once they were out of the City proper, Elva found it colder than he was accustomed to, maybe even colder than he remembered. No one else appeared to notice that he wore his own linen shirt and suede waistcoat under his uniform jacket. He tugged at the knee of his right trouser leg. Come to that, he would have preferred to be wearing his own trousers. The uniforms of the City guards weren't designed for riding. His cartridge belts were slung on his saddle horn and he felt naked without them.

"Does your uniform not fit?" The thing sat its horse awkwardly, as if its last ride had been a long time ago.

Elva brushed away a nonexistent streak of dust. "Not at all, Practitioner. It's just the wrong kind of uniform for riding, that's all. The trousers should be looser here, and here." He indicated thigh and knee. "You see how the material tightens and pulls against itself."

"I do see." It reached across the distance between them and tapped Elva on the knee.

Elva felt a shock pass through him, tightening his grip on the reins. His horse moved forward, picking up its pace as he unintentionally squeezed his knees together. Once he brought the animal under control, he glanced back at the thing, to find it frowning and flexing its right hand. Obviously it had tried some *forran* that hadn't worked.

Its right hand. Not its practitioner's hand. Practitioners often used both hands for a *forran*, but when using one hand only, they always, *always* used their dominant hand, the practitioner's hand. The Godstone had used the wrong hand without noticing. Elva didn't know exactly what the thing had intended, but its frown clearly showed that it didn't know what had gone wrong.

"Practitioner?"

"Hmmm?" It looked up again, straightening in its saddle. Just for a moment its eyes appeared to be a different color, green-hazel, not brown. "You know me," it said. "You don't want to admit it, but you do know me." The strangest smile tilted up the left corner of its mouth. "Go ahead, say it."

Elva pressed his lips tight. *Xandra*?

The Godstone smiled, just as if Elva had spoken aloud.

———

FENRA

The building resembled nothing so much as one of the round towers in the White Court. I had seen towers like this in other Modes, but they usually overlooked harbors. They were named after the engineer who had designed them, but if I had ever known her name, I had long forgotten it. In the Third Mode, towers should be square. From the look of the stone foundation, this one was older than the brick of the upper stories would have you believe, the whole a patchwork of different surfaces, different materials. The sun had dropped considerably closer to the horizon, and I thought I saw it flash off glass in the windows of the upper stories.

"The carriage won't fit," Arlyn said as we approached the arched gateway.

"I will unhitch Terith, and we will leave it outside."

When we got closer I saw a wooden door deep within a stone arch. Thick oak planks, bound with iron strapping that showed streaks of rust. The wood was weathered to silver and the center plank had a crack in it for almost its entire length. Arlyn put his hand on the metal plate set to the right side.

"If there's a lock behind that, it would take a key as big as my hand," I said. Terith snorted and I stroked his flank, hiding my smile.

"Not a lock exactly," he said, his brows drawn down. He pulled his hand back until only his fingertips were resting on the metal. "I forgot . . ."

I waited, but he did not continue. "Does it take a practitioner to open it?" I said as gently as I could. Perhaps it was his recent ordeal, but I had not seen Arlyn as shaken as this before.

"I'm afraid so." He stepped back from the entrance and looked up, blinking, as if he thought he could see the top of the tower from here. "Would you mind?"

I gave Terith a final stroke and rubbed my hands together. "Is it your pattern?" I saw him nod out of the corner of my eye. I brought my hands up in front of my face and blew on them, flexing my cold fingers.

I placed my palms together, concentrated, and drew them slowly apart. The lines of light at first refused to change to Arlyn's pattern, but after several tries, I finally had something I could apply to the lock. With no apparent change to the mechanism, but with a trembling I could feel under my fingers, the lock clicked, and the door swung open the width of my palm.

"Is it safe?" I was not going to insert my fingers into that gap without asking.

"Was the last time I was here." Arlyn answered my question by sliding his own fingers into the opening and curling them around the edge of the door. Not the thickest slab of wood I had ever seen, but very close.

From the age of the wood, and the streaks of rust, I expected the hinges to be stiff and noisy. They opened smoothly and without a sound. The interior, windowless and dark, smelled faintly of earth and stone. Arlyn pulled the door open enough to step inside over the threshold. I followed him through, leading Terith by a grip on his bridle.

"Would you make a light?"

I clapped my hands and a small ball of light rose to the center of the timber ceiling. The inside of the tower was bigger than I had expected. Perhaps the wall was not as thick as it looked from the outside.

"How is it so cold?" I asked, shivering. Considering the dirt floor, the temperature should have been warmer.

"I've never been sure," Arlyn said. "It's always been this way."

This space was set up as a stable, with three proper wooden stalls to the left of the entry, and what would have been the equivalent of a tack room arranged to the right. Along with bits of harness and two large wooden wheels banded with metal, there were two saddles and tools for taking care of horses, brushes, knives to trim their hooves, and so forth. What had looked like a stone shelf proved to be the edging around a small wellhead. A gleam of water showed not too far down. A ladder led to a shallow loft and continued upward through an opening in the roof above. As the space gradually warmed from the heat of our bodies, I could smell fresh hay. *Someone's forran was still working.*

"We go up," Arlyn said, with his hand on the ladder. "And I'll bet the horse doesn't climb ladders."

If Terith could roll his eyes, that would have been the moment. As it was, he tossed his head and snorted.

"It's almost as if he could understand me," Arlyn said.

"Yes, isn't it." I made sure that Terith had plenty of water, hay, and even a scoop of oats before I followed Arlyn up the ladder.

The trapdoor opened into a fine lady's sitting room, from at least the Fifth Mode, judging by the tapestry covering the closed shutters of the window. A hand-woven rug rested on fresh rushes. I could smell lavender. I sneezed. I have never liked lavender. The furniture consisted of heavy wood benches covered with hand-stitched pillows, some filled with straw, some with goose down. The work was exquisite, but I could not help wondering what it and indeed the whole room was doing here, only two Modes away from the City.

No fire in the hearth, but the room felt considerably warmer than the stable space below it, warmer than it should have, considering the stable was empty. Normally, the body heat of the animals helped to warm the floors above. There were two doors in a wall to the left of the ladder. I saw no stairs to an upper story. There was, however, a door in the curved outer wall where sound judgment would say there could be nothing behind it. Just like Xandra's workroom. I left Arlyn struggling with the window shutters and crossed over to the door. Maybe there *were* stairs, if you knew how to look for them. Whatever was behind the door, it generated heat.

ARLYN

Hauling on the shutters, I didn't immediately take in that the sounds I barely heard over the wrenching of the hinges were footsteps. Fenra's footsteps. Going to the inner door. Turned fast, but her hand was already on the latch.

"Fenra, stop!"

Door swung open outward and a fierce wind sucked her off the threshold into the nothingness. Somehow she kept her grip on the latch handle, and I ran to her, hoping it wasn't already too late. Fenra's knuckles whitened under the pressure, and her wrist bent at an impossible angle. Wrapped my hand around it just as her fingers loosened.

Muscles and tendons hard as steel cables under her skin. Braced against the lip of the threshold, threw myself backward with as much force as I could. If I fell and broke something, Fenra could heal me. So long as she was here to do it.

At first nothing happened. The wind continued to suck at the doorway, its deafening howl enough to disorient me. Don't let go. All I had to do was not let go. She wouldn't let go. She wasn't the same kind of fool as her friend Hal. Strained further, brought my practitioner's hand to bear. Still leaning backward like a man rappelling down a steep cliff, moved my right foot a few inches back, dragging Fenra with me. Her hand, wrist, elbow were now inside the doorway. I managed another step. Her whole left arm, her shoulder, and her head jerked into our space and her right hand braced tight against the edge of the opening.

"We have to shut the door!" I had no idea if she'd heard me. I couldn't even hear myself.

Another step. And then another. Fenra's upper body came back into the room. Maybe she could have used a *forran* to pull herself further in, or even to begin shutting the door against the chaos of the void, but I was holding on to her practitioner's hand and couldn't risk letting go. Another six inches. Hands were damp with sweat, could feel her wrist slip ever so slightly. As if she could feel it too, Fenra twisted in midair and grasped my wrist in turn. Sliding stopped. Now she was pulling herself toward me. I had to brace myself even harder.

Finally most of Fenra's torso was on this side of the door. The suction of the void, the noise of it, hadn't diminished. She twisted again, this time bringing one leg in through the doorway. First her knee and then her foot braced against the threshold jamb. With both feet in, she bent and reached for the bottom edge of the door, her arm flailing in her attempt to reach it.

Terrified that one of us might slip, I moved sideways, drawing Fenra closer to the edge of the opening. My shoulders creaked. My elbows, wrists, and knees were on fire. I wasn't sure I could feel my feet.

Fenra hooked the edge of the door with her index and middle finger, inching it forward by will alone, until all her fingers were behind it and she could begin pulling it closed. Took a chance, moved a couple of inches closer, to improve the angle, give her more leverage. I could see her lips moving, though I couldn't hear a sound over the wailing of the wind. The door inched further and further in until finally both of us stood on this side of it, pulling it shut with all our strength and weight.

Latch clicked and the noise stopped so abruptly, I thought I'd gone deaf. Fenra lay collapsed on the floor, her eyes squeezed shut, her shoulders trembling with every shuddering breath. Might have been tears on her face. Might have been some on mine. Twice I opened my mouth to ask her if she was all right. Both times I shut my mouth without speaking.

Finally she rolled over onto her side, grabbed my wrist with her practitioner's hand.

"What . . . was . . . that?"

"A void." Cleared my throat.

"*A* void? Not *the* void?"

"No. I don't know." Pushed my free hand through my hair. "Maybe it's the only one, but I couldn't prove it, one way or the other."

"What's it doing there?" She waved with her free hand, then let it flop back to her side as if it was too heavy to hold up. Perhaps it was. "There shouldn't be anything behind that wall."

"Inside of the building larger than it seems from the outside. Looks like a tower but there's nothing above us." Gestured upward. "The rooms go out, not up."

"But the rooms aren't there any longer." Her grip on my wrist tightened. "When were you planning to tell me?"

"I just thought . . . I thought I'd open the shutters first, get some light. Only that door opens onto . . ." My turn to wave my arm at the door.

She let go of my wrist and rubbed her face with both hands. Could see she was still trembling, as if the room, so warm a moment ago, had become too cold. "Come on." She flinched away when I reached out to help her to her feet. I let my hands drop to my sides. "There'll be firewood, maybe coal in the kitchen." Moved toward the doors in the inner wall. "We can be warmer in there."

"Are you sure *that* door is safe?" Fenra waved me away, obviously intending me to go first. Led the way down the short passage to the kitchen, a wide, high-ceilinged room with a raised fire box against one wall, an open hearth against the other. The fire in the hearth was laid. All I had to do was find the tinderbox, get the thing lit.

"This isn't a Third Mode kitchen," Fenra said. She stood hugging herself, elbows in the opposite hands. "Any more than the other room is a Third Mode sitting room."

"Yes, well. Once upon a time we would have found a better kitchen," I said. "A better sitting room for that matter."

"And now there's only . . . ?" Fenra's gesture took in the two rooms we'd seen. "You had better tell me."

"The Godstone," I said. Fenra kept her eyes steadily on me, waited. "I made the tower as a miniature version of the world, for experimental purposes. Here I could try various ideas without having any impact on the world outside. These few remaining rooms were once the City."

"You made a copy of the world?" Fenra rubbed at her forehead with her practitioner's hand and shut her eyes. Without opening them she asked, "How?"

I shrugged. "How do you think the 'real' world was made?"

"Not by . . . ?"

"No! No, not by me. But I think I've met the one who did. In fact, I think you've met it as well."

"I have?" She sat down on a three-legged stool and leaned back against the wall.

"When you level me, do you ever feel that there's something watching? A presence just on the edge of your vision that you can't quite make out?"

"Yes." Her eyes narrowed.

"Something so big your mind won't hold it." I shook my head, slowly. "When I made the Godstone, I felt it. It terrified me." I moved closer to the fire. "I think *that* was the Maker—not just an outer Mode superstition. I've watched you speak and listen to animals and other natural things, and I think you're really communicating with *it*, experiencing *it*. You don't seem frightened of it."

She nodded, but she was staring off into the middle distance, as if she wasn't really listening to me at all.

"Where does the Godstone fit in all of this?"

I waved around us. "You asked me if I made the world. I knew that *I* hadn't, but I thought *we* had, practitioners. When I opened the gate into the other world, and I found that the practice worked there as well, I thought the first practitioners had taken a stable world like the New Zone, and turned it into what we have now, maybe just for an experiment, or maybe they thought traveling the Modes would increase their power, I don't know. I thought this world was an experiment gone wrong, that it was meant to be stable as well. I thought I could turn it back, and that I could do it all myself. I made this replica to test my theories."

Fenra pointed toward the sitting room, and the closed door beyond it. "And you got that instead?"

I looked away. "I already said I was wrong." I walked over to the window in the southern side of the kitchen and opened the shutter. The sky looked like rain. "Removing the Modes, making them all the same . . ." I shrugged and turned around to face her. "Apparently all that would do is return this world to the chaos it was made from."

"But this is only a copy. What if it isn't an accurate copy?"

"Think I should have taken that chance? I didn't. I don't."

"And that's why you locked the Godstone away."

Ran my hands through my hair. It was getting too long. "I tried to destroy it, but I couldn't. Locking it away was the only thing I could do."

"And now?"

"Now that it's loose . . . now we try again. Or rather, *you* do."

Eleven

ELVANYN

ELVA LOOKED AROUND carefully as they left the City, but though Xandra had told him years ago which Road marker indicated the edge of the First Mode, he hadn't seen anything different as they'd passed. He found himself touching his revolvers, and even his sword, again and again, but they were always the same. Or so he thought. Arlyn had told him that artifacts from the New Zone wouldn't change here, but Elva didn't know whether to believe it.

He knew that the colors of practitioners' clothing—the yellow trousers, the crimson waistcoat and black jacket—never changed, but did Metenari always wear those knee breeches? Those white hose? That long-tailed frocked coat? That powdered wig?

Not for the first time he wished Fenra was with him. At the very least she could tell him whether his guns were still his guns. He couldn't ask anyone he was with; he wasn't supposed to know about the Modes.

In the late afternoon they arrived at the Inn where they were to spend the first night. He knew he should recognize it—and he did, in a way. It seemed to be in the right spot with regard to the Road, and the windmill was still there—but then how and why would anyone have moved it? The vanes were in good repair, however, when he remembered bare skeletal wood, and the sails were evidently brand-new canvas unbleached by the sun. There was even a small wooden cart with a large dog between the shafts waiting with several sacks that could be grain.

None of this meant a change in Mode, he reminded himself. After all, these were really only repairs and renovations; prosperity was a likelier explanation.

"You have a most peculiar look on your face." Predax appeared beside him, leading Metenari's and Noxyn's horses by their reins. It hadn't taken the young apprentice long to adjust to the idea that Elva was someone out of their past. The same couldn't be said for Noxyn.

"When I was here last," Elva said, "this inn was a one-story building, half rough stone and half old timbers, and all of it stuccoed by someone who didn't know how. This courtyard wasn't enclosed, and the stables were just a couple of lean-tos and sheds."

Predax looked around, eyes narrowing as he considered Elva's words. "I've never seen this place before today," he said finally. "But I'm not surprised you see some changes." He cleared his throat. "The world has moved onward since your time." This last was said with the young apprentice looking sideways at Elva, his attempt at a smile offset by the nervous movement of his eyes.

Elva clapped the boy on the shoulder. "Well, we're bound to be more comfortable, in any case," he said with the look he used to encourage his deputies at home. The smile he got in return was stronger.

And that was Metenari's failure as a teacher, he thought. Perhaps Predax wasn't the brightest star in the sky, but he was still a practitioner. To make it so clear that Noxyn was the preferred apprentice, to the virtual exclusion of the other boys, had to be bad for their training in the long run. And worse for their morale.

Not that it mattered much just now.

Elva gave Predax another pat on the shoulder and followed the stable boy leading the horses away. He'd learned over the years always to check on the comfort of his horse himself. Even from the rear, the inn looked brighter and better kept. Stone might still be part of the construction of the lower level, but brick prevailed above, framing each window with decorative patterns. And the windows were glazed. He could see ripples in the glass when he squinted from an angle, which put the workmanship on a level with what he was used to at home.

Home. He'd lived in the New Zone for such a long time—longer than he'd lived here—it shouldn't be odd that he thought of the place as home. Even if he didn't know whether he'd see it again. He wondered what Fenra would make of his house in Dundalk. Would she find it too cluttered?

Entering the inn from the stableyard, Elva found a long hallway running down the center of the building. To his right, judging by the smells, was the kitchen, with wood for the fireplaces and ovens stacked against the hallway wall beside the open door. Beyond this was a dining room, where the ceilings were higher than he remembered, and what had been a roughly furnished room of trestle tables with stools and benches was now a more formal place, with polished plank floors, separate dining tables with individual chairs, and elaborate fire irons on the hearth.

Across the hall from the dining room, in what would have been called a "parlor" in the Dundalk Territory, he found the Godstone, sitting comfortably on a settee to one side of a narrow brick fireplace, facing a similarly dressed but much older man on a matching settee. Elva had rarely seen a practitioner who looked that old. Nor was it just the man himself. The elbows and cuffs of his jacket showed wear, and his cravat was not as white as it could have been. Noxyn sat just behind his master to the left, on a less comfortable chair, holding a cup in his hand.

"I've been stationed here most of my practical life," the older practitioner was saying as Elva hesitated in the doorway. "I've heard rumors from the City, of course. But here I've always been treated with the greatest respect. Thank you, my boy." Predax had come from a sideboard out of Elva's line of sight with a tall, narrow copper pot to refill the elderly man's coffee cup.

The Godstone gestured Elva forward and pointed to a seat not far from its own, but not as close to the fire as Noxyn's. Predax returned to his place at the sideboard. Evidently he didn't merit a seat.

"The chief of my guard," the Godstone said when the older man glanced in Elva's direction with lifted eyebrows.

True, Elva thought, *if only because at the moment there aren't any others*.

"No," it continued. "These rumors are nothing that need concern you. Public opinion ebbs and flows like the tides in the sea, and more so with attitudes toward the practice. I've seen our status—and the public's attitude toward us—change several times. At the moment we're at an ebb, but when things have gone too far we have only to effect some great cure, or some great rescue, preferably of a poor man's child, to swing things round in our favor. These things are easy enough to arrange when needed. Then you'll see the tide turn, and we'll be heroes again until the next time the Red Court feels insecure."

Noxyn looked over as the Godstone spoke, his eyebrows drawing together, and Predax glanced from his mentor to Elva and back again, biting his lower lip. The older man merely raised his eyebrows and then lowered them to a frown, as if he was waiting for clarification. Clarification that never came. Apparently the Godstone wasn't aware that it had said something unusual.

Later, when the Godstone and Noxyn had gone to dine in private with the older practitioner, Predax brought his dish of lentils and chicken to sit across from Elva at one of the dining room tables. Elva smiled. At this point anything was better than being left to his own thoughts.

"Looks like you've got something on your mind," he said, watching the young apprentice stir his food without lifting his spoon from the dish.

"It's what Practitioner Metenari"—the boy never neglected the title—"said about the popularity of the White Court ebbing and flowing with public opinion."

"He's quite right," Elva said. "In the New Zone, the monarchy—like the Red Court," he added when he could see Predax didn't understand, "—and other forms of government go in and out of favor, depending on whether the people have been given the things they were promised."

"But how would the practitioner know this?"

"He'll have studied it, surely. It's something every educated person knows, whether it happens in their lifetimes or not."

Predax shook his head. "*I'm* an educated person. I've never heard of anything like this, and I've studied all the history there is."

"Well, obviously not *all* the history," Elva pointed out.

Predax set his spoon down onto the tabletop. "Why wouldn't the practitioner have told us this before? If it were commonly known, we apprentices would not have to be . . . we wouldn't have to be . . ."

"Afraid all the time? That might be part of your training."

Again the headshake. "But none of the other practitioners have told any of *their* apprentices. We compare notes, you know, and if even one of us had heard of this, we would all know."

That Elva could easily believe. "But what about all this has you scared? I would have thought that you'd be pleased knowing that your status could be secured so easily."

"It isn't that." The boy leaned forward, his sleeve dipping into his bowl. "It's . . . how does Practitioner Metenari know this? And how has he never mentioned it before? The way he said it, the way . . . He speaks as though he's seen this with his own eyes. As if he remembers something he couldn't possibly remember. He's not all that old, you know." Predax must have seen something in Elva's face, because he stopped, mouth hanging open. "You know something," he said finally. "What is it?"

"Remember we talked about how Metenari didn't seem like himself?" Elva watched Predax's eyes widen and his face pale.

FENRA

I was not even aware of sitting down. I had no sense of the stool under me. I only knew that Arlyn had shoved a pillow behind my back when I no longer felt the cold of the wall.

"You are insane," I said. My throat hurt. "Your lowness has turned your head. If you, the great and powerful Xandra Albainil, could not destroy this thing, how can I do it? If the best I can accomplish is to contain it, you are asking me to do what you once did. To use all my power."

"But—"

I lowered my head into my hands. If Arlyn continued speaking I did not hear him. All I felt at this moment was a smothering exhaustion. I did not know what kept me from collapsing to the floor. I welcomed the feeling, tried to sink into it thoroughly, tried to think of nothing except how heavy my bones were, how heavy my head, my hands, my skin.

For an instant I felt a cool, damp breeze on my skin, smelled the sea, but my brain refused to stop circling the image of my facing the Godstone, of sealing it away. Of living without power for the rest of my life. Unable to help anyone, unable to heal them. What would I do? Who would I be? I began to see what might lie behind Arlyn's lowness.

Who would level Arlyn? If the lowness was a result of losing all power, who would level *me*?

These were just surface thoughts. Under them, I knew that I was going to do this, or at least that I would try. I would make the same choice that Xandra Albainil had ultimately made.

"Did you know?" My head felt as heavy as a block of carving stone.

"What, specifically?"

"Did you know what containing the Godstone would do to you?"

I recognized the look he gave me. I had seen him use it on pieces of wood, on tools, even on his own drawings for furniture. Calculation. Weighing of options. Would this serve the purpose? He considered his answer just as carefully.

"Not at first, but it became apparent."

"Would you have done it? If you had known before you started, I mean?"

The shadow of a smile ghosted across his face, and he shrugged. "I knew what I had to do. Nothing could change that. I couldn't let him destroy the world—I couldn't even let him make the attempt."

"Him?"

This he waved away. "I began to think of the Godstone as a part of myself I had to set aside, to put away. The part of me that thought it was worthwhile to risk the lives of every mundane living, just to find out whether I was right about something."

A part of himself. "You risked your own life."

Again the shadow smile. "Of course, that's what makes it fun. And there was always the New Zone. Anyway." He slapped his thighs and straightened to his feet. "That's not what I'm asking you to do. You don't have to contain it, all you have to do is help me shove it through the door." He tilted his head toward the other room. I forced myself to look, the muscles in my neck protesting every degree of movement.

I felt a ghost of a smile touch my own lips. "Shove it through the door."

"Just pushing him won't take anything like the same amount of power as it would to contain him," he pointed out.

"If we can get him to open the door, we will not need to do much pushing," I agreed. "But how will we manage it?"

This time the smile shone in his eyes.

"Arlyn—"

"It's me he wants." He tapped his forehead. "What I have in here. I'm betting he'll do practically anything to get it."

"You will go through the door?" My heart beat faster. Had he seen what I had seen?

"If necessary, of course. Don't look at me that way, Fenra. A minute ago you were gearing up to sacrifice yourself, and I have the feeling you're far more valuable to this world and the people in it than one old cabinet maker who has to be saved from using his own tools to cut his throat."

"I thought you said that practitioners could not kill themselves."

"I did, and you're changing the subject." Now his face hardened, and I saw what Xandra Albainil might have looked like once. "Make no mistake, I will use whatever and whomever I can to rid our world of him once and for all. Me, you, Elva—whatever it takes."

A small part of my mind wondered if I should find the order of names significant.

"That's why you wanted to lure Metenari here, bringing the Godstone with him."

He scrubbed his face with his hands. "I'm hungry. Are you hungry? Let's see if my food spells still work. They won't be as good as the ones Medlyn Tierell used. I'll be sorry I won't get the chance to find out how he did that."

From which I understood that going through the door was actually part of his plan.

It's a pity that none of this idiot's *forrans* are any use to me. I know them, I see how they should work, but they don't. The idiot has a very inflated opinion of his own abilities, his success, and therefore his value to the world at large. In fact, his only value to me is in the information he can provide about the world as it is today—not that the place has changed so very much really. Customs and attitudes can shift a little, as I tried to explain to that old graybeard last night, but mundane nature is mundane nature, and it doesn't seem to have evolved much if at all in the time I was locked away.

How Elva looks at me is far more interesting. Sometimes I'm sure he knows me and is just pretending he doesn't, sometimes I don't think he's on to me at all. It's strange to see him as a stranger might. To watch how he reacts and interacts with people when he thinks I'm not there. I've never had a chance to see this side of him, of course, since I've always been there.

"How much longer to the next Mode?" I ask him.

"Sorry, Practitioner, I don't know what you mean." He goes through that ritual he's not aware of, where he touches his sword and then each pistol in turn.

I watch his face carefully as I explain what the Modes are. I think I'd be able to tell whether he's only pretending not to know, but if there's a sign it's too subtle for me to see.

"So from what you're telling me, I *couldn't* know," he says finally. "I wouldn't even know whether my guns are still the same, or whether they just seem like it to me."

"They're the same," I say. "Do you want me to tell you if and when they change?"

He looks thoughtful for a moment. "I'm not sure," he says, shrugging. "Part of me thinks I'd rather know, and part of me doesn't."

"*You* haven't changed." I laugh aloud. "Always looking at things from different angles. You should learn to make up your mind." When I glance at him, he is looking at me with an examiner's face, as if analyzing what he saw. "Do you know me?" I ask him again.

His eyes narrow. *He knows.* Suddenly I'm sure of it. He has to know. Even the stupider of the two apprentices is starting to be suspicious. Why doesn't Elva say something? Perhaps he thinks that *I* don't know? Could that be it?

"Very well," I say. "Perhaps you can't see where the next Mode begins, but you can tell me how much longer it will take us to get to Xandra's tower." I've never traveled there like this, as a flesh-and-blood person. Physical distances meant nothing to me before.

"We won't get there before nightfall," he says. "We'll want to stop for the night. Though there's a waxing moon tonight if we want to press on."

I think about it. Arriving in the middle of the night would definitely be unexpected. It would catch them both completely off guard. On the other hand, I don't need surprise to help me. Arlyn Albainil's a mundane, and the little girl won't be a problem either. She's barely second level on her good days. She's important to Elva, that's clear, and so long as she stays out of my way he can have her.

"You say there's somewhere for us to stop?" I remember an inn, one story and a thatched roof, but perhaps I'm thinking of somewhere else. If Elva suspects who I am, my asking might make him uncertain.

"Only a Wayfarer's Rest," Elva says. "It can be made comfortable enough for one night."

I think a bit. I'm sure I don't need surprise to help me. "We'll stop then. See to it."

Elva leaves me to ride ahead. Noxyn moves his horse up to take Elva's place. "I've an idea that would help you."

"Do you?" I don't trouble to keep the amusement out of my voice.

"Metenari's an idiot," the boy continues. I almost smile, considering my thoughts of a moment ago. "Accomplished, learned, sure, but a fool in the way the world works."

"Who do you think you're talking to, boy?" When I tilt my head to see him better, he's looking me right in the eye.

"You're the Godstone," he says. "Only a fool wouldn't see it. Is there anything of him left?"

I ignore the question. "And how do you think *you* can help *me*?"

"I'm a better partner than Metenari."

"And how will you prevent me from erasing you, as I've erased him?"

I think I scared him a little, but he only smiles. "First, Metenari didn't know what you are, and I do. I'll be prepared and ready. Second, I know that a practitioner's power can contain you, and I'm much stronger than he ever was."

"Strong enough, do you think?" This is fun.

"There's always ways of getting more power. Doing all this research," he shrugs, "I found a few useful *forrans* I didn't tell Metenari about. Predax is fairly strong. If necessary, I'll just take his power. He'll never be missed."

I decide I like this boy. He keeps his eye on his goal and isn't sidetracked by sentimental nonsense. Ambition is always easy to work with. He's already proven that he is smarter than his mentor ever was. "How do you propose we explain it to Elva?"

"Elva who?" He grins in a way that is unexpectedly familiar. "What do we need him for?"

No sentiment. "He knew Xandra. He could still be useful."

Noxyn shakes his head like an adult indulging a small child. I feel a coil of heat in my guts. "How? Arlyn Albainil is one thing. He has the blood, and maybe even the *forrans*. Why would Xandra have told this mundane anything important?"

Now I don't like him so much after all. Elva is mine. Not someone to be discarded so lightly by someone else.

ARLYN

When I went to help Fenra start a fire in the hearth of the old kitchen, she shooed me away.

"Leave it," she said. "It's my fire, and it might take a dislike to you."

I couldn't be sure she was joking.

We found smoked bacon, bread, and even coffee in the containers I'd left here, and only slightly stale. I'd thought there should be eggs, but we couldn't find any.

"Do you think Elva will be able to convince the Godstone to follow us here?" Fenra sat back on her heels and watched the flames catch.

"I'm certain of it," I said. "I know how he thinks, remember, and that's an advantage. He's going to want the *forrans* I know, and faster than Metenari can find them by research." All of which was true, but only part of my real reason. I couldn't tell how, but I knew the God-stone would follow me anywhere. It wouldn't take much to persuade him through the door. He'd never been afraid of the chaos, he always thought he could overcome it, or that I could, if I really wanted to. "What I'd like to know is whether Elva's managed to be part of the expedition. I hope he has."

"You would want him so close to the Godstone?" She laid out slices of bacon in a cast-iron skillet heating on the fire grate.

"If he's here, that increases the odds that you—that both of you will escape."

The look on her face told me that she didn't believe any of us would escape. Suddenly I wasn't so sure she didn't know exactly what I had in mind. "Will he know Elva, do you think?" she finally said, turning back to the fire and shaking the skillet before turning the bacon slices over with a large fork. The smell made my stomach growl. "If he has any of Xandra in him . . ."

"That would be the only way." I tried to remember how long after the creation of the stone I'd sent Elva away. Had the Godstone already been aware when Elva and I argued about the dangers, the risks in what I'd planned? I'd sent Elva to safety before the disaster with the tower, hadn't I? I shook my head. "I can't be sure. I don't know when he stopped being merely a tool and became aware."

"We will have to hope that there isn't enough of Xandra in the Godstone to recognize Elva," she said.

"If he does, if he thinks he can use Elva to control me somehow, I want you to know now that I can't let that happen."

"I am sorry you thought it necessary to tell me." She sat back after removing the skillet and pushing the coffee pot closer to the center of the fire. "Will the Godstone know *you*? What should we do if he does?"

Fenra asked the question quietly, but it rang in my mind. "I don't know." I was saying that a lot. Maybe too much. "If he doesn't, we should act as though we believe he's Metenari. It won't hurt if he thinks we're stupid."

"Once we have eaten, you should rest. I will take the first watch."

"No need," I said. "They won't be arriving tonight."

She divvied the bacon onto two plates. "How can you be so sure?"

"He'll think he doesn't need surprise on his side."

ELVANYN

The Godstone swung its leg over its horse's head and jumped lightly to the ground. Elva didn't need to see the surprise on Predax's face to know that this was not the way Metenari ever got off a horse. Even his new slimness wouldn't account for this easy movement. Unlike the apprentice, Elva did know who got off his horse that way. He made his own dismount in a less flamboyant manner.

Noxyn appeared not to notice anything unusual. Of course, he and the Godstone had spent much of the previous night talking to each other in undertones, which could definitely account for it.

"What do you think? Hasn't changed much, has it?" The Godstone walked right up to the stone wall and laid its hand against it, almost in a caress. "No, not much at all." It turned to Elva. "Do you remember where the entrance is?"

Elva wasn't sure how to answer. The Godstone wasn't taking much trouble to pretend to be Metenari—maybe there wasn't any of the practitioner left. "It's around the other side, Practitioner. Facing the rising sun." Curved windows ringed the upper stories of the tower. There would be sunlight shining in one window or another no matter the time of day.

"That's right."

The Godstone set off around the tower on foot, and Elva hung back, doing his best not to overtake it, though what he wanted most was to shoot it through the head. His hand even drifted across to touch the gun under his left arm. The sword would certainly be quieter, but he felt the sudden need to see the thing's brains.

Except that wouldn't kill it, however satisfying it might be for a moment. The body would be dead, but the Godstone would just be free. Though he could hold that in reserve . . . He came up on the thing's right side, the better to either shoot or stab, if distraction proved to be necessary. The Godstone stood with its palms on the weathered wood of the door, as if it could sense through this contact what was happening on the other side.

For all Elva knew to the contrary, it could.

"It's worn." An unexpected note of uncertainty underlaid the thing's tone. "It's not this old, surely."

"It might be the Mode, Practitioner." How had Noxyn crept up so silently? A noise from behind them was Predax clearing his throat. Elva kept his eyes on the Godstone. Practitioners or no, the other two didn't matter. As Fenra had said, you could kill apprentices if you took them by surprise. And these two were not suspicious by nature. Predax was too naïve, and Noxyn too blinded by his own cleverness.

"Elva." The Godstone beckoned him closer without looking at him. Elva stifled a grimace just the same. He didn't like the thing using the diminutive of his name, whoever it thought it was. "Is it locked?"

He put his hands on the latch and lifted it, swinging the door open. Inside, a horse nickered at him and tossed its head. *Terith*, he thought. *Fenra is here*. Although he knew he should wish her anywhere else, he felt a flush of warmth, and had to stifle a smile. When he turned around at the sound of the door swinging completely open, he found the Godstone frowning at Terith, its eyes narrowed.

"They'll be upstairs," he told the thing. *Fenra will kill me if I let it hurt Terith*. The thing smiled and Elva gritted his teeth. It had never looked so inhuman as it did at that moment. Then its face relaxed and Metenari's face was back. Except for the eyes.

"After you, by all means," the thing said.

Elva turned toward the ladder, feeling as though he saw the moss on the cobblestones and the crack in the fifth rung for the first time. Well, he'd done his part, he'd brought the thing here. He had to hope that Arlyn and Fenra had come up with a plan.

Standing on the seventh rung, Elva reached for the latch of the trap-door and looked down at the Godstone, still at the foot of the ladder. He raised his eyebrows slightly, as if waiting for orders, and lifted the trap when the thing nodded to him. Trapdoors were a defender's dream. Only one way to use it, and that put your enemy's head right into your line of sight.

Good thing I'm not their enemy.

It was the smell that struck him first. The dried grasses and reeds that covered the floor, the lavender that had been strewn over them. The ashes of the fireplace, the scent of the beeswax candles, and even a faint trace of cooked bacon. And laid over this like a fine mist, almost undetectable, Fenra's scent. One day he'd figure out what it reminded him of. If he had any more days. Only then did he fully take in that Fenra herself was standing off to his left, her hands clasped in front of her. He smiled at her, and drew in a breath when she smiled back. His smile faded when the Godstone came up out of the trap behind him.

Fenra stood close to the kitchen door. She glanced away, and when Elva followed her gaze he saw Arlyn standing next to the door that opened to the other Modes. He understood at once that they wanted the Godstone to go through that door. Elva had used the second door himself, but all he'd ever seen on the other side was another sitting room.

The Godstone stepped in front of him. "Xandra." By its tone it was delighted. "I didn't know you were still alive. These idiots thought you were someone called Arlyn. Why have you been running from me? I'm not angry—oh, maybe I was to start with, but I had a lot of time to think, and I want us to put all that behind us, and start again. Partners. Your knowledge, my strength. There's nothing we couldn't do."

Elva pulled both guns. Here it was, the moment for Arlyn to prove who he really was, and what he really meant to do.

FENRA

"What, no hug?" The Godstone reached out Metenari's arms for Arlyn, and in that moment, with that gesture, I knew that nothing of my classmate remained in the body before us.

"Well, you did leave me for dead, back in the vault," Arlyn said.

The Godstone dropped his arms. "No, that was Metenari. I would never have left you if I'd known you were there."

When he turned to look at me, I saw coldness in his eyes. "Stay out of the way and you won't get hurt."

Yet, is what I read on his face.

The change was so obvious, I could not understand how his apprentices had not known this wasn't their mentor any longer. But no, I realized watching them, they knew. Though Predax was the only one afraid.

I felt Elva's eyes find me and glanced to where he stood beside the still-open trapdoor. He had drawn both guns.

"What's happened to Metenari?" Arlyn said. At that moment it dawned on me that Arlyn had not said what we should do if the Godstone knew who he was.

The Godstone laughed. "Come on! You don't care." The humor left its face. "I'd expected *you* to come for me, not that idiotic waste of space. But you didn't come, did you?"

"I know you realize why I couldn't." The calm in Arlyn's voice astonished me. I could feel my heart thumping in my chest, the dryness of my mouth, and he sounded as if he were in his workshop planing a piece of wood.

The Godstone nodded. "I do now. You knew *how* to reach me, but I see you aren't able to do it yourself. That's why you brought the little practitioner here." He gestured at me and I found myself taking a half a step backward. I could have asserted my own power to hold myself in place, but the whole point of this charade was to keep the thing from knowing exactly how much I was capable of.

Arlyn put the fingers of his practitioner's hand on the latch that had taken us so much strength and trouble to close. I imagined I saw his hand shaking, but it could very easily have been me.

"Let's leave these children here to tend to things," he said. "I've got coffee in the next Mode."

"Oh, coffee! Would you believe I forgot about coffee?" He hesitated. "Through there? Are you sure?"

Arlyn managed to chuckle. "It's the other door that's dangerous."

If I had not already seen that this being was not Metenari, the ease and grace with which he crossed the room would have told me. In everything physical but his actual practice, Metenari had been no better than awkward, and more often clumsy. I wondered whether that accounted for his leaving the City so seldom. I wished I had liked the man better. I wondered why I was thinking of such irrelevant things.

The Godstone stopped arm's length from Arlyn and gestured at him to open the door. Smiling, Arlyn took a firmer grip on the latch. I knew that smile. He had smiled at Elva that way, and now that I thought of it, at me as well. I would have given anything in that moment to be able to stop him. He stumbled, and the Godstone took him by the arm, above the elbow, as if to steady him. I saw Arlyn take a sudden breath, almost a gasp, and I thought the being had hurt him somehow. Arlyn's smile faded, and then grew warmer again.

This was my moment. I moved to one side for a better angle on them. Arlyn acknowledged me with a twitch of an eyebrow. I signaled I was ready. He lifted the latch. Instantly the chaos sucked them off their feet and through the opening into the nothing that lay beyond. Arlyn did as planned, retaining his grip on the door latch while he kicked at the thing clinging to his arm. Bracing myself, I parted the hands I'd kept clasped and drew lines of power around Arlyn to steady him and strengthen his grip on the latch. At the same time, I directed energy against the Godstone, trying at once to loosen his grip and push him farther out the door. In that moment he looked at me, and I saw awareness and understanding in his eyes, and anger. Yes, and fear.

Finally, when I thought I must stop or collapse, their hands slid apart, and Metenari's body flew off into the nothing. I immediately reversed the flow of my power, pulling Arlyn back into the room and slamming the door shut.

For a moment he stood with his forehead and his hands pressed against the surface of the door, his shoulders heaving as he drew in breath after breath.

"Fenra," he said. Without lifting his head, he reached his hand out toward me.

Twelve

ELVANYN

E LVA WAS REACHING for Arlyn when Noxyn shoved him aside, almost knocking him off his feet. Arlyn still braced himself with his practitioner's hand on the door, and when Noxyn took his free arm, Arlyn looked at him as though he didn't know who he was.

"Stop where you are." The snap in Fenra's voice would have been enough to freeze braver men than Noxyn. Elva grinned. "Let him go."

The apprentice whipped around, aiming a look at Fenra that was more than half sneer. "Stay out of this. You're nothing but a third class and lucky to be that," he said.

"And you are an apprentice without a mentor," she said coolly. "Metenari is gone, and I am the ranking practitioner in this room. Let Arlyn go."

Noxyn opened his mouth to argue, but something in Fenra's face made him release Arlyn, who immediately sagged against the wall. The apprentice took a position Elva recognized, feet squared, hands open and rising. Fenra mirrored the stance almost exactly, rubbing her palms together slowly, deliberately. Elva put his left hand on the grip of the gun under his right arm, leaving his right hand free for his sword. Clearly Noxyn, pale with rage, intended to hurt, perhaps even fade Fenra, while she very obviously planned on stopping the apprentice without hurting him. A much dicier proposition. Elva drew his gun.

"You tricked him," Noxyn said, his cold eyes fixed on Fenra's face. Elva wondered which "him" Noxyn meant. "You lured him through that door and now you will get him back." Noxyn held his practitioner's hand over Arlyn's shoulder. "Or I will destroy this worthless piece of—"

Bang!

Elva blew smoke away from the end of his gun barrel. "Well, what do you know. They work fine here, just like at home."

Noxyn stared down at the spreading bloodstain on the front of his shirt, slowly pressing his practitioner's hand against it. Fenra ran forward and caught him just as he fell against the wall, almost dragging Arlyn down with him. She tried to take hold of both of Noxyn's hands, but he wouldn't give her his practitioner's hand. She touched his face, the muscles already gone slack, the eyes empty.

"Elva . . ."

This time Elva wasn't exactly sure what the look on her face meant. "You couldn't have stopped him," he told her. "Not and be certain of saving Arlyn. Trust me, if we'd left him alive, we'd spend all our time watching our backs."

"Elva's right. Nothing we said would have changed his mind." Arlyn's voice sounded as though he'd been screaming all night. He tried to sit down where he stood, his shoulders braced against the wall, but his knees wouldn't unlock. Elva went to help him, but Fenra didn't move aside to let him past. She held his gaze with hers for a moment before she glanced at Predax, standing well back with both hands pressed tight over his mouth, his eyes wide and staring.

"Are you going to shoot that one as well?" Her voice cracked with anger.

"No, I'm not." Elva re-holstered his gun to prove it. "He's not such a poisonous idiot as his schoolmate here."

"He's right," the boy said, lowering his hands. His voice was higher than usual, but steady. "I'm not. This is way over my head. I don't want to know any more about it, I just want to get back to the City and never leave the White Court again in my life."

Fenra flicked her gaze back to Elva. "We can trust him," he said. "Predax was already having second thoughts about Metenari before we ever arrived here."

"And he doesn't want a shot in the heart like his friend." Arlyn cleared his throat and pointed at Noxyn's body with his thumb.

"Noxyn was no friend of mine," Predax said. The tone he used left no one in any doubt that he spoke the truth.

"Fenra, I need to sit down." Before Fenra could move, Elva wrapped Arlyn's arm around his shoulder and led him slowly to the chair nearest the fireplace. He felt Arlyn trembling, and as soon as he was seated Elva covered the man's lap with the heavy wool rug hanging over the arm of the chair.

Arlyn sighed and rested his head against the chair's high back, his eyes closed. "We'll have to get the body out of here if we plan to spend the night," he said.

Elva glanced around to where Fenra still stood over the corpse. She hadn't moved from her position at Noxyn's side, not even to wipe the blood from her hand. Elva turned to her in concern. "Did the killing upset you?"

Her face still stiff, she waved him away. "I have seen death before. I am not usually on the killing side of things, but I know all about having to remove bad tissue so that the good can live. I do not have to like it." She looked at each of them in turn, wiping her hands on a handkerchief she pulled from a pocket of her waistcoat and then folding it carefully to dust off the sleeves of her jacket. Her hands trembled slightly, and it was obvious she was using these ordinary gestures to distance herself from the body at her feet. "I take it we are in agreement that he *was* bad tissue?"

"If I may, Practitioner?" Predax took a short step forward, clearing his throat. "I think Noxyn hoped the Godstone would prefer him to Practitioner Metenari. He, my old mentor, he was ambitious, but a good man. He thought he was helping. I don't think he would have welcomed the Godstone, if he'd known what would happen when he found it, Noxyn . . ." He shrugged. "Well, I heard them talking. Noxyn thought he could control it, even when it was obvious Metenari couldn't. Maybe especially because."

"Just curious, was he? Metenari?" Elva found it ironic that Arlyn would say that.

"Not *just* curiosity, but, well, yes." Predax blinked and looked around at them. "If you tell me where to put him, I'll, uh . . ." Predax gestured at the body.

"Better let me." Elva had rarely seen anyone look so relieved. *He's the one who's never seen death before*, he thought.

"Take him all the way outside, if you would," Fenra said. "Terith is not keen on dead bodies either."

Elva suddenly felt an overwhelming reluctance to leave Fenra with that look still on her face, but he shook it off and bent to grab the corpse by the ankles. "Hold the trapdoor for me, Predax." He pulled the body over to the opening, and saved himself some trouble by pushing it through onto the hard-packed dirt of the floor below. Predax put his free hand over his mouth and turned away. Elva touched him on the shoulder and waited until the boy relaxed before turning to climb down the ladder. The horses snorted and shifted themselves to the other side of the stable. Though Terith moved with the others, he never took his eyes off the body.

"Sorry, Terith," Elva said. "I didn't mean to startle you. Next time I'll call out first." Elva didn't know what surprised him more, that he meant what he'd said, or that Terith seemed to understand him.

The warmth and the light of the early summer evening startled him. He felt like hours had passed. Birds still sang. The sun was just sinking behind the hills. The distant trees had leafed out, but the foliage wasn't the dense canopy it would be later in the season. Elva dragged the body around to the far side of the tower and left it, taking his time to walk back round to the door. It felt odd just to leave it. At home—in the New Zone, he meant—he'd be complaining about wasting time digging a grave, but he'd be digging it. He should have asked Fenra if she had any corpse packets with her; most practitioners traveled with a couple. He'd wished for one often enough when he'd had a shovel in his hands. Just sprinkle the body thoroughly, say a few words of comfort to the bereaved, and walk away. Sunlight did the rest, though some swore by rain. In some ways it was better to be here.

ARLYN

"I have never seen you like this. You are almost glowing with energy."

The woman sounds worried, and looks exhausted, so I take care to stay sitting down and breathing deeply.

Fenra was smarter and more skilled than she let people know. As soon as I had that thought, I stifled it, before he could hear it and get ideas.

"I feel so light," I say. "I didn't realize just how much this was all weighing on me. How relieved I'd be once it finally ended."

"Do Jordy and Konne know where we are?" I asked Predax, getting the question out before I could be stopped.

"No sir, not exactly. Practitioner Metenari felt the fewer who knew exactly what was going on the better." Predax starts at the sound of Elva's feet on the ladder and looks away as Elva pulls himself through the opening. Clearly Predax doesn't even want to think about what Elva has been doing. I might be able to use that.

"I should set wards." The woman, hands on her knees, pushes herself to her feet.

"Is that necessary?" I ask her. "What could possibly be a danger to me here?"

"If we plan to stay overnight it is." She looks at me sideways. "We are far from the Road, and horseflesh is a great temptation, even when indoors." She looks straight at me, frowns.

"Of course, naturally," I say.

Listen, watch, listen, I said to myself, hoping somehow that Fenra would hear me—and that he wouldn't know I was thinking at her.

"Which means I'd better go too." Elva rises to his feet, checks his guns, his sword. "Wouldn't do for some varmint to get her before the wards are set."

I turn my attention inward. What is he trying to say?

FENRA

"Arlyn feels . . . different." I shivered, though the air still held the sun's warmth. My eyes automatically scanned the ground around me for loose rocks, old seedpods or pinecones, and stray twigs. I did not know how much I wanted Elva to disagree with me until he did not answer.

"He's been through a lot," Elva pointed out finally. "Not a surprise he's so shaken."

I breathed on a handful of twigs and set them down in pairs, crossed over each other. "Do *you* know the names of Metenari's junior apprentices?" I asked him.

"Of course I do," he said. "I've met them often enough in the last few days. He was always sending them on errands around the Court."

"I did not," I pointed out. "I did not even know Predax's name until you used it." I looked at him over my shoulder, my practitioner's hand hovering over a likely anchor rock. "When do you suppose Arlyn heard them?"

"Maybe when they were using him to search the vault?"

I hesitated. Arlyn had been very weak, close to death, but after all, if he had heard the names even once . . . practitioners had remarkable memories. I would have felt much better, however, if Elva had sounded more confident, and if the icy spot in my belly would go away. "So when he asks Predax whether Jordy and Konne know where we are, it is only to learn whether they can come after us? Why would that matter?"

He waited too long to answer. I set down the final pinecone before turning to him. "What is it you do not want to say?"

"I watched Metenari say and do things that only Xandra would have said or done. And now you're telling me that Arlyn is showing the same symptoms?"

"If I am right, if he knows things he should not know—"

"That doesn't make him Metenari," Elva interrupted me. "It isn't Metenari that we know can jump into a body. It's Xandra."

"You mean the Godstone."

"The Godstone thought it *was* Xandra. I don't know, maybe it was. All I know is Arlyn isn't—wasn't—much like my old friend Xandra, but the new Metenari was. And now . . ."

"Now what?"

Only long practice of disciplining my physical reactions saved me from jumping or even squeaking aloud at the voice from behind me. Elva apparently had the same kind of experience. I straightened slowly and dusted my hands off on my trousers before answering Arlyn.

"Fenra was just saying that now's a good time, now that the danger is passed, for her to return to the outer Modes, and I'm trying to convince her that a lady of her skills would be very useful back home."

"Home? You're going back there? It's home to you now?"

"Well, yes." Elva leaned against the stone wall, arms crossed. "I've been living there longer than I ever lived here. Except for you, there's nothing left here for me. And you can always come to visit me."

"And how is it, precisely, that I'm to manage that when I have no power and can't operate the *forrans*?" Arlyn smiled and my heart grew cold. "You'd better leave Fenra here with me, don't you think?"

Elva's lips parted, but I could see his mouth was empty of words. "I expect he thinks of the *forran* in the doorway of your vault, the one that opens the dimensional gate—"

"When someone tries to open the vault." Arlyn nodded. "And how am *I* to do that without you, Fenra? Can you explain that? Do I convince one of the other practitioners to lend me their power? Do I give them access to my pattern? Do I reveal who I am?" He looked from me to Elva and back again. "Or is it that I've already done so?"

I do not know what look appeared on our faces, but I was totally unprepared for his sudden laughter.

"Oh, you should see your faces." He scrubbed at his face with both hands, and the smile I saw when he lowered them slowed my heart rate. I knew that smile. It was Arlyn's.

"Yes," he said, grinning. "I'm the same Arlyn that you've known all these years. And also, as Elva seems to have guessed, I'm not that same Arlyn anymore. But it isn't what you fear—stop touching your gun, Elva. The Godstone went into the chaos with Metenari. What jumped from him to me was the part of me the stone stole in the moment I contained it. That part of me that was my, my fire, my . . . my joy of life." He smiled at me in particular, and my heart thudded heavily. "Perhaps even my power."

"And now it's back?" I confess that the muscles in my back eased. Arlyn's regaining the parts of himself he had left behind with the Godstone—especially if that included his power—would explain everything that concerned Elva.

"It's early to be sure, but yes, at least I think so." He lowered his eyes and looked away, but he did not seem uncertain to me, he seemed self-satisfied and clever.

"Then there really can't be any reason for Fenra not to come with me—unless, of course, she doesn't want to." Elva shifted over until he was standing closer to Arlyn than to me. I hoped he had no thoughts of protecting me.

I gave Elva back a smile as good as the one he gave me. I did not know if he was serious, or what my answer might be if he were. "I have a responsibility to the people of the village. I will have to think more before I can make such a decision." I thrust my hands into the pockets of my trousers. Now if the warmth could only stop them from trembling. I needed sleep.

Arlyn clapped his hands together. "The brain needs fuel and rest if there's going to be all this thinking going on. I could use some planning time, myself. I suggest we all go back inside before the rain begins, and see what there is to eat."

He took us each by an elbow, ushering us into the tower. As we stepped into the ground floor stable, I pulled away. For a moment Arlyn's fingers tightened, and I thought he would not release me, but then he relaxed.

"You'll want to see to Terith, I take it."

All my relief vanished as a shiver ran up my spine. In all the years we had known each other, in all the years Arlyn had known Terith, he had only ever called him "the horse." Never by name. He knew Terith's name, but he had never bothered to use it. Arlyn had not been able to see Terith, or other living things for that matter, as persons in themselves.

Be reasonable, I said to myself. After so long being uncertain and frightened, I had fallen into the habit of it. I needed to relax.

As soon as the two men had climbed the ladder into the upper story, Terith began to nuzzle at me, blowing warm air into my face, my neck, my hair.

"You are not helping me." I ran my hands down his neck.

He rubbed his long face against me like a cat, pushing me until I had to brace myself against the side of the stall.

"All right," I told him in a voice pitched for his ears alone. "You disagree. Fine, I will still be careful." Terith's instincts were worth listening to. "Will you be able to free yourself if things go bad for me?" At this he snorted, as I had known he would. Terith believed there was no stable that could hold him. I quietly unlatched the outer door, just in case. Without turning around I told him, "If you feel it necessary, go back to Ginglen's Hotel. I know they will look after you."

I climbed the ladder to the upper floor slowly, buying myself time to think. If I could touch Arlyn even for just a few seconds, I would be able to tell who, or what, we were really dealing with.

"Predax should return to the City," Elva was saying as I opened the door to the upper floor.

"But I'm still just an apprentice. I don't have a mentor anymore," the younger practitioner said. He had evidently given his earlier request more thought. He smiled at me, and I smiled back, stepping off the ladder. "I'll be honest, I don't look forward to explaining to the council what happened to Practitioner Metenari."

I knew exactly what he meant. It was one thing for mentors to lose apprentices—accidents can happen, and carelessness is a characteristic of many, teachers and students alike. But for an apprentice to lose a mentor . . . perhaps it had happened before, but if so I had certainly never heard it.

"Easiest thing in the world," Elva said. "The practitioner disappeared in a puff of smoke while showing you a new *forran*. As an apprentice you could hardly be expected to understand anything about it."

Arlyn's laughter made my skin crawl. "That's just like you, Elva, an answer for everything."

"Things are rarely as complicated as they appear." Was I the only one who caught that note of caution in Elva's voice?

"Excuse me, Practitioner, Captain." Predax cleared his throat, looking from Elva to Arlyn and back again, even sparing a second's glance in my direction. "I was just thinking that if Practitioner Albainil's power *does* return, he doesn't have an apprentice and I don't have a mentor and so . . ." His tone was a combination of diffidence and eagerness.

Arlyn began to speak and Elva cut him off. "Practitioner Albainil doesn't take apprentices. He never has."

"Don't be so hasty, Elva. Times change, and people change with them, even practitioners." Arlyn circled around Elva, reaching out toward Predax. Elva knew better than to try stopping him, but his glance at me had clear concern in it. I shifted to get a better angle on Arlyn's face, and what I saw there made me interpose myself between him and Predax. As Arlyn swerved to go around me, I managed to snag his sleeve between the thumb and fingers of my practitioner's hand.

"Taking an apprentice is a serious step," I said, when he fixed his eyes on me. I steeled myself not to look away. "Even if it's only for making furniture, it's nothing that's done on the spur of the moment. The two people must be well matched, and well suited to one another in style, and in level of potential, examined and tested before a joining can take place. With no disrespect intended to you, Apprentice Predax, no such preparation has been made in this case."

"I am Xandra Albainil. I've forgotten more about the practice than you will ever know. I think I'm capable of choosing my own apprentice." Arlyn pulled his sleeve free of my grasp with a sharp movement and reached out for Predax.

"Fenra!"

I did not need Elva's warning. Simple common caution told me that whoever this was, he could not be allowed to touch Predax. I set my fingertips together, focused on my pattern, pulled my fingers apart, and grabbed Arlyn's wrist with both hands.

The wash of power pushed me to the floor, drowning me. I clung harder. I had expected something, but nothing as great as this. From far away I heard Elva's voice— "Do you want to be consumed? Go! Take a horse and get as far away as you can"—and understood that he was speaking to Predax. I had a minute to think, *If he takes Terith, that will get him back to the City.*

I focused on the fog, the beach, and stepped sideways into the other place, pulling Arlyn with me, and found my hands empty. On what I thought of as Arlyn's rock, someone—something—sat. It flickered, and turned its face toward me. Arlyn's face, younger, sharper, less worn, with eyes shining. A skull glowing bright and anonymous through the skin. I called on the fog and the wind to come to me. Rather than finding my way through them, I hoped to use them to push this person who was not Arlyn away. It was contrary entirely to the vow I had sworn to heal and help whenever I could, but when I considered the whole of the world, I knew I did the right thing. Sometimes it is necessary to excise poisoned tissue. Would it be possible to cut away only the Godstone?

I might have succeeded, if I had not hesitated long enough for that thought. Fog rose up out of the ground around us, wind blowing it toward me rather than Arlyn. A thick mist, a sea mist, cold on the skin and leaving droplets of moisture behind it. A dull blow to my head and I was back in the tower room, my forearms raised to block another strike from Arlyn's fist. We sometimes forget to guard against physical attack, we practitioners. I saw a muzzle flash, though I heard nothing more. *Elva*, I thought. *He is trying to help me.* I did not think he was succeeding.

I felt Arlyn drag me across the floor. I made myself go limp, giving him a dead weight. I dug in my heels, striking at him as well as I could with my right hand, but nothing I did even slowed him down.

"Thank you for teaching me this trick."

I heard the door open, felt the wind of chaos blowing past it, the vacuum it created pulling at my hair and clothes. I clung all the tighter to his arm, concentrating on calling the fog, on drawing power from him, to weaken him. Anything. But the noise and the drag of the wind made concentrating difficult. I felt two hands closing on my ankles, and for a moment I knew what it was like to be the rope in a tug of war. Then all I felt was the noise, the buffeting wind, the scrape of sand, dust, and gravel across my face.

❧

ARLYN

I yank the door shut on them just before I'm sucked through myself. I only meant to frighten her, to make her more compliant. I wasn't afraid of her. Imagine thinking that she could drain power from me.

"She could, though, that's the thing," I said. "She's spent all her time as a practitioner healing—apart from the time she spent as a lorist's apprentice, studying all the history she could find on healing people. Killing someone's just the reverse of healing, you know."

"Spare me your platitudes," I say. "*You* she might be able to affect, but not me. I have all your power as a practitioner and all Metenari's as well. What could some youngster who's barely a second level do to affect me?"

I stayed quiet, trying not to give anything away.

"What?"

"You're right," I said finally. "I don't know what I was thinking."

"You must have been thinking more clearly before," I say. "Or you wouldn't have been able to best me."

"That must have been it," I agreed. It couldn't have been that I felt so tired. Or not tired so much as . . . as . . . run down. No, that wasn't the word. Listless. "I need to sit down," I said to him.

❧

FENRA

At first the noise, and the wind, and the needle-prick pain of the flying grit distracted me so much I could not focus. I choked on the debris, and the wind sucked the breath from my body. But there was more than sand and grit. I had bonded with rock and bricks and mortar as if they were living things, and I felt the same kind of presence around me now. As if I tossed in the lungs of some vast creature, damp, hot, and intolerably noisy.

What felt like a rough hand brushed the hair back from my face. I forced my hands up against the wind, but whatever it was had gone. I tried to cover my mouth and nose. *Hands.* Around my ankles. *Metenari?* He could not have survived. I flexed my foot and stopped immediately as I felt the hand shift. *Elva.* Elva had tried to save me, and he must have followed me in. My first thought a selfish one: I was not alone. My second thought equally selfish: now I had to save Elva as well as myself.

I could not be proud of these thoughts, but on the other hand, it did not occur to me until Elva mentioned it later that I could have simply kicked him away.

Instead I bent at the waist, covering mouth and nose with only my right hand, and reached down to tap his wrist with my practitioner's hand. Immediately he released his hold on that ankle and transferred his grip to my forearm. I was so relieved I could have kissed him. If he had been slower to understand—slower to trust me—it could have killed us both. As soon as my practitioner's hand closed on his forearm, Elva released his hold on my other ankle and, using his new grip as leverage, pulled himself up until we were both oriented in the same direction. Considering there still was no "up" or "down" this gave me a remarkable sense of accomplishment and, strangely, security. Only the change in my middle ears told me we were spinning. Otherwise we might have been floating suspended. If the wind would only die down a little, and take the noise with it, we could almost be comfortable.

Elva kept his grip on the wrist of my practitioner's hand and slipped his other arm around my waist until we were hip-to-hip and one of his revolvers pressed into my breast. I hooked a leg around one of his for greater security. He seemed to be trying to tell me something, his lips to my ear, but I could not hear anything over the clamor around us.

Still, holding fast to Elva I felt anchored and even peaceful. The chaos still swirled around us, over us, through us, but somehow it felt calmer now, as if the impacts striking us were no longer random. I felt, very faintly, the sensation I would get when walking through woods, knowing that some animal watched me. We were not alone, something lived here. Something—no, *someone*—someone was trying to communicate with us. As if in response to that thought, the hand that had brushed my face before touched me on hip, elbow, shoulder, forehead, touched Elva the same way—I felt him flinch—and then traveled across to my ankle, my knee, and back to my hip. What breath I had caught in the back of my throat. This being lived here, in this chaos, perhaps *was* the chaos. My mouth dried and I squeezed my eyes shut, clinging to Elva as strongly as I could.

As the touching began its second cycle, Elva loosened his hold around my waist and bent elbow and wrist awkwardly, reaching for the grip of the pistol on that same side. I hugged him closer, trapping his hand between our bodies. After a moment of resistance, he relaxed again. The being touching us had done us no harm so far.

As soon as Elva relaxed, a warmth spread over us as if we had been slowly lowered into a carefully drawn bath, as if the chaos now approved of us. The wind supported us, the air around us clearing. I felt my tense muscles relax, and I did what I had not had the chance to do before. I reached into the front of my shirt and closed my hand around Medlyn's locket.

ARLYN

I should have been horrified, grief stricken, but to be honest what I felt was a kind of envy. Fenra and Elva—at least it was over for them. They didn't have to keep trying, figuring out what to do next, hiding from him. They could just let it all go. If I hadn't already known it wouldn't work, I'd have been looking for a way to kill myself.

"Elva was my best friend," I reminded him. "That's the second time I've lost him because of you. And now there's Fenra."

"She's the one I don't understand. Why have her with you? She's not what I'd call pretty."

"No." I stifled my thoughts. Easier than I expected. Just let the same thought go round and round. Fenra working as a healer—no, too close, think about the horse, the horse didn't like me.

"Best you could find, out there on the perimeter, I suppose."

"Yes."

"Not as much use to you as you hoped, though."

I stood up. My arms swung back and forth. "As you say, best there was."

"Why didn't you tell them who you were?" Clasped hands went up over my head and my back stretched out.

My head shook. "It seemed like a good idea. I'm not sure it matters anymore."

"You're right there."

Eating next. Not sure how I got there. Didn't remember anything for a while.

It's strange how he still thinks of himself as apart from me. As if we are two different people. Perhaps he went mad while I was gone. The same thoughts seem to circle and circle. Do people do that? Go mad when they lose half of themselves? And the better half too, as far as I can see. If he sighs one more time I'll have to slap myself in the face. I try again.

"I tell you it's because you tried to change the whole system at once. Do you think the world was created like that? It would have been done one Mode at a time. If we did the same, it would work. That's just logic."

"Logic's been wrong before." He tries to draw in another breath but I cough instead. No more sighing. "I don't know that this change is necessary any longer, in any case."

"Really?"

"The City's advanced far enough that even in the outer Modes people live comfortably."

I roll my eyes. "But then I'd never know—*you'd* never know. And stop trying to look at that door. What do you think? Your little class-three friend will find her way back from there? What? What is it?" I think he's about to say something but whatever it is fades away unexpressed. I wish he'd stop sulking.

"Here," I say. "Let's close up that door permanently. I'm surprised you haven't already had someone do it."

I stand up, smiling, and take a deep breath. A relaxed focus, that's the key. I feel the power bubbling up, eager to express itself. His—*my—forrans* float just below the surface of his mind, ready for use like underground springs just waiting for a well. I put my hands together, palm to palm, and bring the tips of my fingers up to touch my lips.

Now. Another deep breath. Concentrate. Feel the power. I lower my hands from my face and pull them slowly apart. Now the pattern would emerge.

Light blooms, solid, bright, almost too much for my eyes to bear without blinking. But it can't blind me, it *is* me. I feel it wash over me. I feel him come to life. I turn my hands around each other, as if I opened a jar. Nothing. The light fades, dissolving away, leaving emptiness behind.

"What are you doing?" I should be able to use these *forrans*. We're using the same brain. We're the same person.

"Nothing."

"Well, stop it!"

"I feel the power, but it slides away. I can't concentrate," he says to me, voice hard.

"What's wrong with you? Metenari was one thing, I can understand why I couldn't get his *forrans* to work, but yours? Your patterns are my patterns. We're the same."

"Are we? Are we the same? You're not Xandra, you know, no matter what you feel. You're just an artifact wearing an imprint you absorbed with my power. Like a mundane wearing a shirt."

"You think because we're in the same body I won't hurt you? Try harder. Concentrate. Focus."

"I'll try, but I can't promise anything. I don't know. I'm tired, I guess. I must be having a low day."

Thirteen

FENRA

THE SILENCE AND the warmth, even the homely smell of the wood, told me we had reached Medlyn's vault. I blinked away tears of relief. Elva held me by the elbows; otherwise I am sure I would have fallen. My brain could only think of irrelevant things. Elva's mustache needed trimming. Had his eyes always been this dark? The bridge models on the shelf behind him weren't dusty. My fingers closed around the leather straps of his gun holsters.

"You're warm," I said. My voice sounded loud and rough in the silence of the room. Unrecognizable.

"I thought that was you." *Elva's* voice I recognized immediately.

To the left of where we stood, the sofa Arlyn had used beckoned and my knees trembled under me. "We could sit down," I said.

"Can you walk?"

"Of course."

"Great, because I'm not so sure I can." A ghost of his usual smile flitted over his face. Still clinging to each other like two people skating on a pond for the first time, we made our way over to the sofa and sat down. My breath left my body in a deep sigh. Elva's hands floated across his pistols, and he frowned when his fingers brushed empty space instead of the hilt of his sword.

"Do not worry," I told him. "If you need one, there are three swords hanging on the wall."

He looked at the spot I pointed to with eyebrows raised. "Too bad I can't see them."

I sighed again, my eyes falling shut. Only parts of my brain seemed to be working. There had been swords in Medlyn's office, I remembered, but were they still there? I was reasonably sure they weren't the kind of memento Practitioner Otwyn would have kept. On the other hand, Medlyn had moved his models into the vault; perhaps he had moved the swords as well. I did not even know why he kept them. "Here." I laid my practitioner's hand on Elva's wrist and concentrated.

He gave the smallest jerk and his eyes narrowed. "Ah. Are they really there, or did you create them?"

"They're really there," I said.

"And will they still be there when you let go of me?"

Even my bones felt too tired to answer him. I forced myself to speak. "When Arlyn held things, they remained visible to him. As soon as I can stand, we will see if that is also true for you." I squeezed my eyes tight. "Metenari—" I began.

"We're *not* going back into that place for him."

I frowned. "He is not there," I said. "What was left of him was absorbed by the chaos. I meant to say the Godstone."

Elva started to get to his feet, struggling to stand at the same time that he pulled out both guns. "It can't come after us? We're safe here?"

My brain seemed to take a great deal of time to find the answer. "There is no 'here,' as the word is commonly used." I closed my hand around the cool bit of gold hanging around my neck. "Without the locket, perhaps there would be no 'here' at all."

"And no one can use the locket but you?" He settled back again. Apparently I was not the only one too tired to stand.

I nodded. Frowned. Shook my head.

"Xandra can't find his way here?"

"No."

"Then let's get some sleep."

"I am too tired to sleep." But I let him scoop up my legs and pull me into his arms as he lay back on the cushions, rested my head against his shoulder, and let the weariness float me away.

It may have been hours or only a moment later when I awoke. Elva's breathing, slow and steady, told me he still slept. A part of me was proud that I had managed—with Medlyn's help—to rescue both of us from a place so unlike any other, but I remembered with clarity the sense of another life, just beyond my mental grasp, a living being in the chaos, watching us. Examining us. I wondered if it was the same being I sometimes felt on the beach. I wondered if the locket would have worked if it had not decided to let us go.

I wondered how I knew it had done exactly that.

Elva's breathing changed and without moving I asked him, "Why did you follow me through the door? Why did you not let go when you saw you could not hold me back?"

"Where I come from a gentleman doesn't let a lady go into danger alone." I could hear a smile in his voice. I almost smiled myself.

"That's nonsense," I said. "And you do not come from there, in any case."

I felt him shrug. "I've lived there longer than I've lived anywhere else."

Before I could think of any response, my stomach growled, and Elva laughed, sitting us both up, and holstering the gun I had not noticed in his right hand. "It's good to know some things don't change. I'm hungry myself. But it looks to me like the cupboard is bare."

I stretched, moving first one shoulder and then the other. I felt as though I had been lifting rocks. "There is food here," I told him. "And drink. But I will have to get it." Elva released me from his arms slowly as I rose, and watched me cross the room to the cupboards against the far wall. He was right—regardless of our own predicament, the world moved on. At least for now.

The selection of food had changed. Chicken stew with parsnips and potatoes in a small stoneware pot, onion buns, fresh apples and plums. I wondered if I had to return the pot to the cupboard when we were done, and what would happen if I did not. I laid everything out on the round table. Elva gave a hopeful smile as he took his seat across from me.

"You look like you're setting a table," he said.

I reached out for his hands. "Here is bread," I said, setting his left hand down on the basket holding the buns. "Here is a spoon, and here a pot of stew, chicken judging by the smell." I set his right hand down on the handle of the pitcher. "And here is wine. At least it is right now."

He closed his right hand around the pitcher's handle, lifted it and inhaled with appreciation. "What do you mean, 'now'?"

I explained the nature of the jug and the baskets. "I would not have expected wine," I said. "But I trust my old mentor, and through him, his pitcher."

"Good enough for me."

Lucky, I thought. I would not have known what to do if Elva had not been able to see the food. At what point would the cup, the plates, and the jug disappear for him, I wondered? When he finished the meal? Or would he always see them now? After a few moments of silent eating, Elva swallowed and cleared his throat.

"If nothing else," he said, "we'll have surprise on our side. It won't dream in a million years that we escaped that chaos. It won't be expecting us at all."

I looked at him over the rim of my wine cup before putting it down on the table between us. "My brain feels like an old rag that's been washed too often. One more effort and it will simply pull apart into loose threads." I massaged my temples, but even my practitioner's hand felt stiff. There was still a question I wanted to ask. "Had you stayed, could you have killed him?"

Elva hesitated long enough that I thought he was not going to answer at all. "That wasn't Arlyn," he said. "Not entirely, anyway. Arlyn would never have tried to kill you."

"Which doesn't answer my question." I picked up my wine cup and set it down without drinking.

"Let me finish. I'm fairly certain I could have killed the body— that's what I mean by 'Arlyn.' But I'm equally sure bullets wouldn't have killed the Godstone. What if it jumped again, with you the only practitioner around?"

"Then it would have gone through the door with me." I decided I was not hungry anymore.

"Then it might be sitting here right now, though I've got the feeling I wouldn't be."

"I am not so sure," I said, pushing my spoon away. "Arlyn knows about the locket, but does Xandra?"

Frowning, Elva moved all the items on the table, lining them up along the edge to his right and then distributing them again. For some reason I thought of cards. "Do you remember when I told you what the Godstone looked like?"

I had to think. "You told me it was like a crystal, about this big." I held my hands a foot apart.

Elva's sad smile took me by surprise. "Did I tell you that when

Xandra was working on it, I couldn't look directly at it? It gave off more light than my eyes could take—or maybe it *was* the light, I was never sure which. Xandra gave me goggles to wear, made specially for me. It did look like a piece of crystal then, a prism, painting colors on the air, but it was still too bright. Even with the goggles I couldn't focus on the shape." He poured himself more wine, still smiling. "I never saw Xandra touch it, but it responded to him somehow. That much I could tell."

"Pure energy," I said aloud. That made sense. Energy could go wherever it wished. That's why we used *forrans*. Without a pattern energy was just . . . chaos. "If somehow Xandra gave a kind of form to pure energy . . . Metenari could not have known what it was—no matter what documents he found. He would have had no way to control it."

"No, it controlled him. At first I think Metenari was still there, but the Godstone wore him away, like erasing the printing on a page. We know now that when he sealed it away Arlyn left an imprint of himself—of Xandra anyway—along with his power. Though I think what it has of Xandra is a crude copy, like when a child copies an adult's drawing. It's Xandra, but certainly not all of him, not the quiet, thoughtful part of him."

"So what we have here is both—no, all three of them? The Godstone, plus the distorted imprint of the Xandra you knew, with both of these imposed on the Arlyn I know?"

"Something like that, yes." He smoothed his mustaches with his thumb and forefinger. "So I think we have to figure that what Arlyn knew, the Godstone knows too. Including the locket."

I stared at the tabletop, turning my spoon over and over in my fingers, finally gripping it until my knuckles grew white. "How long?" I said. "How long until the Godstone erases Arlyn as he did Metenari? How long before he wipes away the part that in the end decided not to risk destroying the world?" I looked up into his bright blue eyes. "We cannot leave this. We will still have to stop him."

"We've been doing great so far." Elva moved his wine cup to one side and leaned forward, putting his hand on top of mine. "Tell me, can you still open Xandra's vault?"

I considered the question. Could I? Practitioners had excellent memories. If I concentrated, I could see Arlyn's pattern as clearly as if I had his drawing in my hand. "I believe so. Why?"

"To get to the New Zone—no, wait, hear me out." His hand closed tighter on mine as I tried to move it away. "What if we can't stop it? Would we be able to at least save ourselves?" He grinned. "I'd hate to lose you."

I would hate to be lost. For a moment I was tempted. I would still be a practitioner, though perhaps I would have to act the part of a physician, or a scientist, as practitioners did in the City. I went so far as to ask him, "Are you certain we have no chance? No chance at all?"

He released my hand and threw himself back in his seat. "Certain? Of course I'm not certain. Anything's possible. But it's not what you'd call probable."

"What of the people here? If there is a chance, however small . . ."

Elva crossed his arms, shrugged, but nodded, reluctantly. "The risk . . . if we fail . . ."

"Consider the outcomes," I said. "One: Xandra tries his new approach and is successful, leaving a world that to mundanes seems exactly the same. Two: he tries and is unsuccessful, destroying the world—"

"A fifty-fifty chance," Elva said. I ignored him.

"Three: if we attempt to stop him and fail, the first two outcomes still remain possible—a fifty-fifty chance, as you say. If we succeed, then the world continues in its present form. The only thing we really risk in trying to stop him is ourselves, against the chance of saving everyone, without changing the nature of the world. Surely that's worth it?"

He shifted and the candlelight flickered on his face. "At home we have a saying, 'When you run out of bullets, use a sword.'" His voice lost its intensity. "Do you have an idea of how we could stop it?"

"We will think of one. At least we know the chaos won't kill us. We may find a solution to *that* instead."

"Good thing I didn't kill the monster, then."

"I do not think it was a monster," I said, thinking of the soft brush against my cheek. "Unlike others we could name, it did not try to kill us."

His teeth flashed. "Where do you think it's gone now?"

I knew he did not mean the monster.

ARLYN

"This body doesn't require more rest. Why are we sitting down?"

Voice didn't sound much like mine, surprising, all things considered. "I'm frustrated." Watched my hands tremble, then stop.

"There's no reason for this not to work. Not now. I can feel the *forrans*. You're hiding something from me."

"How?" That shut him up.

"Then why isn't the *forran* working?"

"I don't know. You tell me. You're in charge."

My practitioner's hand formed a fist tight enough to hurt. "That's just what I am. So you'll tell me why this isn't working."

Took too much energy to laugh. "Why? What'll you do to me?" Keep him distracted, I thought. Don't want to give him the satisfaction.

"You have no fear of death. Why is that? If you can't die by your own hand, I wonder if you could die by mine?"

"I'd love to see you try it. Really, I would."

"Ah. Yes. I see what you mean. I don't think I'd be destroyed myself, but I'd be back at the starting gate again, wouldn't I? Still, there must be something you want."

"Can't think of a thing." Really couldn't.

"What if I could get your friends back?"

Heart skipped a beat, ears buzzed.

"Wow. I guess that answers my question, doesn't it?"

"No. They'll just die along . . . with . . . everyone else."

"We'd wait until I was successful—wait, you've thought of something. What is it?"

Would have laughed out loud. Couldn't kill myself, but I'd be destroyed along with everyone else. Temptation. Saw myself doing it. Would it really be so bad? Everyone dies anyway. Felt like a weight disappeared. Then it was back.

"You're thinking something." My fists pounded my head. "Gahh! It's like an itch. Stop it."

Couldn't face it. Fighting it. How long? Just wait for lowness to be complete. How long had it taken? Couldn't remember. Too long. Too slow. Too difficult. Stall.

"What about start at one end," I said. "Not here, not the middle."

"The end?"

"Or the beginning, if you want to go Mode by Mode."

My body moved to the window and looked out. "Which is closest?" I thought again about not telling him. Problem was, he knew I had an answer. He couldn't see what it was, but he knew one was there.

"I'm fond of you, but don't test me too far."

Tired. What difference would it make? "The City," I said finally.

"So the City's the beginning. Where's the other end?"

"Nowhere. Keeps on going."

He smiled. "As if we create it with our presence. That makes a great deal of sense."

If he thought so.

"We'll go back to the City then, and we'll see which of us was right."

I closed my eyes. Appalled. Was the horse still in the stable? Any horse? Would we have to walk? How long would it take? Forever? Found myself on my feet without any intention of being there.

ELVANYN

He hadn't known why Fenra had stopped him from killing the creature touching them. He just knew that in these matters he trusted her instincts more than he did his own. And he remembered how the horses reacted to her back home. Though to be honest, he'd have felt better if he'd been able to shoot something. Preferably Xandra, even though it wouldn't have killed the Godstone.

Suddenly he felt a floor under his feet and Fenra's weight sagged against him. He still couldn't see anything, but after a moment his hearing adjusted. Somewhere nearby he could hear wind rippling through the leaves of trees. He felt her lips close to his ear.

"Follow me. Do not let go of my hand."

He grinned, safe in the knowledge that she couldn't see him. The last time a woman had said that to him, events had turned out very pleasantly indeed. This time, however, Fenra just led him through what his ears told him was a stone corridor. A familiar size, he thought. There must be dozens like this one in the White Court. He almost spoke, but thought again and shut his mouth. He knew the locket had brought them back to the Court, though evidently not to her old mentor's office. Elva felt he should be able to place these corridors, even in the dark; for years it had been his job to patrol them.

"Where are we?" he said finally.

"Down the corridor from Lorist Medlyn's private rooms in the Watchmaker's Tower."

As if her words were a *forran*, his sight cleared and he saw moonlight coming in a nearby window. Someone had left the shutter open to catch the night breeze. Ah, now he knew where they were.

"If we live through this," Fenra said. "The first thing I do is find a way to go somewhere else, and not just to Medlyn's places. It's like going back to the beginning every time."

"It's a safe beginning, you have to admit that."

"Compared to what? What do you think will happen to us if we're caught here? You're supposed to be with Metenari, what will you tell your Guard captain?"

Don't answer, Elva thought. *These aren't the questions she wants answered.*

They descended one staircase and then another before passing through the left half of a large door swinging wide but silently open into the Moonlight Patio. Suddenly he heard voices and before they could hide—if there had been somewhere to hide—three apprentices walked into the patio from under the arcade on the far side. Elva braced his feet and reached for the sword Fenra had found for him, but just as his hand closed around the hilt, Fenra hissed and jerked him over a small curb into a hedge of thorny bushes.

Though they were no more than ten feet away, across the narrow width of one of the long pools, the apprentices passed them by without reacting at all. Elva would have sworn one of the trio looked right at them. Once they turned a far corner, Elva waited until the sounds of their footfalls died away before asking her.

"How long have you been able to make yourself invisible?"

"Since early in my training. I was always the one sent to steal food in the middle of the night."

"And why haven't I seen this skill before?"

"Because it takes a great deal of energy, and it only works in the middle of the night. And best when my feet are touching actual earth." She stepped back onto the path and slipped her arm through his, now that they were walking on stone. Sucking on a scratch on the back of his free hand, he thought about walking this way with Xandra, holding each other up when they'd had a little too much to drink, or when Xandra was giddy with some revelation or success. Fenra led him down to the far end of the patio, through the beam of moonlight trapped by the lilies in the pool. Fenra's arm stiffened in his, tightening her hold, and when they were back under the concealing darkness of the arcade she drew him further into the shadow.

"We may be too late." Her voice was hardly more than a shifting of air. If they hadn't still been arm in arm, Elva doubted he would have heard her. "The Godstone has already tried something."

"What's gone wrong?"

For an answer she ran her fingers lightly over his cartridge belts, finally resting them on his revolvers. Her breath caught as she drew it in.

"Something wrong with my guns?"

"No, not that." She sounded relieved, but not as relieved as he felt.

"What, then?"

"In the moonlight I could see our clothing. It's changed. We are both wearing knee breeches now, not fitted trousers, and you have clogs where you had boots."

At first Elva had no idea what she was talking about. What was wrong with knee breeches and clogs? Then the significance of her words sank in, and cold shivered his skin.

"But we haven't left the City," he said. "You're saying the Mode is different, but we haven't left . . ."

"The Modes are tied together. If one changes, they all change."

"I know that." Fear made him impatient. "But change comes from the City. And it doesn't move backward, does it?" he added when she didn't respond right away.

"Never before, that I know of. Last time the Godstone was used, things jumped forward, at least that's what Arlyn said."

"What now?" Would this change her mind? Evidently the Godstone had tried changing things, and the world hadn't been destroyed.

"Without checking another Mode, we don't know exactly what's happened. Is it just the White Court precincts? The whole City? The world?"

The guard at the East Bridge merely saluted them as they passed.

Their job wasn't to keep mundanes in. The gate itself stood open, another change. They hesitated in the face of the unexpected darkness of the streets.

"Light your way, good sirs? Oh, sorry, Dom, I didn't see you there."

"Yes, thank you."

A rasp, a spark, and they shielded their eyes.

The lampee proved to be a boy of about thirteen, his grin more confident than his voice. He leaned against his lamp pole, neatly dressed but plainly, open collar showing under a short jacket, neat breeches, worn but decent-looking clogs. The stocking on his right leg sagged, but he didn't seem aware of it. Not so well dressed as someone's private servant would be, not livery. The boy clearly made his living by picking up customers on the streets. He must have just brought someone this way.

"Do you know the way to Ginglen's Inn?"

"Lady," the tone chided her. "It's my job to know that, don't you worry."

"Which wasn't a 'yes,'" Elva murmured in her ear as they set off after the boy.

Fourteen

FENRA

I DUG THROUGH my pockets and finally found a coin for the lampee. He tossed it into the air so that it passed through the golden light of his torch before he caught it, gave me a shallow bow, and saluted us both with a finger to his forehead.

"Any time you need me, ask for Oleander," he said, whistling as he walked away.

Elva had instructed Oleander to drop us at the corner where our street entered the square. It took a few minutes for our eyes to adjust to the relative obscurity of the newly risen moon. We were finally able to make out the outline of roofs against the night sky. The trees and plantings along the walkway were rough masses of darkness, indistinguishable except for their scent, and the whispering of their leaves moving in the breeze.

I stepped toward the circle of lamplight around the hotel's front door and Elva grabbed me by the arm.

"Don't hurt me," he murmured.

Before I could ask him what he meant, he had backed me around into a darker niche between a wall and a yew hedge and taken me in his arms, pressing his cheek against mine. "We're hiding," he breathed against my ear.

"We have done this before," I said.

"And it worked."

For a moment all I could hear was his heart beating in time with mine. Then I heard what had startled Elva in the first place, the opening of the inn's door, and voices. But what had we to be afraid of, here in the street? It wasn't as though we were still in the White Court, and likely to be questioned. And then I realized that we could hear two sets of voices walking down the street toward us, and only one set of footsteps.

This alone would not have been enough to spook a veteran like El-vanyn Karamisk, nor me, for that matter. People have been talking to themselves for the entire history of humankind. However, they do not usually use two different voices. Nor does one of the voices usually belong to Arlyn Albainil.

So I hugged Elva close, burying my face against his neck, trying not to tremble as his breathing moved the hair around my face. The man we had been listening to walked past us without even glancing in our direction. Proof we had disguised ourselves well? Or that he, or they, simply had no interest in others? That would certainly fit what we knew of Xandra. We remained holding one another until we heard the footsteps stop. I peeked around Elva and saw Arlyn's familiar silhouette standing between us and the door lamp.

Arlyn turned slightly, and I froze. "Do not move," I whispered to Elva.

"Wouldn't dream of it," he whispered back.

But Arlyn only looked up to the lamp, as if examining the color of the flame. I had seen that expression on his face before, the muscles slack, the lips hanging just barely open, mouth turning down at the corners. Everyone has a "resting face," an expression their face falls into in moments of abstraction, or when one believes oneself alone. Arlyn had two such faces. One when he was level: alert, interested, and on the verge of a smile, and one when he was low. It was his low face I saw turned up toward the light.

Once more he turned toward us, and this time I thought he would see us, recognize us. Now there was no time for the *forran* that would make us invisible. *Too bad I cannot call the fog here*, I thought. Just as it seemed he had to see us, a cat ran across the road, tagged Arlyn on the ankle, and dashed off down the sidewalk. His head turned to follow it, and he forgot he was going to look in our direction. Instead, he set off once again down the street.

"That was a lucky cat. What are the odds he'd go to the same inn we're heading for?" Elva loosened his grip but kept one arm around me.

"Good, I would have thought," I said. I did not move away from him either. "It's the only hotel he knows. A better question is how long were we in the chaos, that Arlyn is here before us."

We both straightened, but still neither of us let go of the other, as if we needed the human warmth. "Do you know of any other place?" I asked.

"Well." I saw by the light that Elva was smiling. "I know of a place he'd never think to look for us—if ever he doubts we're dead." He turned to me, lifting his eyebrows and tilting his head toward Ginglen's Hotel.

"You are insane," I said.

"Insane, or ingenious?"

"Is there a difference?"

"All right, all right, I admit I was joking. Where else can we go?"

"Wait." I looked into the dark around us. Nothing. I could have sworn I had heard something more than the wind in the hedges. I glanced at the oil lamp at the hotel's entrance. It looked to be burning brighter. Like a beacon. "We need to go in." I set off before Elva could stop me.

"Hang on." He caught up and was beside me by the time I reached the front door. "I was joking, I didn't mean we should go in."

"We are not going to stay," I said. I had never seen the door closed, and it worried me. I hoped it wasn't locked. Catching my mood, Elva pulled out one of his revolvers. When he nodded I turned the handle, the latch clicked, and the door swung slowly open.

At first I saw nothing wrong in the entry hall. Dom Ginglen Locast stood behind the desk, and the footman—Itzen—just off to one side. I waited several seconds before I realized they were not going to acknowledge us. Elva cleared his throat and advanced into the hall, drawing his second gun. Still, the two men did not move.

"Watch the door," I said to Elva. "Let no one in."

"Teach your grandmother," he said under his breath. He swung the door almost completely closed, leaving himself a crack just wide enough to let him see the street outside, and leaned his shoulder against the wall.

As I approached the two men, I could see they were breathing, but shallowly. Though the hall had an oil lamp matching the one outside, both Ginglen and Itzen appeared pale—no, not pale, I decided as I looked more closely, *faded*. In fact, now that I looked around me, the whole hall had the same washed-out look, even the lamplight.

"What's wrong with them?" Elva whispered, without taking his eyes from the door.

"Something has sucked most of the energy from the room and everything in it."

"Can you help them?"

"I can try," I said, rubbing the palms of my hands together. "It's more or less what I do when I'm healing people."

I braced my feet, pulled my hands apart, and smiled when I saw the rose pink lines of light between my fingertips. I twisted and stretched them as though they were taffy, finally breaking my hands apart and flicking the net of light at the two faded men. At first there was no change, and then I saw their faces flush lightly with color.

"That's it?" I could tell Elva was not much impressed.

"From here they should be able to recover on their own, but it would take time they may not have. And you should be watching the door, not me." I examined them more closely. Lips and eyes were just barely open. I often put people into a deep sleep to allow their bodies time and energy to heal before I woke them again. Ginglen and Itzen had that same look. I had given them energy, but they were still asleep.

I repeated my preparation and called a wakening *forran*, humming the sound of bees, and of birdsong at daylight. The light between my fingertips was gold this time, with an undercurrent of green. When I flung the light, Itzen blinked and took a deep breath. He immediately crossed behind the desk and put his arms around the other man. Ginglen took a few minutes more to come completely awake, but this time I waited for nature to finish what I had started.

Finally both men were awake, and breathing normally, though they kept their arms around one another. They looked around as if not sure exactly where they were, but once they saw me Itzen at least relaxed and turned to face me. Ginglen's face remained stiff.

"Your friend was just here," he said, his tone only a little more neutral than his face.

"Not my friend any longer, I'm afraid." I kept my own tone as warm as possible. Out of the corner of my eye I saw Elva holster his guns.

"Not anyone's friend, if it comes to that," Elva said. His calm tone and ready smile did more to relax Ginglen and Itzen than anything I had said. I had the feeling they were giving me the benefit of the doubt only because I had helped the little girl.

"Are they coming back, do you know? Is there somewhere you can go?"

Ginglen shook his head. He turned away from Itzen, but kept one arm around the other man's waist. "I don't remember," he said. "Why is that?"

"He practiced you," I said. "It's hardly ever done on mundanes, but he must have been running low on energy, and he took some of yours. Not all of it, luckily," I added when they both paled again. "If there is any chance he is coming back, you must leave now, and tell any guests you have to go as well. He will not trouble to look for you if you are not here."

The two men exchanged glances and both nodded at the same time. "Shall we take your horse with us?"

"Terith? Is he here?" Could that be why I had felt compelled to enter the hotel?

"In the stable, where we had him before. Just turned up by himself the day before yesterday."

I hesitated. More than anything in the world I wanted to go and see him, but . . . No. I had to clear my throat to speak. "Please, keep him with you. You will keep each other safe." I had no idea where we might go from here, and if there was a chance to keep Terith safe from the Godstone, I had to take it.

"Can you recommend somewhere out of the way?" Elva said from his position at the door. "Preferably where a high-level practitioner wouldn't think to look?"

"You won't be coming with us?"

I shook my head, wishing my answer was different. "We have something we must do."

"There are always rooms to be found closer to the docks," Itzen said. "You'll find plenty of inns among the warehouses. Only apprentices on assignment go there now, and probably not at this time of night."

"Try the Seamaiden's Tail," Ginglen said. He began opening drawers in the desk, taking out papers and small leather bags that chinked faintly as he set them down on the top.

"Thank you," I said. "Go quickly."

If anything, the lighting down by the docks was worse than in the rest of the city, with only riding lights on the tops of masts and here and there candle lanterns in front of shops or ale houses. The place was busy, with a ship at every pier, but there were few people to be seen. Practitioners do not need a great deal of light to see by, but even I was stumbling over uneven cobbles in the streets as we turned away from the water. Finally we reached the area of streets behind the warehouses, where more light spilled out of open doorways, along with warmth, the smell of frying onions and, from one door, singing.

"There, in the next block."

I looked where Elva pointed and saw a sign hanging over the street. The Seamaiden's Tale. *Tale*. There was a story there, I thought. It was a decent enough place, if a little shabbier than I expected.

As soon as we were inside, I saw why Ginglen had recommended it. For the most part the tables held couples, not solitary drinkers. In one corner five students shared a table, a jug of beer, and a bowl of fried potatoes. The couple behind the bar were obviously partners in more than one sense. The place showed wear, certainly, but was scrupulously clean. And there weren't many places that would have welcomed poor students, and allowed them to take up a table for so little custom.

The innkeepers gave us friendly, interested glances, but without excessive warmth, acknowledging that we were strangers, but looked respectable. In their way, they were just like Ginglen and Itzen. They had the same look of being able to size us up quickly and without judgment. Elva led me directly to the bar counter and waited his turn to be served. After setting a bowl of what smelled like oxtail stew down in front of an elderly man not very well balanced on a barstool, the male of the pair turned to us.

"How can I serve you?"

"That stew smells awfully good," Elva said, returning the man's smile with a carefully calculated one of his own. While the man ladled out two bowls of stew from the pot simmering on the side of the fireplace, Elva asked the woman whether they had any rooms available, and stood dickering with her while I carried our bowls to a table as far as possible from everyone else, especially the group of noisy students. I hoped he had money to pay them. I had given Oleander my last coin.

A couple of people examined me out of the corners of their eyes as I settled down in front of my stew, but no one spoke to me. I realized that except for the people who were here together, no one paid any attention to anyone else.

"I do not remember this place from when I was apprenticing," I said to Elva as he sat down across from me, setting a plate of thick slices of coarse bread on the table between us. The innkeeper followed him with two big mugs of ale and Elva waited until the man left us before responding.

"Judging by the way Predax was treated when I escorted him to the Red Court, I'm surprised they let you wander around down here." He tore off a chunk of bread and dipped it into his stew.

"We weren't in any danger then," I said, picking up my spoon.

We chatted about unimportant things, even though no one paid us any attention. Elva asked me about my now-nonexistent limp, and I told him the same story I had told Arlyn, all those weeks ago when we were first on the Road to the City.

"So your leg is not really injured?" he said. Which is more than Arlyn had guessed.

"No," I admitted. "But I learned early that I preferred to be underestimated."

"So what class of practitioner are you really?"

I shrugged. "Hard to say. Medlyn advised me not to take any more of the exams. He could tell I was not happy, and he wanted me to have the option of leaving the City—something that's easier the less important you are."

Elva waited, spoon in hand, until he saw that I had said all I meant to say for now. This small handful of minutes felt like we sat in the eye of a hurricane, knowing that soon the winds would come. Finally we carried a teapot and two cups up to our room, where I could use the leaves—along with two mice I found in the walls—to set wards. For a piece of biscuit each they were happy to wait in the dark corners outside our door. Any practitioner would know the wards were there, but they would not know why, nor would they be able to cross them. Mundanes would avoid our door without being aware of it, and could not hear us speaking.

"Now that we're safely behind wards, what was it that occurred to you while we were hiding outside the hotel?" Elva sprawled with a sigh on the small room's only bed. Big enough for two, but only just. I sat on a low wooden stool, its hard oak seat softened by a worn embroidered cushion.

It took a minute for my slow brain to know what Elva was talking about. I wondered how he had guessed that an idea had come to me. Then I remembered that we had been standing with our arms tight around one another. Apparently I was not as impassive as I thought.

"Arlyn is low," I said. The look on his face invited me to explain. "Not just low in energy like the men in the hotel. That's not what causes it. In fact, it's the other way around. The lowness drains him of energy. It's not unhappiness," I explained. "That's temporary, it passes. If not treated, the lowness is permanent. It develops into a despair that deepens and deepens. In the early phases, Arlyn might fly into a rage, or begin to cry—either or both for little or no reason. Eventually he becomes unable to summon the energy to do the simplest things. Nothing seems worth the effort. Suicide begins to seem reasonable, death welcome."

"How is it he didn't kill himself before now?" Elva rose to his feet and paced in the small space between my stool and the window.

"He tried," I said. "Apparently suicide is not an option for practitioners. Then I came to his village, and I have kept him level ever since." Elva stood still and I watched his face as he sorted through what I had told him. I could almost tell what ideas occurred to him as they passed through his mind.

"Does this mean that the Godstone is low as well?" He sat down on the bed again.

"I do not know, but I believe I can find out."

I described to Elva how I leveled Arlyn. "I've never done it without touching him, but I have treated him so many times it may have created the right kind of bond between us. I should be able to reach him. In any case it's worth a try. If he *is* low, and I can find him in the fog, we might be able to communicate without the Godstone knowing."

"And if you're wrong?"

"Would we be in any worse position than we are right now?"

"Are you joking? He'll know we're alive. That pretty much ensures he'll come after us, don't you think?"

"I see it as a chance I—we—have to take."

He studied my face, his brows drawn together and his lips parted as though he meant to speak. Finally he nodded. "What do you want me to do?"

"Watch over me. Make sure I wake up again. Call my name if I sleep too long."

I wondered how Elva would react to the idea of such a passive role, but I did him an injustice. He had been a guard for years in his first life. Then a soldier and sheriff in his second. Young as he appeared, he was no boy who needed action to prove himself. He left me the bed, and took the room's small stool for himself, sitting with his knees almost touching the rough wool blanket. He set one of his pistols on the edge of the straw-stuffed mattress, near to hand, and made sure the other was loose and ready in its holster.

"Just wake me up, do not shoot me." I was rewarded by a slight quirk to his lips. I lay down in the center of the bed. "A good thing we were not planning to sleep here," I said. "The bed is just barely wide enough for two of us."

"We called it a wedding bed when I was young," he said. "Perfect for those first few months when a married couple find it easy to sleep closely together. Before children and the rest of life's interruptions change their sleeping habits. As a practitioner you probably called it something different."

"I would have called it too small," I said. Smiling, I closed my eyes and began to breathe deeply, tying my breathing to the rhythm of the *forran* I chanted to myself.

It took me so long to find Arlyn that I began to doubt whether my idea would work. When finally I arrived at the beach, I found that instead of the familiar white sand, the beach was now pebbled with small black, cream, and brown stones, rounded smooth by time, water, and each other. They compacted well, and made for comfortable walking, more comfortable, in a way, than dry sand. As much as I looked around me, however, I could not see any fog. To my right lay the sea; the view to my left was blocked by cliffs I had never seen before. A salt line on the rock face showed where the water would be when the tide came in. I would be swimming if I stayed here long enough.

There were no clouds, and yet I could not see the sun. The sky was a dull, even silver-gray as though I stood under a huge pewter bowl.

I followed the cliff around a jumble of black and jagged rocks. In the distance, I saw Arlyn sitting on an upthrust rock, his feet flat on the pebbled beach in front of him, his hands palm down on his thighs. Though we were still some way apart, I could not be wrong in my identification. The leveling *forran* connected me only to him. For the longest time my walking did not bring Arlyn any closer, and I began to worry. Some quality of sea, light, rock, or beach combined to distort my perspective. I had never spent so long a time in this place, and I wondered how much time was passing for Elva.

It grew harder to walk, my feet slipping back slightly with each step I took. Finally I stopped and crouched down, placing my palms flat on the pebbles. They were damp and cool, and for a moment I felt I should lie down. I sang a *forran* of greeting and unity until that feeling passed. A wind arose, blowing my hair to one side and then the other as I straightened to my feet, arms spread wide to the salty air. When I set off once again, my feet moved swiftly, the pebbles no longer shifting under them, until I finally stood next to Arlyn.

He looked older than I had ever seen him, drawn as if he had lost more weight than was healthy. His eyes were closed, but I could see them moving behind the lids, as though I had come upon him dreaming.

"Arlyn, can you hear me?"

At first I thought he could not, or that perhaps he had chosen not to answer me. Or feared to. "You may speak to me," I assured him. "It is Fenra."

"I'm dreaming," he said without opening his eyes. "He's asleep, but I'm dreaming. He doesn't dream, so he can't follow me here."

"Arlyn, can you see me?" I thought he would open his eyes, but he only turned toward the sound of my voice. His face changed, however, as if he saw me.

"Fenra. I'm so sorry." He turned his face down and away, letting his head hang. "I should have stayed away. I thought I could fix it. I always thought I could fix it. Even the things that never needed fixing in the first place." He looked at me again, with his closed eyes. "I meant to use you to fix it. I'm sorry."

"I knew, and I agreed," I reminded him.

"Sacrifice anything," is what he said, as if he had not heard me. "Anyone. The destination is more important than the journey."

"That's contrary to the general philosophy," I said. I sat down cross-legged in front of him on the damp pebbles. They were surprisingly comfortable. I wondered if the tide would stay away while we were here. So much depended on whether "here" was in my mind or in Arlyn's.

"I knew better," he said. "I knew better, I know better, I know best. I don't make mistakes, I didn't make mistakes, I never make mistakes. Others do but not me. I never do."

It seemed he would go on saying this forever, round and round, if I did not stop him. I took his right wrist in my practitioner's hand and held firmly when he tried to pull away. "Arlyn. Listen to me. You know better *now*. You are trying to fix the damage you have done. We must go on trying. That is the important thing."

"But how? You're dead. You can't help me kill him." He sat straighter, the different expression flashing across his face and disappearing again. I had seen that look before. He had thought of something.

"Tell me."

He had slumped again. "I thought for a minute that you could kill me. I think that would kill him as well. But you're already dead, so that won't work."

"I am not dead, Arlyn. I escaped the chaos."

"I know that's what I'd wish for, if I had my wish. This is only a dream. I can't bear it and yet I don't want to wake up."

"In that case I would like you to wish for some drier place." He did not react. "Does he know you are dreaming?"

"This is too much." The words and the tone were such that you would expect the person uttering them to bury their head in their hands, but except for the restless eyes behind their lids, Arlyn never moved.

"Arlyn, you are low. Let me level you."

"No!" So little was I expecting it that he was able to pull his hands away from mine before I could stop him. "No," he said again, more quietly. "I can bear it. It slows him down. He doesn't know it, but it slows him down."

"How is it the lowness came so suddenly?" I doubted he knew the answer, but I had to ask.

A faded version of his grin appeared on his face. "You did it," he said. "When he was pushing you, and you were trying to save yourself.

You tried to make him too weak to hold you, and you made me low instead."

"Arlyn, does he feel the lowness? Does he know what it is?"

"No. I don't know. I think—run! Fenra, run! He's waking up."

ELVANYN

Normally Elva found cleaning his guns soothing, restful. Like a meditation. Now, however, though he'd cleaned both of them, he didn't feel any of the calm he usually did. He thought about doing it again.

"Don't be an idiot," he said aloud. He'd already learned speaking wouldn't break into Fenra's *forran*. He hoped it was working, whatever it was she was doing. He hoped he'd be able to wake her up if she was in there too long. He wished she'd told him how long "too long" actually was. Maybe he should wake her now.

Fenra gasped as if in response to his thought, eyes flying open, panting as though she'd been running. Elva re-holstered the guns he'd drawn in reflex and helped her sit up, holding the cup of water for her until she was steady enough to hold it herself.

"Did you find him?" She nodded without speaking. "And?"

She held up a hand, palm out, and Elva sat back on his heels, waiting as patiently as he could for her breathing to slow to normal. "He believes he is dreaming. He does not know I was really there."

"Where 'there' means a place that doesn't exist."

She cuffed him and he grinned at her. "You know perfectly well what I mean," she said.

"How much did this tire you?" He had never seen her skin so ashy. Even her hair looked limp.

"Does that window open? Fresh air should help."

No one wasted glass on an attic window. Come to think of it, Elva doubted he'd find a glazed window in the whole place. The whole block if it came to that. He turned the wooden toggle holding the shutters closed and tugged at them. Nothing moved. At first he thought the shutters had been nailed shut, but he couldn't find any nail heads.

"The wood's all swollen and warped," Fenra said from behind him. He looked over his shoulder. She sat gripping the edge of the bed with her eyes closed. "It's the damp air coming off the sea," she added, her eyes still shut. "We're very close to the water here."

Elva glanced around the room but saw nothing he could use to pry open the swollen wood. Grimacing, he pulled free the pistol hanging on his right side. He hated to use a gun this way, but he didn't see another option. Reversing his hold, he hammered at the shutters with the grip until he'd chipped off enough wood to force one open.

Fenra was already standing, and he helped her position the stool under the open window so she could put her face directly into the cool breeze that blew through the narrow opening. The air smelled of salt, and fish, and faintly of rot. She held onto the edge of the windowsill with both hands, finally stretching her practitioner's hand out into the air.

"It's raining." She pulled her cupped hand back in and wiped the rain she'd collected onto her face. The air must be doing her good, he thought. She was looking better already.

ARLYN

Phoenix Plaza looks much the same as always, eating places, private clubs, a couple of private homes in the streets immediately around the huge square. Late, but still horses and carriages, private and for hire. I think about taking one, almost raise my hand, but where would I go?

"Here now, fellow, had a bit too much, have we?"

A grip on my arm, holding me up. Voice friendly, half amused. I struggle to get my feet under me, stop myself from brushing him away. There's a darkness, a blank in my memory. I don't remember sitting down. I nod at the man who helps me, convince him I'm well enough to leave. He doesn't want to stay, just to help.

"What was that?" I ask as soon as the other man rejoins his friends.

"Told you, body needs rest."

"And I told you I wanted to see the stars."

"Now you've seen them, can we rest?"

I take a deep breath, look around to check my bearings, freeze.

"It's different," I say.

"It's the City," I said.

"Yes, but a moment ago there were carriages for hire, and gas lamps. Look." I point, he looks up, simultaneously. "There are cords strung across the streets, oil lanterns hanging from them."

"Different," I said, craning upward to see. "I didn't do this."

"No, we didn't."

FENRA

"Elva," I said.

"Yes, I'm getting up, don't worry."

"No, look at me."

He rolled over and swung his feet onto the floor. "You look fine, why?"

I felt like an idiot. Of course I looked "fine" to him now that I was rested. "My clothing has changed again," I told him. "The cut is very similar, but the lapels and cuffs are wider, and look, the waistcoat is single-breasted." I held out the garment for his inspection.

"My guns?" he asked me.

"Unchanged." The look of relief on his face made me touch him on the shoulder. "Your clothes also," I added. "At least, the ones you brought from the New Zone."

"What happened?"

I pulled on my waistcoat and began doing up the—single-breasted— buttons. "The Mode changed sometime in the night," I told him.

"His—*its* doing?"

I sat down on the bed a little more heavily than I planned. This had shaken me more than I realized. "He meant to bring everything up to the same level as the City, not the other way around." He had not de- stroyed the world, but things were getting worse, not better.

The boy had said his name was Oleander, which I doubted very much. He neither dressed nor spoke in such a way to persuade me his parents would have chosen such a name.

"Maybe," he informed me when I asked him about it, "but it's my name now."

As far as I could see, his clothing hadn't changed significantly from the night before, but then, servants' clothing lags behind changes in fashion. He had put his lamp away, now that it wasn't needed, and stood ready to run any message or deliver any parcel—or carry it behind someone while they shopped. He was neatly enough dressed for that.

"He were up early," he said around a huge mouthful of the turnover stuffed with cooked egg and onion we had given him for breakfast. I made a sign that he should swallow before continuing. I did not want him choking on our account. "Wanted a horse and all. He comes out to the stable yard and stands around, but no one comes to serve him, so I goes in and he tells me to fetch him out a horse and saddle it."

"You should never have done that," I said, my heart pounding. "You could have been killed."

"I'm here, ain't I? So calm yourself." The child had the nerve to smile at me. "I ask him if he needs a guide, or a lunch packed for him— see, I was trying to get him to say where he was going."

I shook my head. Only the young were brave—or foolhardy— enough to take this kind of risk.

"Get to the point, ragamuffin, or there's no second breakfast for you."

Oleander shot Elvanyn a grin that showed a bit of egg caught in his teeth. "So he says never mind, he figures he can find his way to the next county by himself." "County" is what mundanes call the Modes.

"Did he mention any particular place?" Elva asked.

The boy shrugged, licked his fingertip and blotted up the crumbs on his plate with it. He looked at Elva with eyebrows raised and he gave a nod to the woman behind the bar counter. Oleander waited until another egg turnover was on its way before continuing. "I figure it's to the town, where else?"

"How do you figure that?" I asked before Elva could.

"No luggage, no food," the boy said smugly. "Not going to sleep rough, is he? 'Gentleman' like him can't be planning to go much further without a change of clothes, so's likely to be heading for a place where he won't need them, or he can buy more."

"Sound enough reasoning," Elva said.

The boy sat back, satisfaction on his face along with a smear of butter from the pastry in his hand. I hoped he was not still hungry; Elva could not have many coins left, and I had no more letters of credit. I thought with regret of the basket in Medlyn's vault.

"You've thought of something," Elva said.

"Someone has to," Oleander murmured under his breath.

"We must go." I got to my feet.

Oleander crammed in the last mouthful of turnover and stood up, dusting the crumbs off his hands onto his trousers.

"We will no longer require your services," I said while he was still trying to swallow.

As the boy choked on his food, Elva thumped him on the back, looking at me with raised eyebrows. I shook my head minutely and touched my cravat. Elva gave the merest shrug and sat the boy down, handing him the dregs of the beer in his mug.

"Don't worry, partner. We need help again, we're not going to use someone else, now are we? We've made an investment in you," he tapped the boy's full belly, "and we're not likely to forget it."

Though I would not have expected it, Elva's gruffness, along with his blatant self-interest, seemed to be what reassured the boy most. It amused me that Oleander would take the word of a man of the street. Each relies most on its own kind.

I opened the locket the moment we were alone.

"Didn't think we'd be back here so soon," Elva said.

"There may be a secret here that will help us get to Arlyn quickly. If it's where I think it is, and if I can use the *forran*." I left him standing in the middle of the room while I went to the section of the shelves that held Medlyn's newest books. These would have been the theories he had worked on after I had left for the outer Modes. I began where the books and papers were strange to me.

Almost immediately I found a book with my own name on the spine. My excitement died away when I saw the names of other apprentices lined up on the same shelf. Not left particularly for me, then.

"What makes you think there's something here to find?" Elva tossed his hat onto the table where we had been eating, sat down, and swung one leg over the arm of his chair.

"Medlyn used to visit me regularly, every month or so. He would come and spend a few days, teach me a new *forran*, help me to make one of mine general so anyone could use it. One visit, a young man was dying. He had gone boar hunting, and had the bad luck to get slashed on the thigh. By the time they got him home, the wound was infected and he was delirious with fever. I took care of the physical wound," I said, "but the delirium would not respond to any *forran* I knew. Medlyn thought he had heard of one that could help, but he had to go to the City for it. Of course I let him go, though he could not possibly return in time."

"What happened?" Elva said when I fell silent remembering.

"Medlyn left and was back the same day."

"The locket, do you think?"

"Perhaps. But I thought he must have finally created a *forran* to move from Mode to Mode without using the Road. When I finished my apprenticeship, he was still in the theoretical stage of the research, but I realized that day he must have finished it."

Elva remained admirably silent, but as I was rereading a particular page I grew aware of his breathing. When I looked round at him, I found him practicing sword movements with his eyes closed. His balance was remarkable, his movements graceful, almost like a dance, if you could overlook the weapon in his hand. As a rule practitioners do not learn the martial arts, but I knew enough to recognize that Elva performed the standard movements and patterns of fighting with a sword, over and over again, with a slight increase in speed at every repetition. As if he could feel my gaze, he drew slowly to a standstill facing me exactly and opened his eyes.

"Did you find it?"

Fifteen

ARLYN

AS LONG AS I didn't know, *he* couldn't know. Lucky. Wanted so much for Fenra to be alive, must have hallucinated speaking with her. Maybe the chaos hadn't consumed her. Maybe she'd used the locket. Maybe not everything I touched got destroyed.

Maybe—maybe what? Mentally I shook myself. Maybe she was coming to my rescue? Maybe we would all live through this? At least with Fenra and Elva gone I could stop worrying about them.

"I can feel you thinking," I say. "Like a buzzing. What are you doing? Why does the body feel so tired?"

"I'm old, you know."

ELVANYN

With one hand resting on the hilt of his sword, Elva examined the whole room again, squinting from time to time, secretly hoping he'd be able to see something where he hadn't before. "Too much to hope for," he said under his breath.

"What was that?" Fenra asked without turning to face him. She looked like she was running her fingers along the spines of books that weren't there.

"I said, 'Is a fire too much to hope for?'"

Now she did turn around. "Are you cold?"

He brushed this off with a wave of his hand. "It might make the place a bit homier, is all." He gestured at the walls around him. "All these empty shelves. Gets a bit depressing after a while."

"Are you hungry?" Without waiting for an answer, she glanced at the basket and pitcher still on the table at the far end of the room. He could see those all right. And every shelf, basket, chest, pitcher, bowl, and plate. He just couldn't see what they contained. When he turned back to her she'd gone back to the shelves as if he wasn't there. His stomach growled, loudly, and she didn't even turn around. Maybe she was concentrating too hard. *Five more minutes, and then I'll ask her to stop and get me something to eat.* Elva sat down on the sofa and spread his arms out along the back. Told himself his guns didn't need cleaning.

Fenra stopped moving, and Elva sat up straight, automatically checking his weapons. He'd seen that frown before. From the look of it, she was reading from a small book, her practitioner's hand half-folded in the air, her right poised to turn a page. She turned it. Her brows lifted, and she began to nod short, shallow movements. When her lips formed a smile, Elva stood up. She'd found it.

She glanced at him and her smile broadened. "I've found it," she said. "But it may take all my strength. We had better eat. I hope you're hungry."

They both laughed when his stomach growled again.

"The difficulty is I cannot tell from the work journal whether this is the final version of Medlyn's *forran.* There must be one, I know he used it, but is it the one I found?" Fenra said, once she'd fetched warm venison stew, soft rolls that smelled of saffron, and apples from one of the cupboards.

"Can you use it anyway?"

She shrugged. "At first the *forran* made no sense, but I finally realized that, unlike most of Medlyn's other *forrans*, this one wasn't designed for general use."

"How do you mean?" Elva asked.

"All *forrans* begin as something unique to their creators," she said as she tore a roll in half. "Then they're reworked and refined until they can be used by others—if that's the creator's intention. Only a few of my healing *forrans* have been reworked, for example, since I have always worked alone, without even an apprentice."

"But you know how to fix this one so you can use it?"

She paused, spoon halfway to her mouth. "If I had not already experienced what the locket can do, not to mention the self-filling containers . . . I was a good apprentice. I spent many hours studying my mentor's *forrans*, and he gave me the key to his vault." She touched the locket through the thin cloth of her shirt. "His pattern recognizes me. So yes, I should be able to adapt this *forran* for my own use."

Now they'd been together so long, Elva thought he could tell what Fenra was thinking, even though her expressions—like all practitioners'—didn't change very much. But there, the sudden obscuring of her right eye as the lid flickered down and up again.

"You've done it?" He wished, not for the first time, that he could help her. She smiled, her teeth ivory white against the darkness of her skin. Her hands moved as if she was shutting the book he couldn't see.

Then she was gone.

His sharp intake of breath sounded as though it came from someone else.

"I'll be damned." He hoped the locket still worked wherever it was she'd disappeared to. He knew that people could last quite a time without food, but he'd feel the lack of water much sooner. He looked around, lips pursed in a silent whistle. Empty shelves. Empty pitcher on the table. Empty baskets. He checked the cupboard where Fenra had returned the stew pot. Empty.

At least Arlyn had abandoned him in a place he could survive.

That thought should have made him feel worse, but instead a smile formed on his lips. Fenra wasn't Arlyn, or Xandra, or anyone else but herself. And Fenra Lowens wouldn't leave him here alone if there was anything she could do about it.

A whoosh of air and Fenra stood dripping on the carpet exactly where she'd disappeared, smelling of lilies and green water. She held up one hand just in time to stop him from hugging her.

"I am soaking wet," she said unnecessarily. "And I stink."

"Swamp?" Elva knew brackish water when he smelled it.

"A round pool, at least that's what it looked like," she said. "Wherever it is, there's no movement to the water. Achhh!" With the very tips of her fingers, Fenra picked up something he couldn't see, shook it carefully, and set it on the nearest flat surface. She then pulled the fronts of her jacket away from her body with a squelching sound. Elva took hold of the collar and pulled it down off her arms, careful not to strain the seams. Cufflinks and narrower sleeves made the shirt trickier.

"Maybe I'd better make sure there's something for me to change into before I go any further," she said, wringing the shirt out as Elva shook out the jacket and hung it carefully over the back of the chair. Water dripped off the sleeves onto the floor. Where it disappeared.

"Perfect." Fenra straightened up from looking into one of the chests against the wall. She had green trousers and a cream-colored shirt in her hands. She tossed the fresh clothing onto the next chest and dove in again, coming up with vest, socks, and cravat. Under other circumstances he'd enjoy watching things appear out of the air.

"Where does everything go, when you're not touching it?" he asked her as she pulled the silk shirt over her head and tucked it into the trousers.

"If I had to guess—and I do—I'd say nowhere," she said. "I think it doesn't exist until it's needed, and then it's always here, like the food. You did not find yourself suddenly hungry when we left the vault, did you? The food didn't disappear from your stomach?"

He waved his hand around the room. "But what about the books, and the other things on the shelves? Why can't I see *them*, even when you have one in your hands? Is it like the changes in the Modes that commoners can't see?"

Her eyes narrowed and she turned around once, scanning the empty shelves. Finally she nodded. "It could be . . . or, unlike the items in containers, the books and artifacts are always here—unchanging," she said. "Their invisibility could be a *forran*, either laid on any mundane who enters, or laid on the contents themselves."

"Can you undo it?" Things would sure be easier if he could see what she saw. He might even be able to help her.

"I have only just now considered the *forran*'s existence. There is no estimating the amount of time it might take to undo it—if I could. As you can see," she indicated the pile of wet clothes, "I have an imperfect grasp of Medlyn's work, even when I have read the *forrans* and notes myself."

"What about laying one of your own *forrans* on me? One that would counteract this one without trying to undo it?"

She tapped her pursed lips with the index finger of her practitioner's hand. "I can try."

FENRA

It hadn't even occurred to me to draw over Medlyn's *forran* with one of my own. That changed the nature of the problem entirely—modify, do not destroy. The endeavor struck me as being similar to some aspects of healing. If I could adapt one of my own *forrans*—say that which allowed me to isolate or enclose foreign matter in an ailing body . . . could my *forran* treat Medlyn's like matter foreign to Elva's body? I began to concentrate.

The basic pattern of my healing *forran* appeared quickly between my hands, and I set it in place to examine it as closely as I had done when I first created it. I walked around it, turned and spun it, examining it from all sides. Not that it had sides, but thinking of it as three-dimensional helped me to come at the thing from a new direction. Eventually I thought I could see how to modify it. I had frequently used this *forran* with good results to isolate infection. As any good healer knows, infections are caused by tiny, minute bits of matter, tangible, but too small for the unaided eye to see, though not too small for the practicing mind. It was only one step from there to using the *forran* to isolate intangible matter—also, in its way, too small for the eye to see.

When I was ready, I opened my eyes and almost lost control over my *forran*.

Medlyn's pattern hovered around Elva. I could see it now, like dust particles in a ray of sunshine. Except that these particles formed a specific shape.

"What is it?" Elva said. "You have a funny look on your face and you're smiling."

"I can see Medlyn's *forran*. Using the locket so frequently must have fine-tuned my sensitivity to his work." I felt lighthearted, as if I had performed some test well and my mentor had praised me.

"So you can take it away?"

I shook my head, lower lip between my teeth. "If I had more time, I think I could. I can see, for example, that it is a *forran* that overlies you, it is not part of you. It's like a garment that, in a way, you step into when you enter the vault. And like the garment, it *can* be removed."

Unlike my own healing *forrans*. When I healed someone, a small part of my pattern remained within them. They could develop some other illness, but never the same one again.

"So it's back to the first plan? Neutralize it somehow?"

I nodded without answering. I walked around Elva and watched carefully as my modified healing *forran* settled around Medlyn's, connecting here and there with tiny flashes of light that made me blink. Finally my *forran* began to shrink, pulling softly at Medlyn's, tucking in here, folding over there, until it became so small that I saw it only because I knew it was there. At this point, in healing, I would reach in and pull my *forran* out, taking the infection with it, leaving a tiny piece of pattern behind. As it was, I felt I could not attempt that now. Medlyn's *forran* was not an illness. What if it would not come out? What if I injured Elva somehow in the attempt?

His lips curled into a smile as he turned from one side to another. Clearly he could now see what I had been seeing all along. Though perhaps not all, I thought. He was so delighted that though his hands moved, they forgot to go through his touching ritual.

"This is marvelous," he said, approaching the shelves holding the models of bridges.

"Do not touch those artifacts!" I moved fast enough to catch him by the elbow before he got any closer. "You do not know what any of them do."

"If they do something." His face told me he was not disagreeing with me. "Anything here could be like the locket, couldn't it? It could take us somewhere else."

"Wait." His words triggered a thought and I laid a finger on the nearest shelf. All of the models were either of fountains or of bridges. Medlyn had started making these in the last year of my apprenticeship—I should have known they weren't just a hobby; he would not have brought them to the vault unless they had more than a personal significance. But none of them opened, or appeared functional in any way.

I turned to Elva, sliding my practitioner's hand from his elbow to his upper arm. Hard muscle tensed and then loosened. "Now that you can see everything, and can find food and drink, it's safe for me to leave you here."

He pulled his arm away. "Leave me? Are you insane?" His right hand actually shifted as though he would pull out one of his guns.

"You would be so much safer." I could not believe he would argue with me. "You could have been lost—I almost lost you in the chaos." I could feel my face grow hot.

"I was trying to help you!"

"Your trying to help me could get both of us killed. And stop yelling at me."

He closed his mouth on what he was about to say and took a breath. In a much calmer and quieter voice he said, "What if you don't come back?"

"Why would I not—" I stopped talking when he raised an index finger to me.

"I know you *would*, but what if you *can't*? I'd be alone here. I won't even know what happened to you. How long do you think I can last?"

"But there would be food and . . ." I subsided again. Evidently I had not thought this through to the logical conclusion.

"And how long before I blow my brains out? Never mind food and drink, how long can I last in here alone?"

I heard the words he did not say aloud. In the other world at least there had been people, animals, things to do and places to go. Crossing my arms, I leaned against the shelves behind me, the wooden edges digging into my shoulder blades. I tried to imagine what it would be like for Elva to be alone *here*. Here there was nothing. How long before the jug that always provided the drink most needed gave Elva poisoned wine? A year? A month? A week?

Finally I raised my eyes to meet his. I had no answer, and he knew it.

"I'm a soldier, a sheriff," he said finally. "I know the world is dangerous, and that my job could get me killed—most likely to, in fact. I can face the danger. What I can't face . . ." He gestured around him with his hands.

I found myself looking into Elva's dark eyes, reading everything in them. "It is better not to be alone," I told him. "For me as well. But," I added as a smile returned to his lips, "we go as partners. You are not my bodyguard and I am not yours. We are each able to protect ourselves, and we will act accordingly."

He shrugged and touched his pistols and sword hilt. "Well, I hope you'll save me, if you should see the need. Practitioner," he added. "We'll be like a posse, with you the sheriff and me the deputy." This time his smile reached his eyes. "You tell me what to shoot, and I'll shoot it. Now what, boss?"

"We eat."

It was a restless meal. Elva kept getting up to examine some item on the shelves that caught his eye, careful always to keep his hands off. There were weapons in one section, another sword, short with a simple hilt; an elaborately inlaid crossbow, so delicate it would surely break if used; an axe that looked more appropriate for cutting wood. Two muskets, seven pistols, and four boxes of shot. Once or twice he stopped to read the titles on the spines of a few of the books.

Nor could I sit as quietly as I should have. I kept being drawn to the exquisite models of fountains and bridges. I smiled when I found one that matched the crude stone fountain where we fetched water in the village. There was one in particular . . .

"Elva." He straightened with alacrity and came to stand at my right elbow. "Does this look familiar to you?" I pointed to a fountain with a wide, flat bowl, surmounted by a smaller one. A column of three human figures rose from the center of the smaller bowl and held yet another, much smaller one above their heads. This smallest bowl had what appeared to be large fruits, showing florets at the blossom ends. It was impossible for me to tell from where the water came, whether it sprang upward from spouts at the edges of the widest level or spilled down from above, pouring bowl to bowl.

Elva drew down his brows, mouth twisted to one side. His face cleared and he nodded, tapping the edge of the shelf with his fingers. "I know it," he said. "I remember it from my earlier life, but I saw it again just lately. It used to be in a public square, in the old part of the City. Now it's been swallowed up by the Red Court."

"It must be from before my time," I told him. "I do not remember anything like that."

"I swear it's the same fountain. What are you thinking?"

I laid a careful right-hand finger on the outermost edge of the model. "I was facing in this direction when I disappeared. An old, disused fountain might well contain water lilies."

"This one did."

I nodded, stepped back, and picked up Medlyn's book from where I had set it on the table. "It might explain something here," I added when Elva looked at me for more. "There's a section where the *forran* appears to describe two separate components. I took it to mean the *forran* itself and the practitioner. Now I wonder whether the two parts are the *forran* and a model, where each model would lead to a separate place." I approached the model fountain, book in hand.

"That book has your name on it."

"No it doesn't . . ." I turned the book over and Elva was right. It did have my name on it. "I swear it had another title when I first picked it up." I looked up and smiled. "Evidently I am doing something right."

"Well, we'd be wrong to try using that fountain again." Elva nodded toward the model. "If you're right, and it's the target, we'll end up smack in the middle of the Red Court. We need to find one less public."

"But if I am right, we know that we can both reach it, and come back from it." Changing only one variable at a time is one of the basics of experimenting in the practice.

"Exactly. If anyone saw you just now—saw anything at all—they'll have set someone to watching. Remember what Predax told me—the tension between the two Courts is high just now, and the sudden appearance of what could only be a practitioner might look like some kind of attack to them."

I try not to be annoyed when I am wrong and someone else is right. I try very hard.

It turned out the trickiest part was finding a model we could both agree on. Partners, that is what I had said, so I could not simply overrule him. There was one bridge I was certain I recognized, but Elva did not know it. Another he swore he had seen, and that Xandra had told him was in the Fourth Mode, but which I had never seen. Between us we tentatively identified seventeen models, though we only had four of these in common. And if we were right, each model corresponded to a different Mode.

"Seventeen. Are there really that many counties? I mean Modes?"

I nodded, still focused on what looked like a plain olivewood bowl, painstakingly carved by hand. Only its position on the shelves told us that it was included in the collection of models. "Probably more, if each of these is from a different Mode. They don't exist anymore, but they may have once. Medlyn was not as old as Arlyn, but he *was* old. There are books, very old books, that describe practitioners reaching a previously unknown area, in effect, a new Mode. No one travels in that fashion anymore, so maybe those Modes disappeared. I am likely the one who has been the farthest out."

"One day I'm going to want to hear all about this, but right now it's not helpful."

I nodded again. I wanted to try the fountain in the village. However, I knew that to be nothing but homesickness, and I could not indulge at this moment.

"We need to know where to go."

That drew my attention. "The last we saw, the Godstone was in the City. We have already agreed that the fountain in the Red Court is too dangerous a destination."

"Maybe it's there, maybe it isn't. But I bet you can find it." Elva sounded assured. "Arlyn said you can find anything natural, and whatever the Godstone is, the body it's in—Arlyn's body—is natural. If there's any of Arlyn left, then you can find *him*. If it turns out he's still in the City, fine, we know how to get there. If we know for sure he isn't, that's the time to figure out which model to use."

I sat down, palms resting on my thighs, and shut my eyes. I knew I could find Arlyn, and only Arlyn, on the beach, if he was asleep. But would that help me find where his actual body was? In truth, I thought it might. I had leveled Arlyn so many times, it should be much easier for me to find him than it had been to find help in Elva's world.

Of course, then I had *been* in Elva's world. And when I had found Arlyn on the beach, we had both been in the City. Would I be able to find him from Medlyn's vault?

I pushed out the breath I was holding and began breathing slowly, regularly, centering myself, feeling for my memory of Arlyn's pattern. I had seen it drawn on a piece of paper, pen and ink. I had given it color. I had seen it alive, so to speak, in Arlyn's workshop. A small part of me that was not engaged in this search felt Elva take my right hand.

"You're pointing. Fenra."

I opened my eyes and saw that I was pointing quite clearly at a model of a bridge.

"That is a bridge over the Daura," Elva said. "I've gone hunting in that county."

"At least we will not get wet," I said.

"So long as we land on the bridge."

ARLYN

"You aren't helping," I say.

I looked out the window. What had been a nice stone bridge, arched across the river just downstream, was now mostly wood. And the three-story hotel we'd gone to sleep in had turned overnight into a single-story inn. We hadn't done this. Not on purpose anyway. Out of our hands. "This was never my idea," I said.

"Still not helping."

I feel his agitation, and I try to control our breathing, but he prevents it. "Let me," I say. "Emotion gets in the way of rational thought and planning." Don't know why I bother.

Maybe I could fall asleep, dream of Fenra again.

"We need to try something else," I say, but I breathe more slowly. Pulse slows down.

"Bad way to experiment," I said. "Stabilize first, then change variables."

I feel a hot rush of anger, more me than him. Surprises me. "Stable is what we're trying to do," I say. "These are *your forruns. You're* the one who said start in the City and where did that get us?"

"Uh-huh." Wondered was he getting low. "Now that we're here, any new ideas?"

ELVANYN

Fenra gave him a look that plainly told him she wasn't amused. He thought she looked tired, but then, after all they'd been through, had Elva ever seen her well-rested? She was thinner, he thought, than when they'd met. How long ago had that been? He rubbed at his eyes. Maybe he was the tired one.

Or maybe he was just trying to stall. Now that he could see everything in the vault, the prospect of staying here felt like a good idea. It was hard to think about leaving what felt like the safest place on earth. Fenra stood with her arms crossed, index finger of her practitioner's hand tapping against her right upper arm. She glanced at him.

"How soon can you be ready?" she asked.

So much for stalling. "What's to prepare?" he asked her. "I'm already carrying all the weapons and ammunition I've got."

"There's nothing here you can use?" She tilted her head toward the shelves holding weapons.

"That's all shot for pistols, no bullets." He paused. "I've taken a second knife. Maybe you want one too."

He waited for her to tell him she wouldn't need it, but instead, after a moment for thought, she picked one of the knives off the shelf and studied it, eyebrows drawn down.

"Not that one," he said, taking it from her hand. Her fingers were icy. "Here, this one's got a better grip, and a belt." He strapped it around her waist and adjusted the knife to hang handy for her practitioner's hand. She smelled of almonds. Toasted almonds.

"This ruins the hang of my trousers," she said.

"Yes, that's what's important right now." But she had made him smile.

She took a few more minutes to prepare, touching the locket lying between her breasts over and over, as if she communicated with her dead mentor. For all Elva knew to the contrary, she did. It was well known—at least it had been in his time—that different practitioners had different powers, and that some were stronger than others. He'd seen Fenra do things he'd never seen anybody else do, even Xandra.

He hoped that was going to be enough.

Finally she finished whatever meditations she'd been doing and gestured him to come closer to her.

"Put your arms around me," she said, waving him even closer. "Hug me as tight as you can, grip your own wrists. Whatever happens, do not let go of me."

"Not a chance."

This time he thought he could feel the transition, a sudden wash of icy cold, then burning hot, then a sensation as if tiny insects crawled over his skin. Each took fragments of seconds to happen, and finally his heels and feet struck heavily down on what felt like a wooden floor, with a sound like a blow to an empty barrel. This bridge wasn't stone.

His arms were empty.

"What happened to not letting go?"

FENRA

I could smell the sea. I saw nothing, heard nothing. But I smelled the sea. Slow as a sunrise, the darkness around me grew lighter, and I was standing on the beach once more. It was then I realized Elva was not with me.

My breath caught in my throat and my chest tightened until I had to force myself to inhale, deliberately drawing in air and pushing it back out. He must have been left in the vault. Reaching into the top of my collar, I pulled out the chain that held the locket. My fingers shook as I fumbled at the catch, but not as much as they did a moment later.

Nothing happened. *Wait*, I thought. Was I even here? Perhaps my body had remained in the vault? When I reached into this place to level Arlyn, neither of us was physically "here," if that word had any concrete meaning. Our bodies remained in the real world. If those words had any concrete meaning.

"Sometimes other people go away, but I'm always right here," I said aloud. My voice sounded exactly as it should sound in the open air. My friend Hal used to say that when Medlyn was trying to teach us moving *forrans*—which actually meant finding out if we could. It's a rarely found power, and my old mentor was the only practitioner who taught it. Hal's observation was his way of saying that you had to keep focused on yourself, to not lose yourself in the move. Some did get lost, we heard, though we were never told what happened to them.

Me? I could use someone else's moving *forran*, but I could not write one myself. Healing was my strong power. Ironic, considering the twisted leg that had brought me to the City in the first place.

I began walking with the sea in the direction I always took when looking for Arlyn, alert to any change in light or wind, expecting the fog at any moment. After walking until my legs and back were tired and sore, I sat down. The nearby arm of the bay appeared no nearer. "I must really be here." My voice still sounded normal, and I wanted something to listen to besides the sounds of the water. "Otherwise how could my back ache?" There was a flaw in that logic, but I was too tired to find it.

Ione Miller, back in the village, used to sing when she was bored, or doing some repetitive task. It gave her something to do, she used to say, along with letting her feel that time was passing. No one wanted to hear me sing. Not even me.

After a while I began to feel I was not alone. I reached out with the power I use to find or call animals—and help, and Arlyn—but I could detect nothing living. Not even in the sea. Yet something *was* here, I knew it. I shot fast glances over my shoulders, hoping to catch some movement, anything that could explain the feeling that I was being watched.

"Would you show yourself, please? Do you need my help?"

As if in answer, a wave rose up the beach and didn't return to the sea. In fact, a portion of it came further up than the old water line, rattling pebbles and pushing worn empty shells and other debris before it. I scrambled to my feet, though the water stopped at least two feet away. As I watched, it filled a small hollow, pooling without draining away through the pebbles.

The surface of the pool bulged upward, as if air was blowing into it from below. A large bubble formed, a shimmering iridescence, not unlike blown glass, marking where it shivered in the air. It moved abruptly, roughly, as if it wanted to be a different shape. I took a few slow steps toward it, my hands raised. The bubble immediately disappeared, though the water remained. I retraced my steps. Perhaps my movement had frightened it.

As I had this thought, the beach before me erupted into the air. I threw up my arms to protect my face. Water, sand, shells, and pebbles swirled as though caught up in a waterspout. The force of the wind increased and I fell to my knees, covering my head as best I could. It had been a little like this in the chaos, I thought, but there I had not been alone. Finally the noise and wind eased, and I was able to peer between my forearms and look. What had been a shapeless cloud of debris had solidified somewhat, but it was lumpy and grotesque, as though a child had been trying to make a sculpture of an animal she had never seen. Though I could not imagine the size of the child who could play with what I saw before me.

Movement slowed, and quickened. Just as I thought I could see a shape, it was gone, replaced with another. Finally, moving as though it was stiff and in pain, the debris began to take a human shape, deformed,

I thought, until I realized the lower half of the being remained below the surface of the beach. Though it seemed to be moving still.

I scrabbled backward like a crab, pushing myself with hands and heels. The being placed the palms of its hands on the beach in front of it, as if it leaned forward on a table.

"Please don't," I whispered, unsure what I asked for. In my imagination I saw the being pushing itself upward, and pulling its unseen lower torso and legs out of the beach. Was it human-shaped all the way down? And if it wasn't? Which would be worse? I covered my mouth with my right hand to stop my teeth chattering and I believe I may have lost consciousness. When I became aware again, I was lying on my side, still with my hand over my mouth, and the being was still in front of me.

It was looking at me now. Before it hadn't been, but it was now. I lowered my hands and pushed myself into a sitting position.

"Are you the Maker?" I asked. I wished I had not made fun of Arlyn when he suggested that one existed.

The being tilted its head as if considering my words.

"No." Its mouth did not open, its lips never moved. Its voice was the sound of rock rubbing against rock, of waves, of drifting sand, and, strangely, of the wind in the trees.

My own mouth was dry. "May I ask what—who you are?" A living being, definitely, though without gender.

As if it had heard and understood my thoughts, this time the mouth opened and the lips formed words. "I am here. This is me."

"You are . . ." I cleared my throat again to loosen my vocal chords. "Was it you who examined us in the chaos?" I gestured around me with my right hand. "Is that where home is for you?" Perhaps it had pulled this place from my mind.

"No," it said again.

I squeezed my eyes tight shut. I had not spoken aloud. "Not my mind," I said, determined to speak my thoughts aloud rather than have them plucked from my brain. Ah, it would have heard that. "Is this place in your mind?"

"All places. Reach you. Difficult. My child."

"You look for your child?" What kind of child could it possibly have?

"No. You. My child."

All the fine hairs on my arms stood up. "Me? I am your child?"

"All."

"All people?"

"No. My people," and it held up its left hand.

The air froze in my lungs. "Practitioners," I breathed.

"Yes."

The air carried an overwhelming feeling of relief, that I understood.

"Yes. Slow. It will come. Patience."

I hoped it meant explanations were coming.

"Yes."

At first I thought it had difficulty speaking our language, but I realized suddenly that it was both more simple and more complicated than that. It had never before created a part of itself that could speak. It had reshaped the pebbles of the beach into a sort of mouth and tongue.

"Is there something I can do for you?" A weird thought occurred. "Are you hurt? Do you need healing?" How long had it been trying to attract my attention? Years?

"Yes."

Then it opened its eyes. "Come," it said, holding out a hand large enough for me to step onto.

I could hear my heart thumping in my ears, feel it hammer in my chest. "You want me to come with you?" I repeated.

"Be with me, join me." It held out both hands.

But I was already a part of it, isn't that what it had said? If I went—or joined—would I come back? *What about Elva?* I pushed that thought away.

"Safe," it said. "Both. Come."

It said it was injured. I had offered to help, to heal it. I had to at least try.

"Only way," it said, once again not quite answering a question I had not quite asked.

I took a deep breath and let it out slowly, trying to regain control over the beating of my heart. I knew that I would help it. I did not understand what was happening, and this might be a way to find out.

"Yes!" it said, with obvious enthusiasm.

Licking my lips, I approached it with my hands held out. Its form became more solid and I lay both hands on the index finger of its practitioner's hand. I blacked out again.

When I came to, I was sitting cross-legged with my hands resting on my knees. I sat on the same beach, but it wasn't the same. A blue sky dotted with a few white clouds arched overhead, seagulls called to each other, and the sun was warm on my face.

"Can you hear me better now?"

I blinked, and it was sitting cross-legged in front of me. Now there were feathers and tendrils of seaweed mixed into its body of pebbles and sand and shell. It was smaller, though still much larger than I, and less daunting. The bits and scraps that made it up still moved, but smoothly, calmly.

"I can," I finally managed to croak. I cleared my throat. "How?"

"This is real space, not the space between. At first, I could only reach you there. Once there, here becomes easier."

If you say so, I thought. "What it is you need?" I asked. I could see no injury.

"I am the World," it said.

"You . . ." I swallowed. "You are the world. Our world." I wanted to ask about the other dimension, but this wasn't the time. "Everything, the Modes, the people, this place," I waved around me, "are all parts of you?"

"Everything you experience, even yourself, even Elva."

Everything around me tilted and began to spin. For a moment I thought I would faint, and I put my hands down on the pebbles to prop myself up. I took several deep breaths, waiting until everything settled down again. I was speaking with the world of which I myself was a tiny part. So small, so unimportant that it had never bothered to bring itself to my notice before.

What could it want from me?

"I must be made whole."

A shiver swept through the being, as if it would crumble apart, and for the first time I feared for it. I reached out, but before I could touch it the shivering stopped, and every bit of debris settled back into place. I rubbed at my face. It waited until I looked up to continue with its explanation, and though its language skills had improved, I still could not fully believe what I was hearing.

"Let me see if I understand you," I said. It was not impatient with me, yet I felt a sense of urgency. "You are the incarnation—if that's not too pointed a word—of the world I live in."

"Yes, an incarnation. But not where you live. You are a part of me. Every who, every where, every when, all are parts."

My stomach gurgled with nausea. "Everything that exists?"

"Yes. You know healing. You know how the body works. Many parts make the whole."

A basic principle of the practice. The whole can be greater than the sum of its parts. Like a practitioner's pattern.

The being nodded. "Yes. You are the pattern. Practitioners are. My pattern."

I waited for it to say something more and then realized that the last two statements were in fact one. "*We* are your pattern?" It nodded again. "We are what you use to make *forrans* for the world—for yourself?"

"You travel through me, Mode to Mode, and I grow, I heal, I move. Injured now. I grow weaker and weaker. You, my children, no longer strengthen the pattern. I no longer heal and grow."

"The White Court," I said. "Most practitioners do not leave it anymore. Now only apprentices travel, and even then as little as possible. You need practitioners to travel again."

"Yes," it agreed. "But there is more."

The Godstone. "Is it the Godstone? Is it injuring you?"

"Yes. No. Yes."

That was helpful.

It was apparent that the being understood even unvoiced sarcasm. "Yes, it is Godstone. No, it is not injuring me. It *is* the injury."

I held my forehead between my palms, speechless as the meaning of the words suddenly struck me. "The Godstone's a part of *you* like everything else. When Xandra Albainil thought he made it, he really tore it from you." When I looked up, I was relieved to see the being had shut its eyes. "Trying to stabilize the Modes, is *that* injuring you?"

"Yes. Must be movement, growth, change. Practitioners have the power to maintain, to clean, to fix, to heal."

"We're the *forrans* that keep you healthy." I nodded again. "That's why we can see the Modes. When Xandra broke off the Godstone piece, and tried to stop the Modes from changing . . . you have been ill, all this time?"

"Since the Godstone's time. Yes."

How horrible. All my life the world had been injured, suffering, and I had not even noticed.

"You noticed. You did not understand. Bring all of me here."

"You want me to bring the Godstone here? To, uh, re-attach it?"

"Bring it here. I will do the rest."

"How do I get it here?" I was not even sure I knew how I had arrived here myself.

"You know. Take its power."

"That I most definitely do not know how to do."

"Yes. You do know. Make it low."

Sixteen

ELVANYN

"HOW DID YOU get here?"

Elva spun around, both guns out and cocked. The God-stone stood at the near end of the bridge, one foot braced on the parapet, leaning forward over its knee. It looked more like Xandra now, leaner, sharper of feature and with darker hair than the Arlyn he'd first met with Fenra. Even the mouth was flatter, as if it should have been smiling, but wasn't.

"I don't know."

"Are you really here?"

"Don't you see me?"

"You were swallowed by the chaos beyond the door, you and that little third-class girl." The Godstone straightened and looked around. "Is she here too? I could use her."

"I don't know where she is." Elva had never been happier to tell the truth. At least the hollow empty space in his chest distracted him from his fear. "I'm not even sure I know where I am."

The timbers under his feet shivered. He looked down but didn't see anything unusual. The river was quiet.

"Didn't see that, did you? Or did you?"

"See what? I felt a little tremor, that's all."

"Ah, that's interesting." Just for a second Elva saw an expression he'd seen a lot in the old days, Xandra wondering about something. Then it was gone. "The Mode just changed," it said. "You know what I mean, even if you can't see it."

Elva glanced around, but no, everything looked just the same as it had moments before.

"Did you do it?" he asked.

"Well, yes, in a manner of speaking." Elva winced. He'd never seen that toothy smile before, and with luck he'd never see it again. "At least I started it all going."

"And is it working?" Elva walked toward the Godstone. He wasn't any safer further away, and closer he'd have a better chance of killing it before it could move. "What you wanted to do. Is everything becoming the same now?"

All expression left its face suddenly, and just as suddenly the teeth gritted in anger. "You don't understand anything about it. You wouldn't even know what questions to ask."

Elva hoped his smile didn't look as fake as it felt. "I never did. That's what you liked most about me." Absolutely true, Elva realized as he followed the Godstone off the bridge across a cobbled yard and into the inn's wide main room. Xandra had been too suspicious of his fellow practitioners. At least he could be sure Elva wasn't trying to steal any ideas.

"Well, I can ask this question, at least. What happens next?"

The Godstone stood in front of the fireplace in the west wall of the room, staring down into the flames. What could it see in there? "You see? A stupid question. What do you think I'll do? Continue, of course, finish the experiment."

"But if it's out of your control—"

It took a step toward him, the fireplace poker suddenly in its practitioner's hand, raised in a white-knuckled fist. "Who said that? Who told you that? Was it that third-class girl? Is she here too? What do you know? Tell me!" It spun around and flung the poker across the room, narrowly missing a servant girl who dropped the tray of dishes she carried into the room.

Elva jerked his head at her, though she didn't need any encouragement to run, and drew his left-hand pistol, aiming for the back of the Godstone's head.

"Hold still a minute, would you?"

FENRA

At first I could see nothing, and I feared I had somehow lost my sight between the beach where I spoke with the World and this new place.

A chill breeze touched my cheek, lifting the hair that had slipped free from my braid. At this touch all my senses snapped awake. Above me the sky full of stars, around me the sounds of the wind in trees and bushes, an owl hooting nearby, the rustle of small animals through grass and hedges. The air smelled of recent rain, and the sound of moving water was beneath my feet.

A bridge. I stood on a bridge. Though made from timbers, and not the stone I expected, I knew it for the bridge over the Daura that Elva and I had tried to reach together. I had no doubt that he had arrived without me; the World had said he was safe. It had diverted only me. But what had happened to him in the meantime? How long had I been on the beach? Hours? Days? I took myself firmly in hand. I had been given my instructions, and I could not allow thoughts of Elva to distract me.

Even so, I kept thinking of the time he insisted that he stay behind in Medlyn's old office. He had wanted to make sure that the guard, Rontin, was well, and would not be punished for failing in her duty by allowing us to escape. I had told him then we had larger concerns, and he had told me there were no larger concerns.

I pressed the palms of my hands together and felt for the connection I had with Arlyn. After a moment it led me toward the bank of the river on my left. Despite the trees, there remained sufficient light for me to make out the dark outline of buildings with here and there the soft light of lamps or candles showing through cracks in the shutters. Walking as softly as I could on the uneven timbers, I reached the end of the bridge and stepped off onto a roughly cobbled stretch of roadway. There were buildings on both sides, but the larger one on the right, where the lights were, looked most like an inn. In the City, even in a town, I would have expected a lamp at the entrance, but no one here would waste an outside light after the household had gone to bed. The silence told me that the night was well advanced. Earlier, and there would have been people still in the common room.

This inn was not on the Road, though if I followed the Daura downstream it would bring me there. It seemed that all of Xandra's hideyholes were off the Road.

I approached the building, humming softly in the back of my throat to calm any animal within hearing. I walked through the stableyard without disturbing the horses, the dog, or the boy asleep on a pallet near the wall under a pile of old horse blankets. I needed to find Arlyn and get him out of the inn, into some open space where there would be less danger to others, and I could be in direct contact with the World. It struck me that my *forrans* had always worked better when I was standing on earth. In the village I was often barefoot. I would have to remember that if I had a future to remember it in.

I pushed the main door open and stopped in the doorway of the first room, my hand lightly on the latch. I could smell gunpowder. Someone had fired a gun nearby, and not too long before. At what stage was this Mode? Would there be guns? Muskets were known in all but the outermost Modes, but this did not smell like that kind of powder to me. There was a more metallic tang to the odor.

Elva, I said to myself. *He has shot someone.* Perhaps more than once.

ARLYN

As soon as the bullet hit me I was completely myself again, as though the pain and the heat drove him out. I wish I'd known that before. My knees hit the wooden floor with a crack and I thought, *That's going to be a bad bruise*, and just before I stopped having any thoughts at all, I felt familiar hands on my shoulders, heard a familiar voice.

"What have you done?" Fenra. That was Fenra.

"I shot him. I should have done it ages ago. Hey, don't fix him!"

"Really? A bullet will only kill Arlyn, remember? Only the body." I'd heard Fenra use that tone before. Nice it was being aimed at someone else for a change. "It will free the Godstone, unharmed and capable of transiting to someone else."

"The someone else being you, since you're the only practitioner here? I'll just point out that you weren't here when I pulled the trigger and I had no reason to think you would be." The tension drained out of his voice. He'd been so afraid he had lost her, I realized, and that fear had turned to relief, and the relief to anger. "Where were you?"

Would she answer him? Had she heard the question?

"I met the World." I heard the truth in her voice. I didn't understand it, but I heard it.

"You'd better explain," Elva said.

Good idea, I thought. I blinked. I could see again. I wasn't alone. I could feel him now, but he was stunned.

"This—" her hand left my shoulder to gesture, "—all of this is a living being, the City, the Road, the Modes, all parts of its body."

"And we're the fleas?" Elva. Skepticism.

"Not at all. We are like blood, circulating through the body, moving needed elements from place to place. Arlyn did not create the Godstone, as he believed—"

My ears perk(ed) up a little here.

"He broke off a piece of the World."

"That's why it can affect the Modes?"

"Exactly, and the World wants it back. Needs it back, in fact, to return to perfect health. Without it, it is going to die. If it dies, we die with it."

"Not true." Wasn't me speaking. I didn't know. Might be so, or not. More wrong than I thought, then. "I'm not a part of anything," I say. "What are you trying to do? It wasn't the world you spoke to, silly girl. It's the Maker of the world."

"No," I said. "Listen to her." I struggled and Fenra helped me sit up. "This is part of her special connection. She knows."

I wave this away. "What difference does it make?" I ask. "Look how long it's taken to get to the state we're in now. How much longer do you think it would take for the thing you met to actually die?" I cough to clear my throat. I'm breathing easier now. "Generations. More. You'll all be long gone. But I'll still be here. I'm not a part of it!"

I felt hot, could feel the blood in my face. Not *my* rage, though. I sighed. "Lowness will get you long before the World dies."

Felt Fenra looking at me. Me, not him. Careful not to meet her eyes.

"Is that what's happening to you? Why the body is so tired?"

"Yes. And it's happening to you too, apparently."

"If this lowness is so strong that it can affect you even when I'm here, why hasn't it killed you before now?"

That was my opening. "Fenra." I tried to shrug, show my indifference. Couldn't manage. Made sure he noticed. Weak.

"What? *She's* been keeping you alive?"

Managed to nod. "Yes." Explanation took too long.

"So she's good for something." Now I turned to look at her. I hoped she understood. Not like I could kill myself. He reached out for her, hesitated. "*She* doesn't have this lowness?"

"Shall I shoot him again?" Elva asks. I hear a metallic click.

"You have said it yourself," Fenra says. I know she's speaking to me. "I could not carry you any more than this mundane could. I am not strong enough. It must be a first-class practitioner. Like Xandra. I may have a talent for healing, but I am only a third-class practitioner."

"Then fix him. Fix us now."

FENRA

"Fenra, don't do this." Elva gripped my wrist. I knew why he protested. He believed that if I did not level Arlyn he would eventually shut down, trapping the Godstone. But the World had not given me "eventually" as an option. The World wanted him now. This might be our only chance.

"I must keep him alive," I said. "While he is, the World still has hope."

"Ironic, isn't it?" The Godstone laughed. "You need me, so you have to keep me alive, even though my being alive is the last thing you need."

"You know, Xandra, you didn't used to be so much of a shit. I used to like you."

"I could make you like me again, don't think I can't."

Elva's face lost color and his lips pressed into a thin line.

"He does not mean what he appears to mean," I assured him. I hated to see even the touch of fear on Elva's face. I had not hated the Godstone before now. "This temper and nastiness is part of the lowness," I said. I helped the Godstone to its feet. "Come, sit here and let us begin."

I lowered him into one of the armchairs and moved the other closer, facing it squarely. When I was settled, I held out my hands. He hesitated, watching me through slitted eyelids, but finally put his hands in mine.

"Relax," I said. "I am the one doing all the work."

At first I could not concentrate. I hoped he could not feel my rapid pulse. My hands felt like ice. I hoped he would think that normal. I thought I heard Elva speak again, felt his hand on my shoulder. I wanted him to stand well back from us, beyond the effect of the transition. I wished I had been able to warn him without alerting the Godstone.

Once I began, the familiarity of what I did relaxed me. In the blink of an eye, we were standing on the beach, surrounded by fog.

"Relax," I told the Godstone again. "This is how I level him. Ask if you do not believe me." It must have done so, as I felt the tension in its hands release. I began by feeding him a little trickle of power, enough to convince the Godstone that I was doing what he expected me to do, not so much as to make him doubt I was a third class. Under the mask of this little outgoing stream of power, however, I began to pull a larger stream back to me, further hiding what I did with images of the tide, and the wind in the trees. The fog began to clear just around us and I relaxed even further, until I realized that we were in the wrong place. This was not the beach of pebbles where I had met with the World, but the sandy beach where I usually found Arlyn. Which meant that our physical bodies were still in the sitting room of the inn. Had I used too much of my own power when I had healed Arlyn's bullet wound?

Dismayed, I clenched my hands, unable to stifle the movement before the Godstone noticed it.

"What are you doing?" He dragged me to my feet by my upper arms and shook me. His chair fell backward to the floor. We were still touching, and so the power continued to drain from him to me, but now the covering, outgoing trickle had stopped. The Godstone raised his practitioner's hand into the air above my head, fingers stiffened, ready to strike me with a *forran*.

Suddenly Elva seized the thing's wrist, pulling its arm down and twisting the hand to one side. I swept aside my horror that he touched the thing and took advantage of the Godstone's distraction to pull more power from him, faster than I had before. Good, I was one step closer, but I had no time to rejoice. As much as I had taken, there was still more, perhaps even enough to overwhelm me. Xandra Albainil was much stronger than I had guessed—unless what I felt now was the Godstone itself.

"Hang on, Fenra!"

I wanted Elva to be somewhere, anywhere, else, but his voice, his touch, his presence, anchored me. I clenched my teeth and forced an image of my pattern into my mind, making it brighter, stronger, drawing as much power out of the Godstone as I could. In the space of a breath I felt lighter, larger, steadier. I had reached the turning point, and while the Godstone was still dangerous, our power levels were almost even.

I felt his sudden surge of triumph as he pulled at me. Had he tricked me? I saw the shadowy image of the inn room, felt the warmth of the fire. I remembered something Medlyn had taught me, early in my apprenticeship. Use your opponent's strength against him. I stopped resisting the pull of the Godstone's power and instead thrust mine at him like a spear, using his own exertion to stab deeper than I could have on my own. I dragged my pattern with me, felt him falter and waver under my surprise attack.

Suddenly the beach rose up around us again and this time I saw the pebbled sand, the rock, smelled the sea and the faint air of rot. The image stabilized, the inn room disappeared, and we stood alone on the beach. Or, not exactly alone. Elva was there, one hand still clinging to the Godstone's wrist, the other hand wrapped around mine.

But there was no sign of the World. It needed more time to form, I thought, remembering how long it had taken before. *Hurry.* I was afraid I would not be able to keep the Godstone distracted very much longer.

"What were you hoping to accomplish with this?" The Godstone tore itself free of me and I rubbed my hands together, preparing for his attack. Instead, he turned on Elva, swinging his clenched fist and striking him like a hammer on the side of the head. Elva dropped like a bird shot from the sky and my heart stopped beating. There was no blood, but I could see a dent in his skull.

"What are you going to do now, little practitioner? You've siphoned off enough of my power to put up a good fight—too bad you don't know how. And even if you did, your friend will be dead by the time you finish with me. Or maybe you'd like to leave me be, and save your friend instead? Which shall it be?"

The image of Hal falling away from me flashed across my mind. I could not have saved him, but I had wanted to. What had Elva said? *There are no larger things.* If I let my friend die to save the World, could I live with myself? The World had told me that every part was important to the whole. I laid my fingers along the ridge of bone in Elva's skull. It was not too late. Limp as he was, there was still life here.

The Godstone's laughter had triumph in it, but I blocked it away. I forced myself to concentrate on Elva, on the way the bones of his skull felt, and how they should truly be. On the nerves and fibers and sinews of his neck, shoulders, and spine. How each and every part of him formed the body that both was and was not Elvanyn Karamisk.

The World was like this, I thought, as I straightened pieces of bone, smoothed muscles in the scalp and face. This is what it had meant.

The Godstone had stopped laughing, but I did not turn toward him until I felt that Elva was out of danger. I straightened to my feet, hands trembling.

"So, that's it for you then, little practitioner. Look at your hands! It's like you've got a palsy. I'll bet if you tried to walk you'd be lame. And for what? For a mundane, someone of no importance whatsoever."

Behind him, a wave had collected into a pool and began to take form. The Godstone had kept its attention on me as I healed Elva, and the World had taken advantage. I kept the Godstone's focus on me, as the avatar of the World took shape behind him. I took a step forward, holding my trembling hands in front of me, as if reaching for his help. I looked into eyes that held nothing of Arlyn Albainil in them. Nothing of the furniture maker, the craftsman, the man the children liked.

"You are wrong," I said. I did not step closer, but I knew that I could. And without limping either. "Each and every part of the World is important. The loss of any piece out of its time diminishes it. Even so small a piece as Elva. Or me. Or you." My fingers shook with the need to seize him and suck away his power, to kill this thing.

Perhaps something in my face alerted him, for it spun around, freezing into stillness when he saw the World. His expression spoke of arrogance and frustration, but there was no fear. No awe and no wonder. Not even when the World held out its massive hand.

"What's this? Am I supposed to be frightened?"

It was then I understood that the World could not re-absorb him automatically. Perhaps they had been separated too long. Perhaps I had not drained enough of its power.

"Fenra."

Elva looked up at me with a pale face. He let me pull him to his feet. "What happened?"

"No time," I said.

"Go," he said, pushing me away.

I ran to where the World and the Godstone were reaching out to one another. Even though the World was so much larger, I felt with horror that the contest might be an even one. It was damaged and hurting—perhaps more so due to the activity of the Godstone in recent days. It had been injured for so long, and this might very well be its last chance to heal.

I held back, waiting for some sign. They touched, and the beach around us shimmered, the colors fading and deepening several times and then stabilizing once again. I would have been reassured, but I thought the World looked smaller. A moment later I was sure of it. The Godstone wasn't being absorbed. The World was not winning. The part was too diseased to rejoin the whole—it would only poison it.

ELVANYN

Fenra ran from him toward the Godstone and the creature that could only be the World, but he stayed where he was, swaying just a little, a little off balance. There wasn't anything he could do to help. His mouth was dry, but he blinked back what felt like tears. He was no practitioner, but he could see the cloud of sand and pebbles and shells and air whirling around the Godstone—sometimes clearly a human shape, sometimes just a dizzying swirl of movement—and he could feel power as if the air was saturated with it. He had an itchy spot on his scalp just above his left ear and he rubbed at it, wincing at what felt like a bad bruise. Something must have knocked him on the head. Gritting his teeth against the unexpected pain, he pulled both guns free of their holsters. He'd always felt more relaxed with them in his hands.

Fenra had reached the combatants and grabbed the Godstone by the shoulders from behind. It shrugged her off and she staggered back, lunged at it again, this time managing to get an arm around its neck. She couldn't bring up her other hand; it was too tall for her to get leverage for a chokehold. Elva moved until he flanked them, looking for the best angle to take a shot without hitting Fenra. That was the most important thing. He wasn't worried about hitting the World, he was sure the bullets would pass right through.

He wished that Fenra hadn't saved Arlyn, though he knew why she thought she had to. He knew all about the big picture; it just wasn't the picture he cared about right now.

Fenra was growing pale, her breath coming short. She looked like she might faint, and while the Godstone was pretty shaky looking itself, Elva figured it was by no means down for the count. He edged closer. Fenra must have caught sight of him out of the corner of her eye, because she turned and looked straight at him. She smiled, pulling stiff lips away from clenched teeth, and moved her head sharply up and down.

Elva thought he understood, and lifted his guns.

In the last moment the thing moved and he missed the head shot, catching Fenra under the right armpit instead. The missile must have passed right through her—the Godstone cried out and its hands dropped from their grip on her forearm. Elva holstered one gun to have a hand free and ran to where Fenra lay on the pebbles, biting her lower lip in pain.

"Wait, wait," she gasped. The wound in her side had stopped bleeding. The Godstone dropped to its knees beside her, and reached for her. Elva took aim. He couldn't read the look on its face. The World stretched out and took hold of the thing with its hand of pebbles and water and sand and air and the Godstone began to scream.

Elva took Fenra into his arms, pressing as hard as he could against her wounds. The screaming continued as the Godstone was torn free of its host and reabsorbed by the World. The towering image brightened unbearably, light making every scrap of stone and shell stand out sharp and clear. Elva curled over Fenra, raising his arm to shield his eyes. One of the great hands reached down toward him and he turned Fenra away. The back of his neck stung with sand and he shivered.

Abruptly everything stopped. He could hear the movement of the water as the waves ran up the beach, but that was all. He opened his eyes and without looking behind him felt for Fenra's wound.

"I am not hurt," she said, but she clung to him all the same. Her hair smelled of clean salt spray.

"Arlyn," she said. Elva helped her to her feet and together they staggered over to the body. Blood trailed from its eyes and ears.

"Is he there?"

Still holding tight to his arm, Fenra lowered herself to her knees, cradled Arlyn's head in her lap, and wiped off the blood with the edge of her hand. "Arlyn," she said, but not as if she was asking. Elva sank to his haunches and took hold of his old friend's right hand.

"He's gone?" Just a whisper of sound out of barely parted lips, almost drowned out by the wind and the waves.

"He is, Arlyn, he is gone."

"Fenra, can you . . ." Elva had to ask, even though he was sure he knew the answer.

She shook her head, her face looking just like the World's did, calm and impassive.

"Elva." Arlyn changed the angle of his eyes so that he was looking at Elva, the corner of his mouth twitching as if he tried to smile. "A person gets too far gone," he said, "until they can't be saved. Not even if Fenra uses all her strength, and then you'd be stuck here, wouldn't you?" This time the smile formed, showing blood on his teeth. "Leave me. I'm part of the World now, just like him. Just like you."

<hr />

FENRA

Of course we did not leave him.

"He is part of the World now," I said, repeating his words as I brushed Arlyn's hair back from his face, watching the tiny movements that showed he still breathed. "He was not before, but he is now."

"Sure, but he was a part of us, too. Both of us." Elva slipped his arm around my shoulders and I leaned against him. "Turned out I didn't like Xandra very much after all. But I kind of liked Arlyn. I think we could have been friends."

I nodded. "You would have liked watching him work. He will be a great loss to the village, in more ways than one." I smiled, remembering Ione Miller's last words to me. "The children liked him."

We stayed until we knew that our friend was gone, and then we stayed some more, unable to tear ourselves away. Twice I touched the locket under my shirt. I was sure that it would work, now that the World no longer needed me.

"The tide is coming in," Elva said finally, lifting me to my feet. "What about Arlyn?"

"We'll leave him here," I said, still looking down. "Where he belongs."

After the beach, Medlyn's vault—mine now, I supposed—felt like home, warm, comfortable, smelling of wood polish, fresh-baked biscuits, wine and cheese. I was pleased to learn that the *forran* I had used on Elva to allow him to see the contents of the vault still worked. A small victory compared with what we had done, but I felt it pointed us forward, telling me we had somewhere to go from here.

"What will you do now?" Elva asked. He picked up the pitcher and refilled our wineglasses. He was trying not to look at me.

"I will have to go back to the village," I said. "At least for a while, once I have collected Terith. I left unfinished business there, and now, without Arlyn, I may need to do something more for the place, to make sure it doesn't fade away."

"And then?"

I shrugged. "Now that the World is healed, there will be great changes—possibly even the creation of new Modes. I would love to see how that happens." I picked up a cherry, turned it over in my fingers, and popped it into my mouth. I spat the seed back into my hand. "First, however, I should go to the White Court. I have a great deal to tell them, and they will have changes to make once they understand what the World requires of them."

Elva took a swallow of wine and set down his glass. Turning it in his fingers, he finally looked up and met my eyes. "Why don't I come with you? Always good to have a witness."

I was momentarily speechless, my mouth dry. "You said you would go back to the New Zone. You would live longer there," I added when he did not respond.

"Maybe." His smile warmed me right through. "I'd go back if you came with me."

For a moment I could see the lives we would lead. Elva would be the High Sheriff again, I would be the town doctor. We would live together in the same house. But it was only a dream. "I must stay here."

Elva reached across the table and took my hand. "If you stay, I stay."